RADIO SIGNAL FINDING

Jim Sinclair

McGraw-Hill

New York San Francisco Washington, D.C. Auckland Bogotá
Caracas Lisbon London Madrid Mexico City Milan
Montreal New Delhi San Juan Singapore
Sydney Tokyo Toronto

Library of Congress Cataloging-in-Publication Data

Sinclair, Jim.
Radio signal finding / Jim Sinclair.
p. cm.
Includes index.
ISBN 0-07-137191-5
1. Radio direction finders. 2. Radio—Receivers and reception. 3. Television—Receivers and reception. 4. Shortwave radio. I. Title

TK6565.D5 S56 2000
621.384—dc21 00-063828

McGraw-Hill

A Division of The ***McGraw-Hill*** *Companies*

1 2 3 4 5 6 7 8 9 0 DOC/DOC 0 6 5 4 3 2 1 0

ISBN 0-07-137191-5

The sponsoring editor for this book was Scott Grillo and the production supervisor was Sherri Souffrance. It was set in Melior by York Production Services.

Printed and bound by R. R. Donnelley & Sons Company.

This book was printed on recycled, acid-free paper containing a minimum of 50% recycled, de-inked fiber.

CONTENTS

ACKNOWLEDGMENTS

Most of the information in this book is collected from equipment that I have worked on and from people I have worked with over the years. I am grateful to the following groups and individuals for letting me learn from their experience:

All the active members of the South East Radio Group and later the Wireless Institute of Australia, S.A. section.

All the people I worked with as an employee of Telecom Australia.

All the users of HF radiotelephones in central Australia and Northern S.A.

The radio staff of the Alice Springs and Port Augusta Royal Flying Doctor Service bases.

All the many people who entrusted me with their equipment when I worked as a sole trader.

All the officers and volunteers of the Country Fire Service stations at Clarendon, Kangarilla, Happy Valley, Coromandel Valley, and Cherry Gardens.

There is one particular section of the book which was borrowed from the experience of others. The technical information on satellite TV was provided by Ron Hooper of Southern Satellite Communications Pty. Ltd. and some of the basic information on satellite mobile telephones came from Chris Joseland of Australian Satellite Services.

Even this list has some bits left out, particularly the book, *Foundations of Wireless* by A.L.M. Sowerby and M.G. Scroogie, and the contribution of John Sheard, a radio proprietor in Mount Gambier who very patiently answered all the silly questions when I and others were very new to the game.

Jim Sinclair

CHAPTER 1

The Scope of This Book

1.1 Subject Outline

This is a book about receiving radio signals; specifically, how to set up receiving systems that work as well as possible under difficult conditions. I assume you are someone whose task is to install or operate a receiver and aerial system so as to obtain maximum performance in a fringe area. To make best use of this manual you will need to have some background understanding of electrical principles such as would be normal for a small business operator working as an electrician or auto electrician in a country town or a pastoralist who does mechanical and electrical repair work on your own station's vehicles.

I assume you have access to at least a multimeter and some basic tools; that you know the meanings of basic electrical terms such as volts, amps, ohms, decibels, and kiloHertz and can read and understand a schematic diagram; and that you understand the distinction between resistance, reactance, and impedance. There will be some references to more advanced test equipment but I will try to provide an alternative (which may be more "fiddly") means of performing each test. In most cases this will use some function of the receiver itself so you will need at least a circuit diagram or preferably a complete service manual if possible.

There will also be some references to operations in the office environment of a big city. These will assume the same level of technical understanding because in most cases if you are working in that way you will probably be part of a team and somebody on the team will have the required technical proficiency.

There will be no great depth of theoretical information about why radio signals behave in a particular way; that is already covered in a companion book, *How Radio Signals Work,* published by McGraw-Hill (ISBN 0-07-470329-3), and where needed references will be made to particular sections of that book. This is a book of practicalities about real radio systems in fringe-area situations and real people working in isolation with limited facilities.

For all installations, however, you will need to bear in mind some "rule of thumb" theoretical principles as follows:

For all signals at every location there will be a "microclimate" of particular spots of strong and weak signal due to standing wave patterns from local reflective objects. There will often be a difference of 6 to 8 dB between the strongest and weakest signals at a particular locality. Section 5.6 in *How Radio Signals Work* illustrates this process.

For all terrestrial signals there will be a general tendency for stronger signals to be available at higher elevations. In many cases, however, the rate of change of signal strength with height due to this cause is smaller than the variation due to the microclimate effects mentioned above.

In all cases the sensitivity of a receiver varies due to the bandwidth of the output signal (audio treble response or data bit rate). The receiver becomes more sensitive when output bandwidth is reduced (the receiver is made more selective).

For correction of apparently poor performance on your system you will need to learn the difference between a receiver output due to weak signal and that due to interference of many different types.

You may notice as you read through that the techniques described in this book resolve into two broad classes; there are some directed to making the signal collection system big enough and efficient enough to collect very weak signals and there are other measures designed to minimize collection of unwanted or interfering signals and radiated electrical noise. In a sense these two aims are partly in opposition but you will need to keep both in mind at all times. The overall aim is to make the best possible ratio between the wanted signal and all the others.

This is a book specifically about receivers and receiving systems. There is a distinction (indicated in Figure 1.1) between the interests of a transmitter operator and those of a receiving station operator. At a transmitter station there is only one signal of any real interest—that is the one you are generating and your interest is served by ensuring it leaves you with the correct power level, with the carrier on the correct frequency and as free as possible from harmonic and other spurious signals, and with the modulation undistorted. The receiver has no choice but to deal with many signals, some of which may be dozens of decibels stronger than the one that is of interest. The one signal you want must share the big wide world with all others and may be tossed about by many circumstances somewhat like a small boat in a wild sea. The receiving system must accept the existence of all these disturbances and interferences and step by step either remove them or change conditions so that their effect is irrelevant.

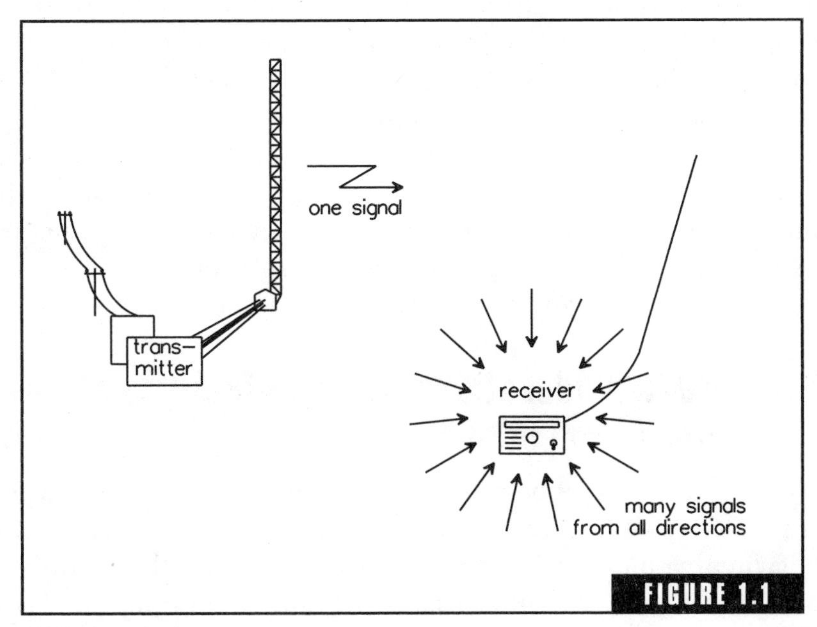

FIGURE 1.1

Transmitter and receiver.

The uses of radio may be divided into three broad classifications. A radio signal may be used for *broadcasting, communications,* or *power transfer.* In the broadcasting process a program (originally generated either by someone talking into a microphone or from the output of a video camera or computer) is transmitted to all who may listen, without any direct expectation of a reply. The broadcaster usually hopes for an audience of many thousands but does not know exactly how big the audience is or which people they are. The greatest number of receivers produced are those which are designed for reception of a broadcast of some sort so equipment for receiving mass media broadcasts is considered first in this text.

The term *radio communications* usually refers to uses of radio systems in which one person speaks into a microphone and transmits a signal expressly intended for another particular person to hear and reply to. It may also be used to describe the sending of messages for a small group of people to hear and acknowledge in turn. Communications in modern times may also include links where a computer originates a message and transmits it to another particular computer and the data transmitted could have the significance of any of the whole range of purposes for which computers use electronic data (transmitting text or a picture or switching a machine on or off, for instance). Consideration of radio communications systems is referred to in passing in the early chapters of this book but is the major subject of consideration from Chapter 9 onward.

The use of radio signals for power transfer includes all the RF heating, diathermy, and microwave oven class of uses. This book is mainly about receivers which are not generally used in those services, so those uses of radio will be ignored except in relation to their potential to interfere with the transmission of intelligent messages.

1.2 Some Considerations Common to All Systems—Planning

The whole project may be as simple as buying what you think is an appropriate type of receiver, placing it in a convenient spot, plugging it in, switching it on, and connecting a sufficient length of wire to get a good clear signal; or it may be as significant as building a dedicated structure (often called a "shack") at a carefully selected low-noise location to house this and perhaps other receivers. In many of those larger installations the whole project includes a transmitter and receiver, often at separate sites with landline or a radio link for communication between them. The transmitter is usually the part of such a project that costs the most and requires the major part of the effort required for the whole project. The effect of the receiver on overall performance of the completed installation is, however, potentially equal to that of the transmitter, so any economizing applied to the project which may affect performance capabilities of the station should be done first to the transmitter. For example, a 3-dB increase in sensitivity of the receiver, which may sometimes be gained by careful tuning and noise matching, is equivalent to doubling the transmitter power, which could require spending another million dollars.

Money (and other resources) is actually the first consideration. In a lot of cases weak-signal problems can be solved by the brute-force method of throwing enough money at them, but what is really of more practical importance is the minimum amount of money needed to give a good-enough signal. Before anyone does detailed planning work of any sort you should make a rough estimate of how much it will cost to achieve your objective and whether you have the resources available to complete the job.

The grade of service required is another factor that greatly influences the overall cost of the project. There is a vast difference between a service that works but is noticeably affected by propagation disturbances and one that offers 99.9% reliability so that there is a workable signal received even when propagation is disturbed. Fading may reduce the signal level by a figure which is commonly in the range of 25–40 dB, so signal-strength figures must allow for that loss. In some cases a single channel cannot provide satisfactory reliability no matter how much allowance for fading is built in, so other channels with a frequency-diversity system (or day and night channels on HF circuits) must be designed for.

Short-term fading is a random process, so a graph of the instantaneous signal level tends to show a Gaussian distribution. For high reliability the conditions must be workable well into the tail of the distribution, so there is not much improvement in reliability for a quite large increase in average signal strength. To upgrade reliability from 99 to 99.9% may require a 10- to 15-dB increase in average signal level.

Other basic questions to ask are:

- What type of signal is wanted and what intelligent modulation does it carry?
- Is it the strongest signal in the spectrum or does it exist in the presence of other stronger signals?
- From which direction will the main signal be received?
- Is the expected signal a groundwave or skywave or will it arrive by refraction or reflection? (Chapter 5 in *How Radio Signals Work* explains these terms.)
- Will the signal have vertical, horizontal, circular, or mixed polarization?

After you have worked through the implications of all these questions there is another that should be asked before any construction work is done:

- Is it economically feasible to place the receiver and a smaller aerial at a position where a stronger or clearer signal is available and send the receiver output to the required location by a cable or landline connection?

Early in the planning you can influence such factors as aerial and equipment selection, tower height and positioning, and in some cases, performance objectives of the receiver. If these basics have already been decided there may be limits to what you can achieve.

If the equipment provided is typical of what several close neighbors are using for the same purpose (for instance, television in a country town) then the job becomes to obtain at least as good a signal as is common for the neighborhood. At the other end of the scale, if you are selecting equipment for a specialized or rarely used purpose for which no local comparable installations are available, it may be worthwhile to ask a communications engineer to calculate an expected figure for field strength, aerial gain, and directivity required and then work to those figures as specifications.

In many cases the planning for most economic performance involves a tradeoff. For instance, with terrestrial signals the quadrupling of tower height will often have the effect of doubling the strength of signal available; also, for screen-backed arrays doubling the size of the array will also double the signal power collected. For a given situation a sufficient signal may be received from a 10-m tower with a 100-element aerial array on it or possibly with a 200-m tower with only a few elements. Neither of these is the most economic combination, the optimum is probably going to be in the range between a 30-m tower with about 60 elements in the aerial and a 60-m tower with about 40 elements. In all cases where a trade-off function occurs there will be an overall best figure, which in the case of Figure 1.2 and for all costing functions is a minimum. In the case of some technical performance functions the aim may be to maximize a particular resultant value for best performance.

Sections in this book which specifically consider practical planning of particular types of service are 2.2, 2.4, 2.6, and 2.7 in Chapter 2; 3.2, 3.3, 3.4, and 3.9 to 3.12 in Chapter 3; and 4.5, 4.8 to 4.10, and 4.12 in Chapter 4;

1.3 When the Installation Has Worked Previously

If your task is to make good an existing installation, the first question becomes "Has it ever worked properly?" If the answer is "No, it has never really been good," then the appropriate first step may be to check the original design particularly with respect to the amount allowed for fading margin. If, however, there was ever a time when operation was satisfactory, you should forget the possibility of redesign and ask "What's changed?" and that will resolve into questions such as:

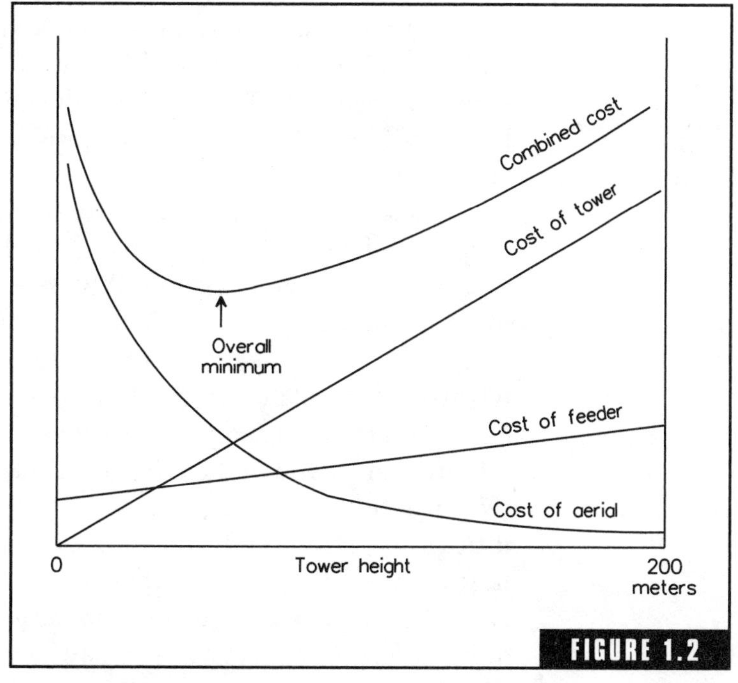

Cost minimizing.

Has the aerial been damaged or shifted out of position or have the joints corroded?

Has the feed line been damaged?

Has the receiver lost sensitivity due to an electronic fault or has the tuning drifted?

Has the transmitted signal changed?

Has a new piece of equipment been installed nearby that is causing interference?

This process of fault localization can be taken to further stages, but in most cases from then on the factors to be investigated will be specific to the type of equipment and the signal you are attempting to receive. Those questions will be asked again in different ways in later chapters.

1.4 Layout of the Rest of the Book

In the modern world there are a vast number of uses of radio with great variations between different types of service. I included specific information on a number of example situations in the hope that you will find at least one example that is close enough to your specific case to be relevant. I placed first the general interest services such as broadcasting and television and present information about more specialized services later.

In most cases whatever is written about other services for a similar frequency band will be relevant to the particular service you are presently interested in, possibly with some modifications due to different operating conditions. For instance, many of the parts of Chapter 5 on the subject of television interference that deal with VHF television are related to the FM broadcasting (Chapter 7) and radiocommunications (Chapter 9) services that also operate on VHF and the low-frequency end of the UHF band. There will also be a similarity between all the information on television systems and the operating conditions for broadband data links if due allowance is made for the higher carrier frequency of the data links. The point is that the underlying technical conditions of operation of a number of apparently quite different types of service may be very similar if carrier frequency and operating bandwidth are close to the same.

This book does not generally give information of sufficient detail for component-level fault location of particular items of equipment. If you propose to use this information for design of the installation of a particular receiver you will also need an owners/operators handbook for the particular make and model of the receiver. For electronic fault-location work on the receiver you should have a manufacturer's service manual available. That service manual should have at least the following:

- safety information about all hazards that are exposed by removing covers
- a schematic diagram (which may be on several sheets of a drawing)
- a memory map and flow chart of the program if any digital control system is used
- an explanation of the operation of each major section of the circuit
- the location of all test points and the expected signals and/or voltages at each one
- fault location information or a suggested test sequence

There is one broad class of services that I have ignored totally and that is the vast number of very short range, very low power services such as cordless telephones, headset communicators, garage door controllers, and so on. In most cases the difference between a weak signal and a strong one is only a few paces or a few seconds driving and the detail differences between all the various services is such that a lot of paper could be used explaining technical solutions for something that is not really a major problem. If you are faced with a "fringe area" type problem with any of these services, the preferred solution is to shift either the transmitter or receiver to a place where a clear signal path is available.

CHAPTER

2

Long Range on the Broadcast Band

2.1 When All Signals Are Weak and Distant

A typical situation where this applies is someone on a Pacific island who has a specific need for information (for instance, news broadcasts or weather reports) from broadcasting stations on the nearby continental land masses. Other relevant places could be agricultural, pastoral, mining, or timber settlements on any of the continents well away from the nearest town.

There are no powerful local signals to overload the receiver so it can be made as sensitive as possible and the aerial can be made as large as required. The overall aim in resolving the weakest possible signals is best achieved by considering each part of the system separately, as is done in the next few sections. The receiver electronics box is one section, the feedline is another, the earthing connection is a third, and the aerial itself is yet another and, finally, when all these are as good as you can get them, you will need then to think about operating procedures and times. Note that all these components of the system have a multiplying effect on each other and all must be correct before any part of the system will work properly.

This chapter deals with the MF broadcasting band (frequencies in the approximate range of 520–1600 kHz; see Figure 2.1) using amplitude-modulated signals. In most countries of the world there

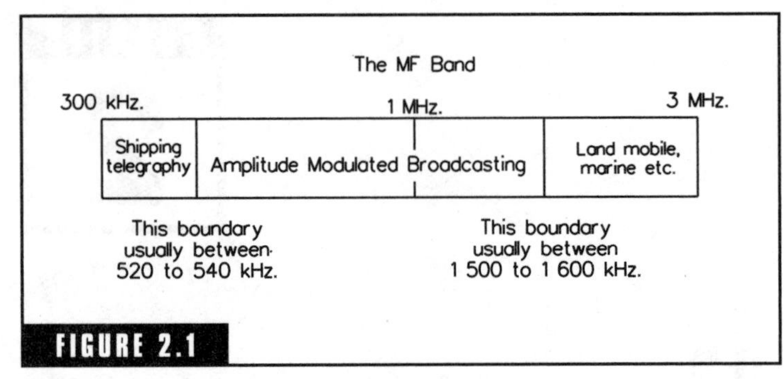

FIGURE 2.1

The relevant portion of the spectrum.

is some allocation in this range for broadcasting either by governments or large commercial organizations using amplitude modulation with transmitter power that may be anywhere in the range from 25 W (for purely local services) up to as much as 1 MW. These services are intended to be used in combination with relatively low-sensitivity receivers in any location where the signal strength is sufficient to give reasonably noise-free reception.

With higher sensitivity receiving systems reception of these signals is often possible from locations well outside the designated primary service area. This chapter is intended to describe the techniques that can be used to gain maximum sensitivity in a receiving system for that purpose. Some of the techniques may be useful for other bands but may need scaling or other modification to adapt from the MF broadcasting band. Also for this chapter I assume that the receiver provides an audible output to a loudspeaker or set of headphones and that you are able to locate it in a quiet-enough spot to be able to hear the loudspeaker clearly. If you intend to use the receiver output for something other than an audible signal you will need to read this chapter in conjunction with Chapter 12 ("Data, Codes, and Selcalls") and if soundwave interference is a problem, there may be some helpful hints in Chapter 10 on mobile communications.

Working conditions for amateur operators on the 160-m amateur band (in those countries where that band exists) are very closely related to operations for maximum sensitivity on the MF broadcasting band. If you are an amateur operator wishing to use these techniques you will normally expect to use the equipment for both transmitting and receiving. Matching for power transfer from a transmitter must be much more critically adjusted than is required for receiving a signal, so you may find that some designs of aerial which work well for receiving are not easy to match to a transmitter.

For reception of signals of the minimum workable strength on the MF broadcasting band all of the following factors are significant:

- received bandwidth minimum necessary
- receiver sensitivity sufficient
- receiver does not pick up spurious signals from other sources
- receiver and feeder shielding sufficient to ensure that limiting noise level is as provided by the signal from the aerial
- aerial and earth system configured to efficiently couple signal into the feeder
- any aerial directional effects that occur are aligned to favor the wanted signal
- time of day and seasonal conditions are not adverse to reception of the wanted signal

The first three of these factors are treated in detail in the next section, most of the others are dealt with in Chapter 3, and the last one is described in Section 2.3.

Note that many MF broadcasters add information to their transmissions in multiplexed or subcarrier form. Section 2.6 deals with multiplexing for anyone who has a specific need to receive that portion of the signal; for all the other techniques described, the multiplexed signal, if it exists, can be ignored.

2.2 Maximum-Sensitivity MF Receivers

For the receiver itself the limit of sensitivity on the MF broadcasting band will be atmospheric noise picked up by the aerial. Gain before the detector stage must be sufficient to reach the electrical noise level, but amplification more than that in the front end is of no value. A receiver that is good for Dx (originally an abbreviation used by telegraphists to indicate "distance unknown, but usually very great") will give some thermal noise when operated at maximum gain with no aerial connected, and the noise output will increase as soon as an aerial of any sort is added.

There are some receivers designed for use in the primary service area of a local transmitter which are made with very little gain before the detector and engineered for interference-free reception no matter how strong the local signal is. Millions of sets of that type are manufactured each year as personal portable or mantle receivers; often the receiver function is included as part of another function such as a clock or as the tuner section of a high-fidelity audio system. Figure 2.2 shows the block schematic diagrams of a couple of typical circuits. If one of these is tuned to a channel with no signal present it will usually give very little output due to atmospheric noise; receivers of that type are not much use for long-range reception.

Selectivity will make a difference; most types of noise are scattered broadly over the spectrum while the wanted signal is concentrated in a definite band. The best signal-to-noise ratio will be gained by making the IF bandwidth only just wide enough to fit the required signal. A receiver designed for high-fidelity reception will be less sensitive than one designed particularly for voice communications because music requires a wider bandwidth than is needed for the intelligence components of speech. Skirt selectivity can also make a difference. A

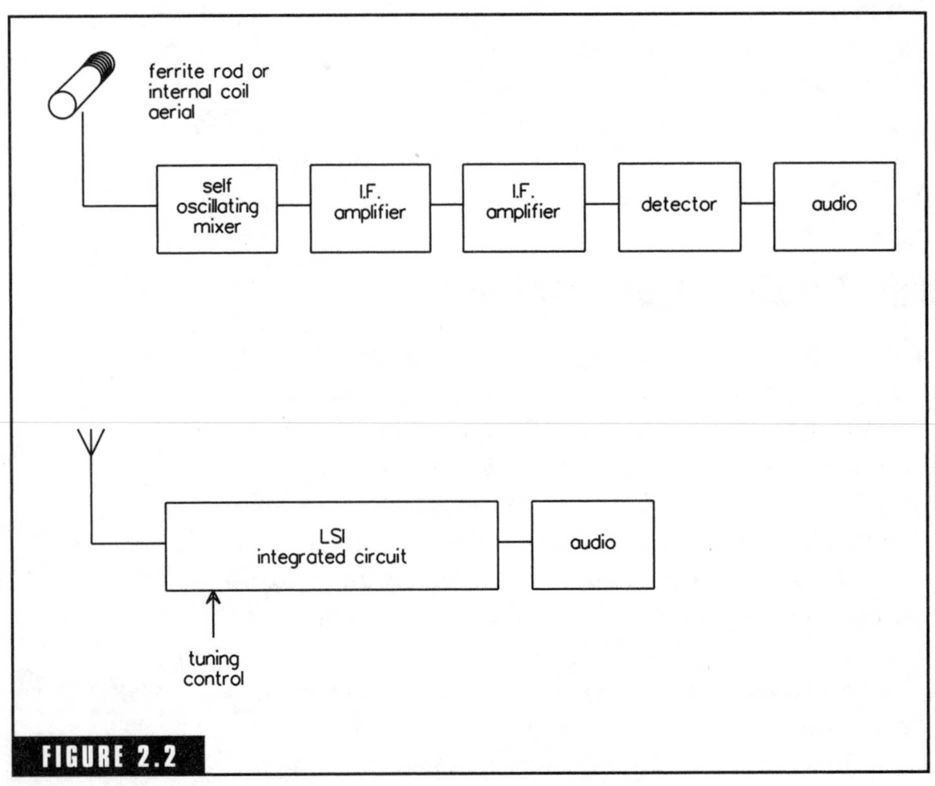

FIGURE 2.2

Typical low-sensitivity receivers.

receiver that includes a filter with steeply sloping sides on its response curve will do better than one that just uses a series of tuned circuits. See Section 9.2 in *How Radio Signals Work* for background and diagrams of tuned circuit and filter response curves.

The overload performance of a receiver and its responses to out-of-band signals may also be considerations. Cross-modulation, intermodulation, and image responses are all mechanisms that can involve signals at frequencies well separated from the broadcasting band. Overloading of the receiver may not be noticed in the desert island or remote area situation but should be borne in mind as a possible source of odd stray signals if the settlement or camp uses any radiocommunication or telemetry transmitters. In the environment considered in Section 2.5 overloading may be possible in some cases and will be the limiting factor on sensitivity in relation to Section 2.6. Overloading by out-of-band signals is more likely to be noticed in dual conversion sets, as shown in the second example of Figure 2.3, compared to simpler circuits due to the relatively large number of amplifying stages before the signal is limited to its final bandwidth by the adjacent channel filter.

One trick that may be worthwhile for reception of voices is to use a receiver designed for high-fidelity (HF) single-sideband (SSB) communications, which has the narrowest bandwidth possible for speech reception. Fiddly adjustment will be required because the SSB receiver has a carrier insertion oscillator, which will show the signal as a very loud heterodyne whistle until it is exactly

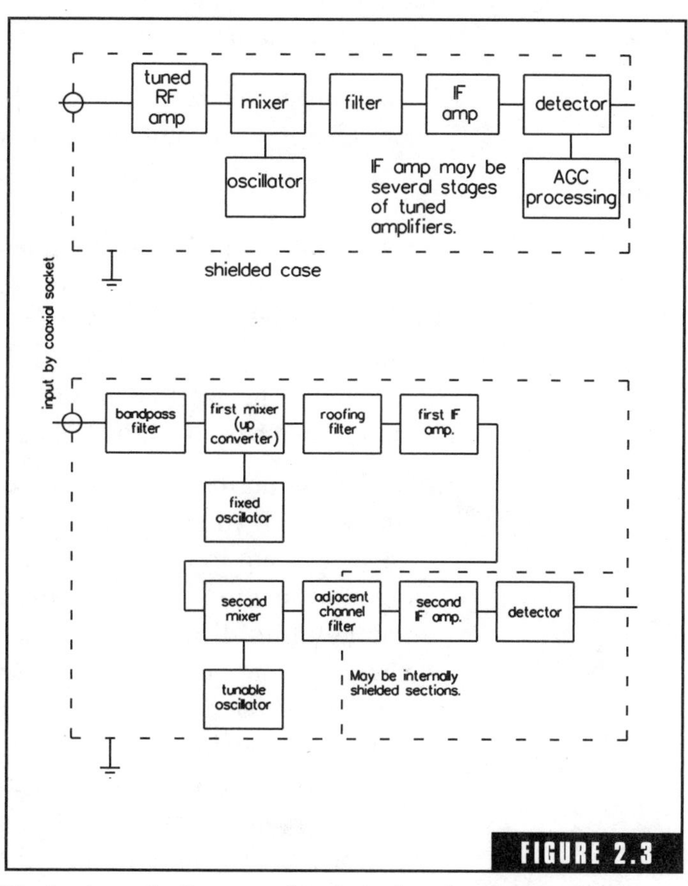

FIGURE 2.3

Block schematic diagrams of typical rebroadcasting receivers.

tuned, but once tuned correctly it will give the best possible resolution of a weak voice transmission. When you tune through an AM signal with an SSB receiver you will hear a whistle which decreases in pitch as the tuning is made nearer correct. When you are almost there the whistle becomes a low-pitched growl and then changes to a "growly" timbre on the voices. At the last stage before exact tuning is reached, when you are within about 5 to 10 Hz, the growly timbre changes to a fluttering very like the musical effect of vibrato. Exact tuning is reached when that flutter is reduced to the slowest possible rate; a steady voice is ideal. An SSB receiver also has the advantage that it is receiving only one sideband at a time so if you can select either upper or lower sideband you can pick whichever is clearest. If you need repeated access to one particular broadcasting station it would be an advantage if the local oscillator and CIO could be crystal-locked for long-term stability.

Two major sources of equipment for this type of service are:

HF transceivers for marine or land mobile use. These are all crystal-controlled but in some of the newer ones the operating channel can be programmed via a frequency synthesizer which can be tuned to include the broadcast band. If you are looking at equipment of this type, check that it can be programmed to receive both upper and lower sidebands on the broadcast band. There are also receivers made by the manufacturers of these transceivers which are identical to the transceivers but with the transmit components left out; they may be marginally cheaper than a complete transceiver and also have the advantage that a transmitting license is not needed for their operation.

Transceivers and receivers intended for use on the HF amateur bands. These are all tunable and all can be set to both upper and lower sidebands, but in many cases they are built for amateur bands only and modification to include the range of the MF broadcasting band may be quite difficult and expensive. Have a look at the service manual before you buy; in general, those in which the RF circuits of the receiver front end are simply switched from band to band with no provision for exact channel tuning will be difficult to change to the broadly tunable form of operation required to cover the full broadcasting range. If, however, you have a need to receive only one or two stations at particular frequencies, this may not be a problem.

If you are specifically interested in high-fidelity reception you will have to accept the limitation on sensitivity imposed by the wider bandwidth. In these cases a receiver that offers a selection of IF bandwidths will be an advantage so that you can make the best compromise between background noise and frequency response.

One source of good-quality equipment for high-fidelity service is receivers described as either "rebroadcasting" or "broadcast monitor" receivers. For the private operator new rebroadcasting receivers are ruinously expensive; they are professional pieces of equipment used when broadcasting stations want to receive a program off-air and use it as the program input for their own transmitter. But sometimes stations dispose of secondhand equipment which is still basically workable at prices not much higher than you would pay for an advanced version of a transistor portable. Check the front-end performance of any equipment offered as a "monitor receiver"; it could be an instrument more properly called a "modulation monitor," which is designed for direct connection to a sample of the transmitter output. These instruments have no amplification at all before the detector and almost no selectivity and are therefore not suitable for long-range use.

With regard to front-panel indicators, a display of tuned frequency or a logging scale which can be read to the nearest 1 or 2 kHz ***and is accurate*** will be an advantage. There may be times when you will have to set your receiver to a particular frequency without the signal being audible and then make other adjustments to maximize sensitivity or you must wait for changing propagation conditions to bring it in. At these times being sure of which channel you are listening on is a great advantage. A signal-strength meter is also an advantage but to be really useful it must either be an analog meter or a bargraph with at least 50 steps in the display. There will be times when to get the best out of your system you will need to make several trimming adjustments, each of which has only a marginal effect on the signal but all combined become very significant. The ability to see the effect of small changes is a great help.

A comparison between Figures 2.2 and 2.3 will indicate the factors that make the difference between a "local station" and a "Dx" receiver. The translation between these block diagrams and a detailed schematic circuit may not always be obvious but you should also be able to pick out differences in a list of specifications. Specifications for a Dx (voice communications only) receiver should be at least as good as:

Sensitivity better than 0.5 mV for a 10-dB signal-to-noise ratio

Selectivity (bandwidth)

Amplitude modulation: 6 kHz for 6-dB rejection, less than 20 kHz for 60-dB rejection

Single sideband: 3.4 kHz for 6-dB rejection, less than 10 kHz for 60-dB rejection

Intermodulation, cross-modulation, and spurious signals better than 70 dB below desired signal for all inputs more than 50 kHz different from the desired signal.

When you first get your receiver to its intended operating site do an initial check by just connecting it to the power supply and listening with no other connections. As you tune across the band you should be able to hear thermal noise which sounds like wind rushing or a waterfall. If you can hear anything else, such as clicks, plops, or whistles, that noise is interference which is either being conducted into the set via the power supply or its connecting wires or arriving by radiation from nearby power wiring. (It may be an indication of defective shielding in the receiver itself.) You will need to identify and remove all such interference before you proceed to the next stage of setting up. Section A4 of the Appendix, titled "Clean Power Supply," is devoted to that activity.

2.3 Times and Seasons

Once you have the receiver itself working correctly on site the next requirement to complete the installation is an effective aerial and earth system. The whole subject of aerials, feeders, and earthing is large enough for a chapter of its own and is the subject of Chapter 3. After installation of a maximum-sensitivity receiver and a correctly directed aerial in the right location, the next factor to consider is ***time of day.***

The time-of-day factor is related to discriminating between groundwaves and skywaves and to separating out the skywaves from different locations. Each day there is a horizon line of sunrises which sweeps over the Earth's surface at the speed of its rotation. Figure 2.4 shows how this works. There is also a similar horizon line of sunsets about halfway around the globe. Skywaves will propagate anywhere in the dark hemisphere but be quickly absorbed if they cross one of

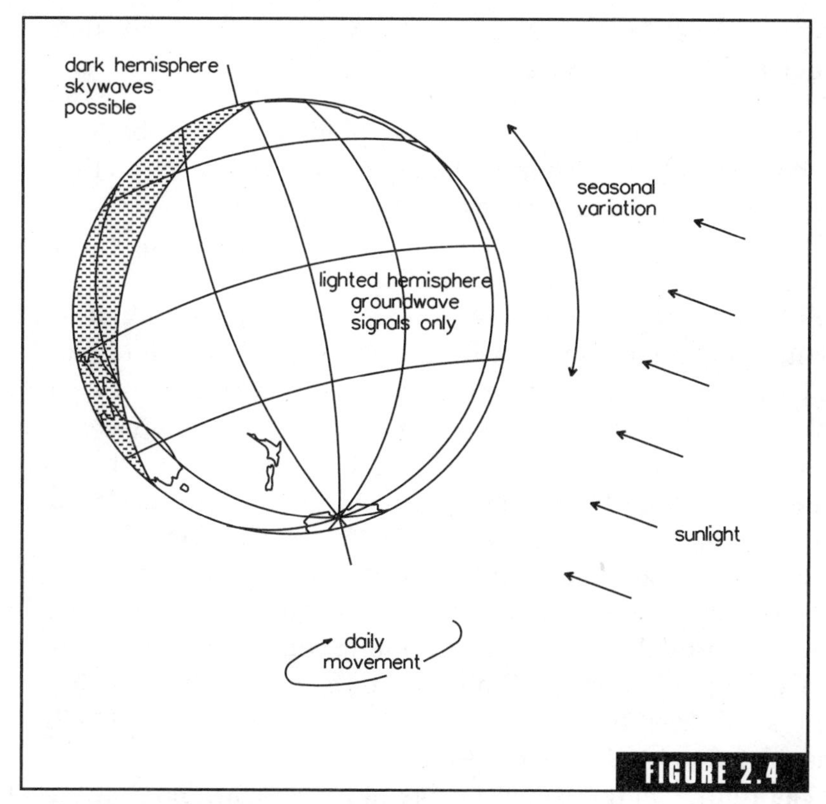

Earth with lighted hemisphere shown.

these horizons into the daylight. This will result in very obvious variations in strength and clarity of signals on a daily basis which can be used both to enhance the wanted signal and also at times to shade out interfering signals. The seasonal factors are:

The cyclic yearly movement of the lighted hemisphere as it moves back and forth over the Earth's surface between summer and winter. In some months of the year times of availability of particular stations and times of strong interference from other particular stations will vary in a regular pattern.

Variations in propagation and noise related to seasonal and weather conditions in the region local to the receiver. The band is always more noisy on summer afternoons (or, in the Tropics, during the build-up to the wet season) due to regional weather conditions which are always more conducive to convective clouds (i.e., thunderstorms) than at other times.

You may be able to use these factors to predict that your best chance of receiving signals from a particular region will be in a particular month of the year at a particular time of the day.

If your aim is simply to have a radio program to not really listen very closely to (this form of operation has been described as "audible wallpaper") and you are not fussy about just which station it is, in the daytime you are looking for groundwave signals and you will simply tune across the band until you find the strongest clearest signal and stick to that. At night the same station may suffer interference from the skywaves of more distant stations. Good performance is achieved with an aerial designed for groundwave reception, but it may need a few modifications of design to ***reject*** skywaves at night. On the other hand, you may be someone a long way from home who has an interest in news from a particular region. In that case you will probably be seeking a particular station and trimming your equipment for that purpose and trying to reject all others. You may find that the signal you want is only clearly heard for a few hours each day and possibly at a time related to local sunrise or sunset.

The timing and seasonal factors must be considered at the planning stage, particularly in association with any directional properties that are planned for the aerial system.

2.4 Dealing with Strong Local Signals

A typical situation where this applies is when someone lives in a country town or provincial city with one or two local broadcasting stations and they wish to hear stations from a distant capital city or they live in one major city and have an interest in signals from another one. If you are in this situation your degree of success will be largely controlled by a specification of the receiver that is often not well-documented. It is the range between strongest and weakest signals your receiver can deal with; its technical name is *dynamic range.*

The previous section dealt at length with receiving the weakest possible signals; this section presents the mechanisms that limit a receiver's ability to handle strong signals. That limit is set by the ability of amplifying and rectifying devices to handle large signals without distortion. In a tuned RF or IF amplifier distortion has two major effects: it may be seen as *cross-modulation* or there may be extra signals generated by *intermodulation*. The difficulty with specification is that because the receiver is tuned, the effect of strong unwanted signals depends on how far off tune they are as well as on the actual number of millivolts or volts applied to the aerial terminal.

For cross-modulation, a typical way of specifying performance may look a bit like a figure for distortion expressed as a percentage of modulation (between 3 and 10% is typical) due to an unwanted signal of a certain level that is a specified number of kiloHertz (commonly about 50) off tune. The level of the interfering signal may either be expressed as an absolute number of millivolts or as a decibel ratio between wanted and unwanted signals. Cross modulation is heard as if it was another station on the same frequency interfering with the one you want to hear but it is not another transmission because it disappears if the wanted station goes off air.

Intermodulation is the production of sum and difference frequencies by pairs of strong local signals. A receiver's sensitivity to intermodulation also depends on the frequency relationship between the channel it is tuned to and the frequencies of the strong signals. Intermod products can appear in blank channels; there does not need to be a desired signal to be compared with so the specification will read as no signals over a certain level (usually a small number of microvolts) for any combination of signals up to a specified level a certain number of kiloHertz off-tune. Intermod products can be heard as extra stations but not clearly because they always carry the programs of both the original stations, usually with fairly severe distortion of the modulation.

For both cross-modulation and intermodulation the performance of the system is determined by selectivity of passive components before the first active components (diodes and amplifiers) of the receiver. A receiver that has only one tuned circuit which serves double duty as an input coupling transformer and feeds signal straight into a mixer as shown in the schematic diagram of Figure 2.5 will never be very impressive when dealing with strong local signals.

At the other end of the complexity scale, if there are several tuned circuits with only loose coupling between them (for maximum selectivity) feeding to an RF amplifier with another tuned circuit in the coupling to the mixer, that receiver may give very good rejection of

mixer (transistor or i.c.)
output to IF amp
aerial
earth
local oscillator

FIGURE 2.5

Schematic diagram of the front end of a basic receiver.

strong signals. For an illustration of this compare the circuit shown in Figure 2.5 with that of Figure 2.6. If, however, those tuned circuits are electronically tuned with circuits that use varactor diodes then those diodes are themselves active devices which will limit the strong signal performance of the set.

In reference to Figure 2.6 the design rules for efficient transfer of energy through coupled tuned circuits, as shown here, are given in Appendix Section A5. In this circuit the capacitive coupling should be less than critical to gain immunity from overloading.

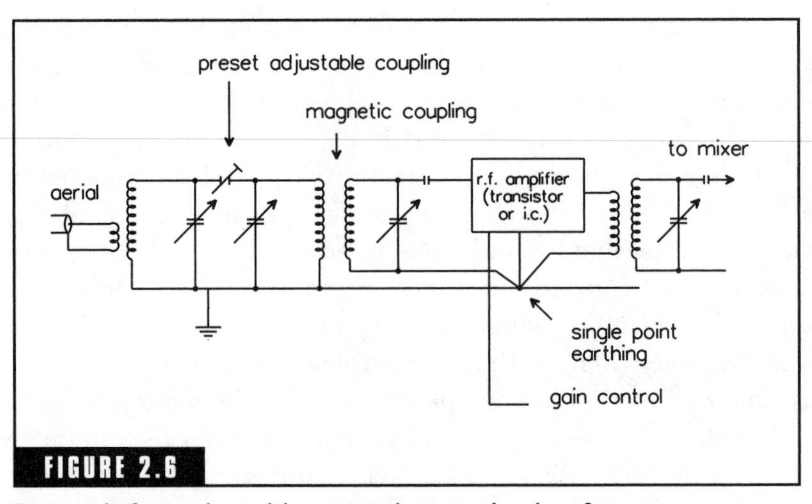

Front end of a receiver with very good strong signal performance.

Unless you live obviously very close to a powerful transmitter you will probably find that strong signal problems are something you become aware of after you have gotten your system running close to its maximum sensitivity. People commonly buy a good receiver and initially put up a small aerial and have clear but weak signals; they then try a larger aerial and the trouble starts. It is tempting to simply reduce the signal input on the basis that the weak signal limit is set by atmospheric noise which is many decibels stronger than the internal noise of the set, and if you reduce everything (perhaps with a series resistor or potentiometer, as shown in Figure 2.7) until the atmospheric noise matches the level of the set noise you lose nothing at the weak signal end of the range and gain immunity from the strong signals.

In some cases a minor change in weak signal performance corresponds to a dramatic reduction of interference; the response to a distorting signal is nonlinear so a 1-dB change in input level often will correspond to a change of 3 to 4 dB in output of the intermodulation product. This is the basic ploy used by motor vehicle manufacturers in design of broadcast band car radio systems and, in a few cases where the overloading is only marginal, a measure as simple as that may be all that is needed.

There are, however, many more effective ways of using the headroom between atmospheric and internal noise to gain immunity from strong signals. The basic aim is to use it up as insertion loss of a filter or preselection tuning system so that a reduction of, say, 10 dB in the wanted signal may correspond to reducing the strong signal by 20 to 50 dB.

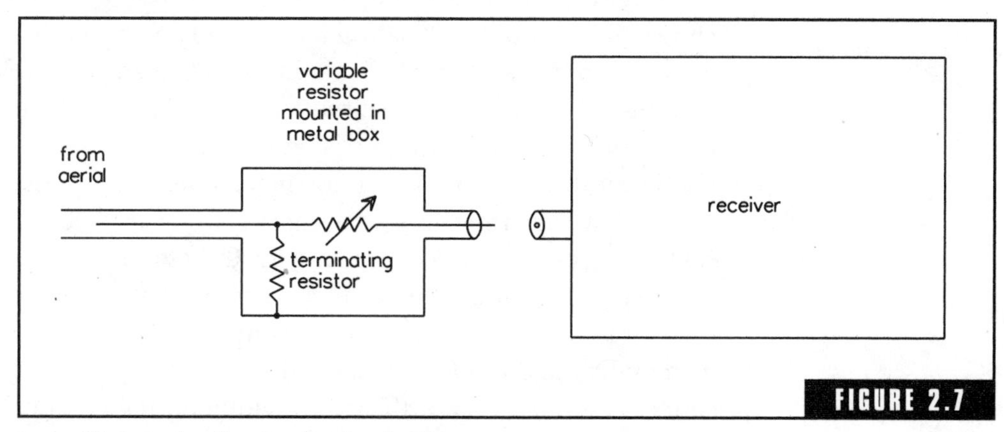

Controlling overloading by signal reduction.

Before you start too much detailed design it is advisable to think a bit about how much reduction of overloading you actually need to achieve interference-free reception. A single tuned circuit cannot be expected to give more than about 30 dB rejection, even for signals well off-tune (i.e., half or double the frequency), and its rejection of signals close to its tuned frequency will depend on the "Q" factor of the circuit. If you calculate you need about 15 dB rejection of a signal 45 kHz away in the spectrum, you could probably achieve it with a single tuned circuit with loose coupling. (Tight coupling will lower the working Q of the circuit; one of the respectable ways of gaining a marginal improvement in selectivity is to design for loose coupling to raise the working Q of each circuit.) If, however, 25 to 30 dB rejection of the local signal is needed, you will certainly need two tuned circuits and for over 30 dB you should design for three circuits and so on.

The Q factor is the ratio between reactive and resistive components of the circuit, and the same number also is a direct measure of the relative bandwidth of the response. A circuit with a Q factor of 100 will give 6 dB rejection of signals 1% off-tune and a circuit with a Q of 10 will give 6 dB rejection of signals 10% off-tune. Almost all tuned circuits will give a selectivity response curve the same shape as the curve shown in Figure 2.8; the effect of changes of Q factor is to make the curve wider or narrower. For receiving system circuits a good practical figure to aim for is a Q factor of about 50, which for a transmission of 1 MHz will give 6 dB reduction of signals 20 kHz off-tune and 14.4 dB reduction of signals 50 kHz away. Figure 2.8 illustrates this selectivity curve in a graph.

In Chapter 3 a tuned matching transformer is described in Section 3.3 and shown in Figure 3.8. It is a good way of incorporating one extra tuned circuit in a system that is marginally affected by overloading. If you need the effect of two tuned circuits, an extension of the tuned transformer function, as in the circuit of Figure 2.9, can be used and the coupling between the two circuits can be made adjustable to maximize the separation between strong and weak signals. In the circuits of Figures 2.9 and 3.8 loading can be reduced (raising the Q factor of the tuned circuits) by constructing the aerial coil with less turns compared to the tuned circuit coil, by reducing the value of the coupling capacitor in Figure 2.9, and by moving the output tap toward the earthed end of its tuning coil. Appendix Section A5 has further information on the practical design of filters using coupled tuned circuits in this way.

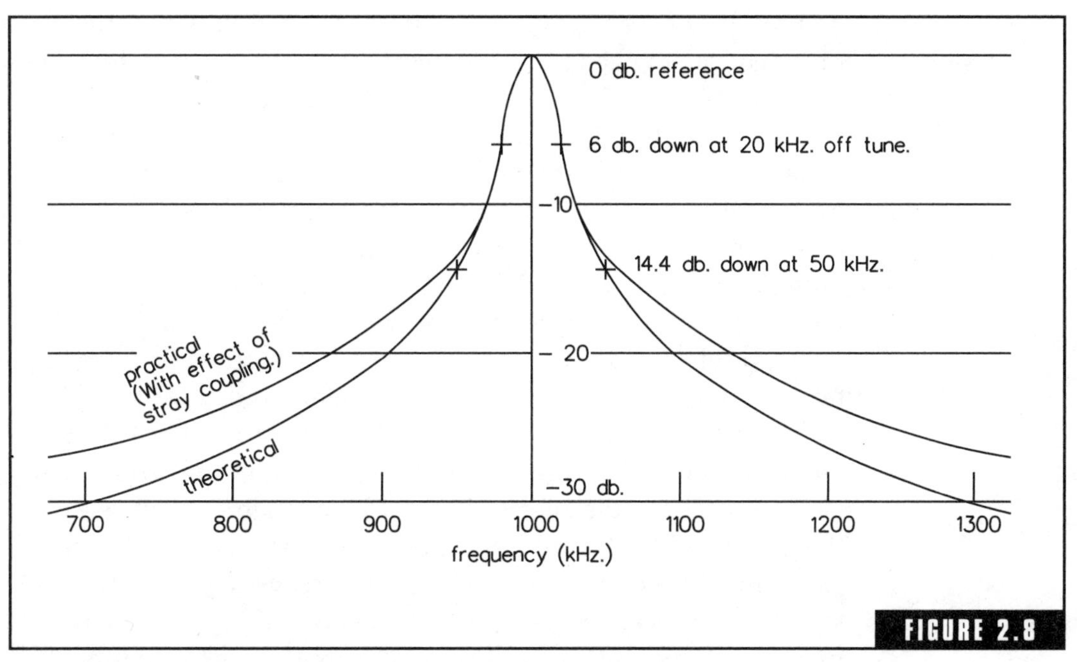

FIGURE 2.8

Graph of response of a single circuit with a Q of 50 tuned to 1 MHz.

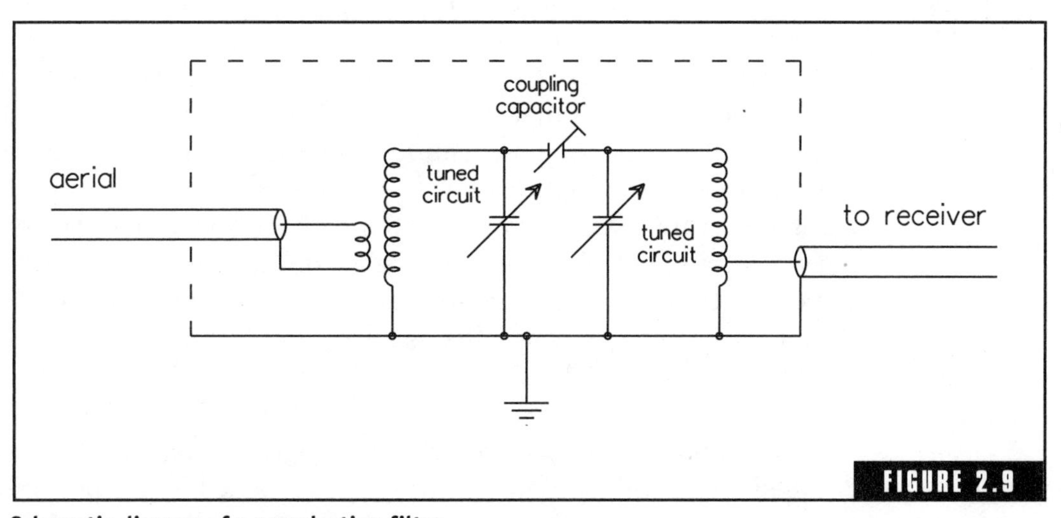

FIGURE 2.9

Schematic diagram of a preselection filter.

Preselectors can also be made in the form of PI (Π) couplers. On the MF broadcasting band a PI coupler will not have a very high Q if constructed using tuning capacitors in the normally available range and will be difficult to tune across the band if constructed using fixed capacitors. PI couplers are most useful for matching impedances. For accurate adjustment, some measurement of either directional power flow or standing wave ratio is needed; they are regularly used in transmitters where the output power is available to drive the power-metering circuit. More detail is provided in Chapter 3.

A preselector, as shown in Figure 2.9, which really is a simple filter will not work well unless it is constructed in a metal box with coaxial cable input and output, and the coils need to be either individually shielded or aligned for minimum coupling. When two tuned circuits are used the selectivity will be enough to reduce a weak station to inaudibility until the preselector is close to exactly tuned. Tuning will become a three- or four-step process; you will first have to set the main receiver dial to a close position then peak the preselector using band noise as the indicator of tuning. Step 3 is to tune the main receiver for the wanted signal and, finally, if needed, recheck the preselector for exact maximum sensitivity.

If you need the selectivity of a third tuned circuit before the receiver input, it would pay to look for some selectivity in the aerial itself rather than try to build too much selectivity into one box. This is because of signal leakage around that box. If the box contains two tuned circuits each of which can offer an ultimate figure for selectivity of 30 dB, you expect when you combine them to have 60 dB rejection of signals if they are far enough away in the spectrum. Rejection of 60 dB means that the power allowed out of the filter is 1 millionth of what is at the input. There are many ways (some are shown in Figure 2.10) in which small fractions of the signal may be coupled, even within a shielded box, and the result may be that instead of the expected 60 dB of ultimate rejection, there may only be 40 to 50 dB, and because the leakage signal is being coupled around the tuned circuits, the addition of extra circuits will provide no improvement at all.

Tuned loop aerials have selectivity and also offer the advantage of a rejection notch in the horizontal polar diagram. There is a diagram of a tuned loop aerial in Figure 3.12. They are loaded by the power drawn off by the coaxial cable, so the Q factor may be as low as 10; this means

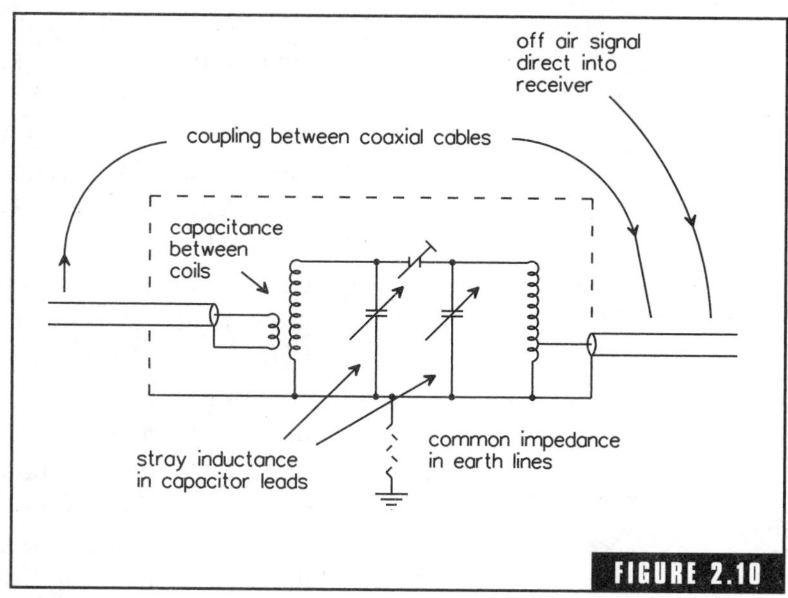

Circuit of a prefilter showing leakage paths.

they will have little effect on signals close in frequency to the desired one, but for signals well enough separated in the spectrum they will give the 30 or so dB ultimate rejection of a single tuned circuit.

In cases where the overloading results from one station or two which are close in the spectrum there may be opportunity for a trap circuit to be permanently included in the coupling components between the aerial and the feeder. This can either be in the form of a parallel tuned circuit in series with the feedline arranged, as shown in Figure 2.11, or as a "T" notch filter, which with cunning design could be formed to have an impedance-matching function in addition to the overload rejection.

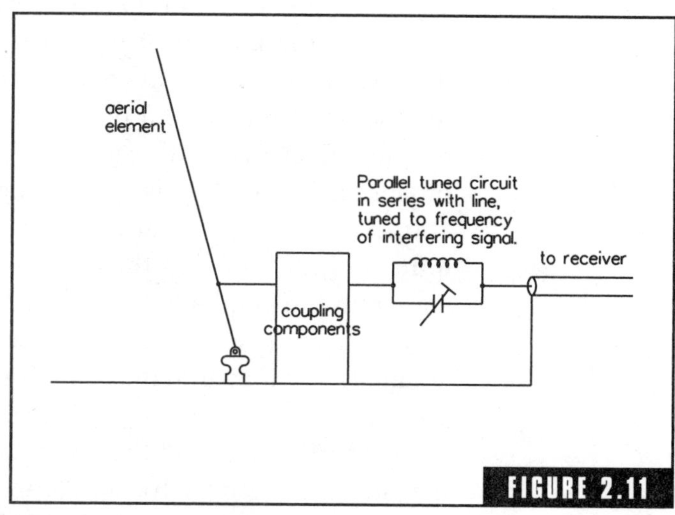

A trap in the aerial lead.

The T notch circuit of Figure 2.12 works by splitting the signal

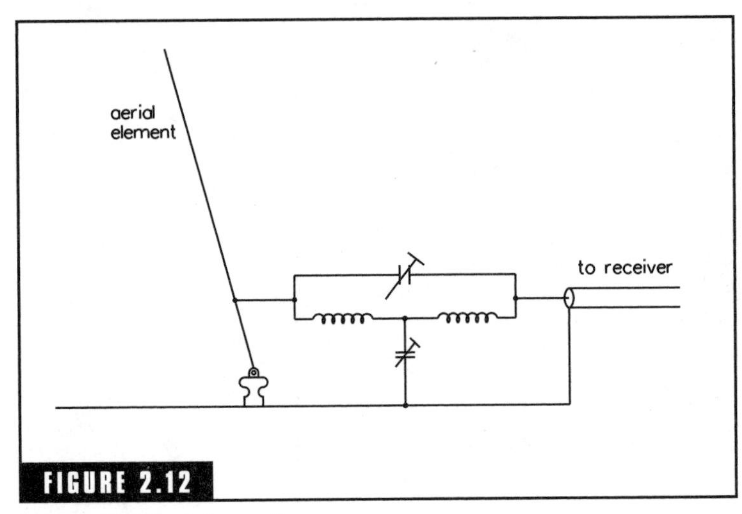

FIGURE 2.12

Circuit of a T notch coupling.

into two components and recombining them in such a way that, given the appropriate impedance conditions, at one particular frequency they will be exactly equal and of opposite phase, which in theory means complete rejection of that one frequency. A T notch may be able to give greater reduction of a particular frequency than can a tuned circuit trap, but its response is so sharply defined in frequency that on the MF broadcasting band it may only give rejection of the carrier but not the sidebands or perhaps maybe one sideband but not the other.

If you require to receive a station close in the spectrum to the rejection frequency of a T notch using a receiver which has a filter (flat-topped frequency response) as the main component for adjacent channel selectivity in the IF amplifier, you may find it is capable of causing a form of distortion to your signal which may be hard to trace if you are not aware of its cause. Although the rejection notch is very narrow at the bottom of its dip it has a skirt of selectivity which is quite wide, and if you are able to do signal-level checks while switching the notch in and out of circuit you would probably be able to detect its effect over a range of spectrum equivalent to about half the broadcast band. For signals in the range from approximately 20 to 100 kHz separated from the notch frequency the filter will create a quite definite slope on the overall selectivity curve of the receiver, as shown in the graphs of Figure 2.13.

If you set the receiver to the "correct" tuning point for one of those channels you may find that signals transmitted with high-level (close to 100%) modulation in the treble register can have the level of the carrier reduced by the slope of the notch selectivity so that the relative level of the sidebands is over 100%. These sidebands are severely distorted by the detector; the distortion sounds like overloading of the audio stages

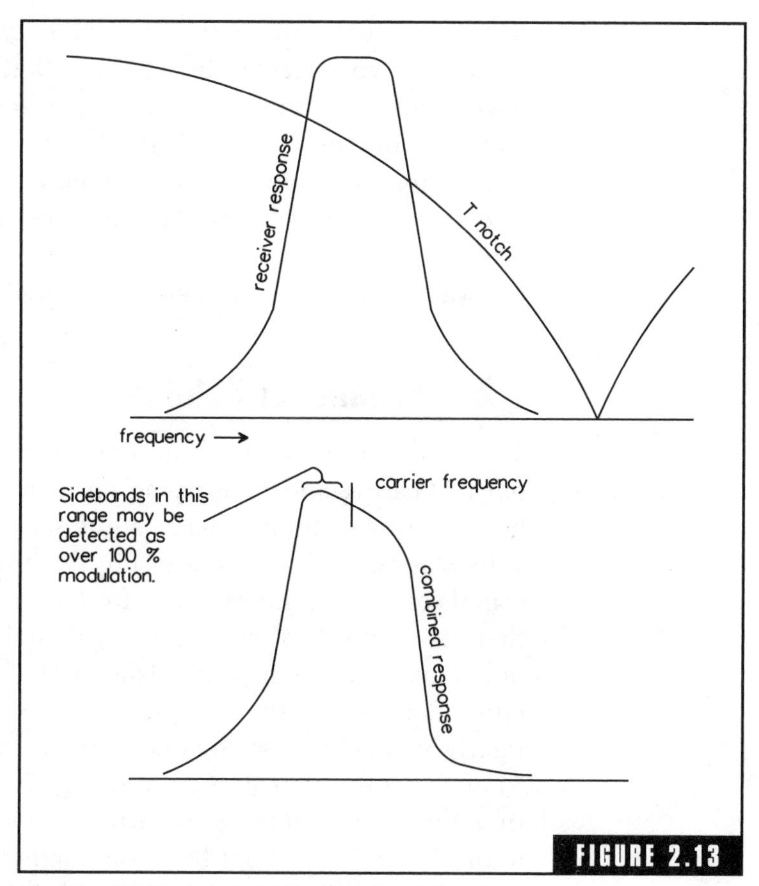

FIGURE 2.13

How the "sideband shriek" effect is produced.

but is not cured by reducing the audio volume. This effect is called "sideband shriek." It is most noticeable when the receiver has a filter in the IF amplifier with a flat-topped passband and steeply falling sides for maximum adjacent channel selectivity and is less obvious when selectivity is provided by a series of tuned circuits which give a more rounded response. It is not seen at all when the carrier is locally generated by a "carrier insertion oscillator" (as in an SSB transceiver).

If your operating site is very close to a high-power transmitter station you will gain some immunity from it by placing your aerial at the point on your block furthest from the aerial of the transmitter. The field strength close to that aerial is high enough to induce substantial power flow into metal objects of all sorts and if any of them have rusty joints the rust will rectify the radio signal to produce both audio- and radio-frequency intermodulation products. It is not at all uncommon for the operators of high-power broadcasting transmitters to go outside on nights which have no wind and a little bit of dew forming and be able to hear rusty barbed-wire fences whispering sweet nothings to anyone who will listen.

These intermods must be reradiated before they can affect your receiver, and in most cases the objects are not shaped to radiate efficiently, so the signal only goes a short distance. But there are many hundreds of sources of them, so the most effective way to avoid their effect is to select a site which is far enough away so that they are not a problem. If you select a site which initially works well but later on you have problems with intermodulation, check first for rusty metal objects close to your own aerial. The physics of the process of transferring power from one aerial to an intermod generator and then inducing it into another aerial means that the most likely place for the intermod generator is close to one or other of the aerials; objects which are halfway between the two aerials are less likely to be sources of trouble.

2.5 Broadcast Band Dx

There are people scattered around the world who have made a hobby out of hearing and identifying far away (i.e., from other continents) broadcasting stations. These people seek signals from the furthest possible stations, which usually implies they are receiving the weakest signals often with interference from stronger, closer stations; however, there is no need for concentrated listening for hours at a time—their only need is to identify a callsign and a short segment of the program either side of it. Other people live as expatriates or have friends and relatives in another state or country and want to hear news broadcasts from that distant region or are sporting enthusiasts who want real-time on-the-spot reports of particular events. Their needs are similar to the Dx enthusiast but in a less intense form. If you belong to any of these groups of people you will definitely be interested in the intelligence content of the program, and you will probably only be interested in relatively short (perhaps half-hour or so) segments. All the considerations of directional response and timing mentioned in Section 2.1 and explained in more detail in Section 3.11 will apply equally well to you. For the expatriate or sporting enthusiast there is one additional factor in your favor. In most cases broadcasting stations are part of a chain and you may be able to find the same program at other spots on the dial. It is the program you are seeking not the individual transmission, so you have the option of trying all of them and picking whichever is clearest.

Broadcast band Dx-ing is a specialized hobby and there are clubs devoted to it; if a club is available to you then it will greatly increase your chance of success to be an active member of it. If you must work on your own, however, the steps to success are:

Initially collect as much information as you can; listen on the band to see what is around and find a few stations that you think might be very distant.

Contact stations by letter and ask for a program guide which includes operating frequencies and standard times when callsigns are transmitted. When the guide arrives check that what you are hearing matches with the advertised program.

Adjust your receiver and aerial system for maximum sensitivity in the direction of the target station using the techniques described in Sections 2.1 and 2.2 and then eliminate overloading using the techniques of Section 2.4.

Use time of day for maximum discrimination of the wanted station (refer to Section 2.3).

Listen for a long-enough period to train your ears to the announcer's voices.

When you can identify the callsign, send a signal report that is useful to the station management. If possible include a short recording of the off-air signal.

This should be rewarded with an acknowledgment that confirms what you heard. Your quest is only successful if they reply with confirmation, and that may depend on whether you can present your letter to them as a worthwhile contact. To achieve that you need to include in your report some information that is valuable to them. The basics, of course, are your location and the time and date you heard their callsign, but also give them a report on the strength and clarity of their signal, whether there was interference, and, if possible, the callsign and location of the interfering transmitter. It is an aid to authenticity if you can give them some detail about the programs either side of the callsign; ideally with some information you have heard which was not advertised in the program guide.

A person with no particular training in listening will find they just become aware of signal degradation when the signal-to-noise ratio is about 40 dB or distortion amounts to about 1% and lose the intelligibility of the message if the signal-to-noise ratio is worse than about 10 dB or if the combination of harmonic distortion and intermodulation products is greater than about 25 to 30% of the signal. (In both these cases the figures represent about the same total power of nonintelligent sound in the final audible signal.)

By concentrating our attention on particular aspects of the sound we can train the information-processing function of our ear–brain combination for much greater performance in either one of two directions. In one direction the people whose job is to monitor a broadcasting transmitter have no interest in the content of the program but are vitally interested in the distortion and noise content of the transmitted program. Those people can, with experience, become so sensitive to these factors that they can totally ignore what is being said and have no memory of the content of the program but be instantly aware of new forms of distortion, intermodulation, or noise at levels which correspond to distortion meter readings of much less than 1%. In the other direction people such as ships' radio operators or aircraft pilots, whose job is to listen for the intelligent message in a noisy signal, can train themselves to be almost unaware of noise and distortion even as high as 3 to 4% of the signal and to be able to follow the intelligence of the message when the signal is so weak that noise and interference are of equal level (3-dB signal-to-noise ratio).

The mental process is similar to that of being in a crowded room and choosing which conversation to hear and which to ignore. When you have trained your ears to listen for message content you will find you have made a semipermanent reduction in your ability to hear distortion and intermodulation; we each have a range of sensitivity to either intelligence of the message or to distortion, and you can, through training, shift your range along the scale in either direction.

2.6 Diversity, ISB, and Multiplexing

If you are a transmitter station manager and your interest in receiving programs off-air is in order to rebroadcast them on your own transmitter, most of the techniques for gaining maximum sensitivity of a weak signal (except possibly for frequency response limitation) will also work to give the best possible signal-to-noise ratio of a stronger signal. There is also another technique you can use which will not normally be practical for the hobbyist or Dx enthusiast, and that is diversity reception. This may take any one of several forms and different forms of diversity can be used in combination. The basic principle is that you arrange for the signal to be received by several different paths and then select the strongest and clearest signal from all those available. The following list shows some of the major ways in which diversity reception can be arranged:

If the same program is available from two transmitters on different frequencies then two identical receivers can be used with selection for the strongest signal. Each receiver's AGC line gives an indication of signal strength and these can be compared using an electronic switching function.

An amplitude-modulated signal contains identical information in each of two sidebands and either one contains all the information needed for the receiver to reproduce the original audio program. In cases where the signal suffers interference from a narrow-band noise source or another program there may be a significant difference between the clarity of programs from each of the sidebands. In those cases it is worthwhile to receive each of the two sidebands independently using single sideband techniques and then select the clearest signal from the two available.

If any form of multipath fading is to be avoided then receivers placed at different physical locations will show fading interruptions at different times. Separation between sites needs to be at least a couple of wavelengths for good effect. With identical receivers the outputs can be selected either for signal strength by comparison of AGC voltages or for signal-to-noise ratio.

A complex system can be arranged which combines any or all of the above methods along the lines shown in Figure 2.14.

In a diversity reception suite two (or more) receivers and an electronic switch are used, arranged as shown in Figure 2.14. If the receivers have the facility for ISB reception, then the audio output of both sidebands is extended to the diversity switch. Indication of each receiver's signal strength is provided to the diversity switch by analog DC signals, which usually can be derived from the AGC line of each receiver.

If you are assessing a receiver for rebroadcasting or any similar high-priority use and have any thought in the back of your mind that one day you may wish to use it as part of a diversity suite, the facilities you need to check for are:

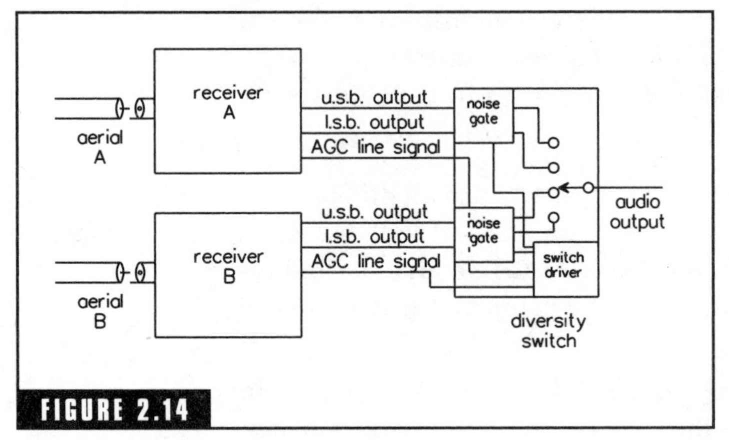

FIGURE 2.14

Diversity switching general arrangement.

- an external connection which carries a voltage indicating either AGC line conditions or signal strength
- if the receiver has the capability of ISB reception extension of the audio output of both sidebands to an external socket

Neither of these functions are particularly expensive to add to a receiver at the design stage.

The cost of including independent sideband capability may be significantly higher. SSB reception is well established and works well for communications-quality voice circuits but is much more difficult and expensive to implement with higher quality audio as is required for a rebroadcasting program in an MF or HF broadcasting relay station. For communications receivers the carrier insertion frequency needs to be close but there is several-Hertz tolerance and the message can be passed even if the CIO is misaligned up to about 50 to 100 Hz. For rebroadcasting the tolerance is much smaller. In theory the carrier is available to be used and would give an exact reference frequency, but any change of relative phase between carrier and sidebands will cause distortion of the audio output. To reduce the effect of these factors the amplitude-modulated signal can be translated to an IF in the 50- to 100-kHz range before sideband demodulation. That step means that frequencies can be much more closely controlled and filters made with more tailored response curves than is possible at higher frequencies.

The economics of diversity and/or ISB reception will depend partly on what use you intend to make of the final signal. If your interest is in short segments, as, for instance, a sporting program that draws program inputs for particular events off-air from a variety of sources, you will probably find your audience will tolerate some degradation of audio quality for the sake of rapid access to on-the-spot reporting; on the other hand, if you are setting up a station for long-term use as a repeater, then the required standard for audio quality will be much higher and a more complex diversity scheme will be justified.

AM broadcasting signals may have extra information added to them in the form of multiplexed signals. One process that is common is to phase-modulate the carrier with only a few-Hertz deviation to provide a "difference" signal for transmission of stereo effects. The spectrum space available for deviation is only very low, so in general it cannot be used for a completely separate signal, but it can be used to carry the information required to provide a worthwhile degree of stereo separation. AM stereo broadcasting requires a stronger signal for clear demodulation than the minimum needed for reception of the signal in monophonic form.

This form of multiplexing and ISB reception cannot be used in the same receiver at the same time. If ISB is referenced to the carrier frequency it will either prune-off the stereo information or attempt to follow the short-term variations in carrier frequency and cause that information to be intermodulated onto the AM signal in a very objectionable form.

Another form of multiplexing that is occasionally used is to add a very low-level subcarrier at the high-frequency edge of the passband of the modulating signal. That subcarrier can be either amplitude- or frequency-modulated with a few bits per second of data which can be used as control signals for a relay transmitter or other remotely sited installation. The relative level between the main carrier and the subcarrier may be from −20 to −40 dB depending on the exact frequency of the subcarrier relative to the edge of the passband of the transmitted signal. A level of −40 dB would be appropriate for a tone clearly within the passband which is expected to be transmitted and received without losses. The higher levels would be appropriate for tones which are sited in the spectrum on the slope of the frequency-response curve where some losses are expected in the transmission process. Signals in that form can be used in conjunction with ISB reception.

2.7 When the Band Is Totally Congested

This is liable to be the situation that applies in the middle of big cities in places such as multistory apartment buildings where the neighbors all use computers of a variety of different types or in large industrial plants where there are many different types of electrical machinery in use. In most cases the congestion of the band will be due to electrical (including digital data) noise pollution rather than a multiplicity of intelligent purposeful signals.

Is Another Location Possible?

If you require to receive a distant broadcasting transmitter from a location of that type, recognize that you can't make a silk purse out of a sow's ear, and first of all ask that question. If your purpose is not absolutely committed to that location it may be possible to install a fixed tuned receiver at a remote site and bring the program to the site as an audio signal.

Would Another Frequency Do as Well?

If the answer to the first question is quite definitely "No," then you should also question the purpose of your attempt to receive that station. If you only seek the program and not the particular transmitter, contact the network managers and get a complete listing of all the stations that carry the same program and try them all. Don't forget that in many big cities there is also a VHF FM service which will almost certainly give better results in the severely noisy environment of a city center.

If after that you really are sure that you need to be able to hear that particular transmitter from this particular location (you may, for instance, be the station manager wanting to verify transmitter output and modulation from a city center office without the involvement of telecommunications landlines), then directional aerials and very severe restriction of bandwidth are the final ultimate weapons available to you. (Aerial directivity in particular relation to the MF broadcasting band is dealt with in the next chapter.)

Proving the Carrier

The existence of a carrier can be verified with an AM receiver set to less than 100 Hz bandwidth. If the transmitter is close enough so that the groundwave signal will always predominate, then it is possible to keep a fairly reliable watch on transmitter output power with a receiver of this bandwidth with manual or preset gain control (i.e., with AGC disabled). An output lead carrying a DC indication from such a device could be taken to a contact voltmeter and used as a backup for the carrier fail and aerial fail alarms. Note that an off-air signal such as that is not reliable enough to ***substitute*** for any of the alarms from the transmitter site and should not be connected to any after-hours call-out system, but it does offer a second opinion which checks the alarm sensors and any telemetry systems in use.

Proving Modulation

You can prove the existence of modulation using the single-sideband principle with audio bandwidth of only a few hundred Hertz. SSB transceivers designed for maximum intelligibility in severely noisy conditions are arranged so that the IF is filtered to give audio response from 300 to 2400 Hz and equipment of that type is a good starting point for this class of service. You will need to ensure (by prefiltering or front-end attenuation) that such a receiver can be used at your location without suffering from intermodulation or cross-modulation. If you are sure of that and require to restrict the audio bandwidth more severely than for a normal SSB channel, you could process the signal using a separate audio low-pass filter with an adjustable cut-off point. There is probably not much to be gained by changing the bass end of the response from 300 Hz, but there is a definite gain possible at the treble end.

Restriction to less than 2400 Hz will affect the intelligibility of the signal itself, but under the worst noise conditions the upper frequency could be made as low as about 800 Hz and still give a signal that is good enough to identify a callsign or a particular piece of theme music. The bandwidth that gives the best compromise between intelligibility and noise will vary with the conditions that apply at the time, so the audio low-pass filter needs to be made with an adjustable cut-off frequency.

Noise Reduction

Whatever definable characteristics the noise has can be used to reduce its effect. Impulsive noise can be reduced with a noise limiter in the IF amplifier of the receiver; the chopper types which switch off a tiny section of the program each time a noise pulse is detected work better than those which allow the noise pulses to go through and simply limit the audio output volume to a predetermined level (there is an explanation of this in Chapter 4 referenced to Figure 4.4). Noise that is produced by power mains-operated machinery is liable to be closely associated with the local mains frequency and its harmonics and can be reduced with audio notch filters. If the signal includes a heterodyne from another transmission, that also can be reduced with a notch filter on the audio output. Noise that comes from a multitude of unrelated sources, however, will show most of the characteristics of thermal noise, so there will be an unavoidable limit to how much you can achieve in any of these directions.

Because there are probably not many other people who want exactly the facilities you require for the ultimate in performance at your site, you will probably find you cannot buy exactly what you need off the shelf. The best you will be able to do is buy some of the functions as components of a complete system but the experimenter's old adage that you can build better than you can buy will definitely apply to operations of this type. You will almost never be able to design something that works perfectly just by theory alone; your first design will probably work but there are a lot of minor adjustments and trimmings that will improve the final result. You will need to mould and trim your system to your particular purpose, and that trimming will be a process that goes on as long as the system is in service. In many cases the measures suggested in this chapter will only give a minor effect on the signal. At the same time as you are trimming and testing there are also natural variations going on which will be superimposed on the results of your tests. Unless the change you are testing has made a gross change in your system's performance, you will need to ensure that you carry on the test for long enough to average out the natural variations and isolate any changes as the result of what you have done. If you are planning a significant reconstruction it would be an advantage to have the old and the new working side by side arranged so that you can switch between the two for direct comparison.

The system for ultimate possible performance under extremely noisy conditions will probably take a form similar to the functional block diagram of Figure 2.15. Setting up a perfect system can take considerable time. In truth, you will probably never have a perfect system; your approach to perfection will be something like a time-constant graph where as time goes on smaller and smaller changes will take longer and longer to achieve.

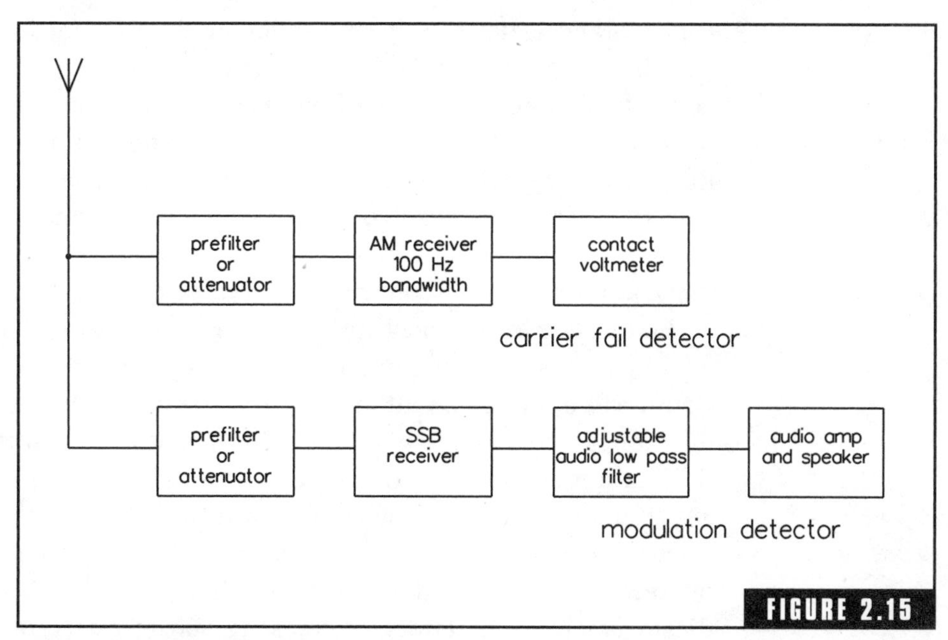

Office-based off-air transmitter monitor.

2.8 Repairing Working Systems

This section is intended for radio technicians or other service personnel and is written on the assumption that you are called to fix a fault as a commercial transaction on equipment which is owned by another person. However, the depth of technical knowledge required is not beyond the scope of equipment owners with a hobby interest in electronics.

If you operate in a country area and a regular client of yours buys a new, fancy receiver, try to obtain a copy of the service manual for it as soon as possible. You will need it sooner or later and it will be easier to get now rather than in some years time when it has become an archival item. Better even than buying the manual yourself, encourage your clients to make inclusion of a service manual part of the original purchase. Even if you do not intend to do work on the equipment itself at the component-replacement level, there are many items on the market these days for which you require test procedures which are detailed in the service manual just to do proof-of-performance tests.

When you receive a service call to a high-performance receiver installation try to establish first of all whether the complete station has ever worked satisfactorily and whether the present fault is intermittent. If it has never worked as well as was expected, the next step may be to check it against some of the design principles mentioned earlier in this chapter. If the station worked as expected at any time do not think about redesign; your task is to find out what has changed and correct it. If the report suggests a possible intermittent, the next step is to try to establish the conditions under which the fault is present—you cannot fix a fault you can't observe.

Before you disconnect anything for detailed tests try to establish whether you should reasonably expect a signal to be there at that time. Could the operator be attempting to receive a skywave signal in the daytime, for instance, or could it be that the transmitter is off-air—does tuning across the dial bring in other stations? Take close note of any reported effects of weather conditions; faults that clear themselves during rain and gradually return after a day or so of dry weather often indicate corroded joints; faults that appear during rain often indicate cracked insulators or tracking across dusty ones; faults that gradually get worse during long dry spells suggest problems with earthing. If initial observations of this type don't lead you to something obvious, then you will need to disconnect parts of the system to isolate the fault to a particular section.

Start at the point of connection of aerial and earth. Disconnect them and operate the receiver at maximum sensitivity—you should hear some thermal noise, which is a fairly steady roaring noise a bit like rushing wind or a waterfall at a middle distance. Repeat the initial setting up test described in Section 2.2, but this time listen for atmospheric noise. If you are not sure you are seeing any effect, connect a multimeter set to a low voltage AC range across the loudspeaker terminals. When the sensitivity of the receiver is close to its original specifications the thermal noise should give a fairly steady reading, and with a couple of meters of wire connected to the aerial terminal the meter will show appreciably higher and the needle will move around more. For that test make sure the receiver does not have any form of muting or squelch function in operation and if the receiver has a noise limiter, switch it off for maximum response. That test works best if you use an analog multimeter.

If you hear no noise at all or a very quiet whisper, suspect the receiver or its power supply. Measure the power supply voltage with the receiver switched off and again when it is switched on; if it is within 10% or so of the specified figure and doesn't change by more than a few percentages when the load of the receiver is connected, that indicates the power supply is correct and therefore the fault is in the receiver. If you hear thermal noise at about the correct level but there is no sign of atmospheric noise when you connect a test aerial, that indicates a fault in the very early stages of the receiver and may be as basic as a broken wire at the back of the aerial connector. With modern well-made electronic equipment the incidence of random faults in internal circuits can be so low that the most likely causes of failures are things that have mechanical movement such as switches, dial mechanisms, potentiometers, and plugs and sockets.

At this point of time if the above tests suggest a fault in the receiver itself, a detailed check of manufacturer's specifications (probably requiring removal to the workshop bench) should confirm that indication and possibly give you some clues about which section of the receiver is faulty. Rectification of these electronic defects should cure the problem.

If you suspect the earth connection, you may be able to check it with an ohmmeter if you can find something to compare it with. The public switched telephone network and mains power supply grids are both systems that rely on low-resistance earth connections; they usually aim for a figure of about 5 Ω. If you disconnect the radio earth from everything else and test between it and the earthed metal of a device (not if it is double insulated) which you know is connected to either the power mains earth or a telephone earth point, you should be able to read a low resistance. The earth connections are in series for this test so the actual resistance of the radio earth is approximately the meter reading minus 5 Ω. That test only works reliably with an earth connection that has known characteristics.

If you are called to a station in a remote area which uses a motor/alternator set to generate electrical power and has a satellite or radio telephone, you will not be able to rely on that test. In those cases if you can split the earth system into two identical sections you can measure the resistance between them. (With the arrangement shown in Figure 3.3, for instance, you could disconnect each stake from the building and measure between any pair of them.) Be aware that the two connections are in series and when the station is operating they will be in parallel, so the resistance you actually measure is four times the final value. If that is not possible and it is still imperative that you measure the resistance of the earth connection there are purpose-built meters available which are used by power and telecommunications installation crews for measurement of all aspects of earthing. To be sure of the whole earth connection you will also need to check the continuity of the wiring between the earth stake and the actual connection to the receiver and also run your eye over the wiring to make sure that its path is still as originally installed. Remember that a multimeter will only check for DC resistance and the RF impedance is the combination of that resistance with the reactance of the series inductance due to the length of wire.

In those cases where the receiver and its power supply pass their initial check and you can prove a workable earth connection, the indication is for a fault in the aerial and checks you do on that will depend on its design. However, for installations that have worked properly before, you will usually find that visual inspection combined with multimeter checks for continuity and/or insulation of suspect parts will be enough to isolate the fault. For those installations where a design modification is required, be sure you understand the original aim before you change anything. There are some cases where aerials for entirely different purposes can look very similar and the difference in performance is only due to relatively minor differences in the length of some of the sections of conductor.

If the reported fault is described as interference, there are two possibilities. One is that a newly installed piece of electronic equipment is radiating a signal on the frequency of the station you are trying to receive; the other is that some intentional transmission is being received by your station with a high-enough power level to overload the front end of the receiver and cause either intermodulation or cross-modulation. Resolving which of these applies in your case can be very tricky, but in those cases where the source is from equipment owned by a neighbor it is vital to be sure before any accusations are made. The important point is that if the interference is due to a spurious emission it is the responsibility of the owner of the source to prevent it, but if the cause is due to overloading of the receiver then the receiver owner is responsible for curing the effect. One test that can give a good indication is to place a calibrated attenuator in the receiver input line (this normally only works well with a coaxial cable feeder) and adjust signal levels a few decibels at a time. If changing the level of the wanted signal also changes the level of the interference at the same rate, that is an indication that its source is outside the receiver, but if a few decibels reduction in signal level causes a change in the ratio between signal and interference, then overloading is indicated. If overloading is the cause, then Section 2.4 gives some helpful suggestions for its cure.

There is a possibility of faults due to drift of value in any matching or coupling components or traps used at the base of the aerial. In the case of a trap drifting off frequency, this will show as an increase of level of the particular signal being rejected and can be simply corrected by retuning for maximum rejection. Drift in other components may give no more than a somewhat vague loss of sensitivity but should be suspected whenever loss of sensitivity occurs over only part of the band. In cases where aerial base components are tuned, drift will be seen as a change in the dial position required for correct tuning and may mean that the aerial tuning is out of range at one end of the band.

For single receiver installations, there are quite a range of other possible fault indications, but in almost all cases they will be found to be due to electronic faults within the receiver itself and can be isolated by bench testing and repair.

When the receiver is part of a diversity system or there is other equipment (such as data detection and processing) using the output of the receiver faults which are proved to be within the electronics can be localized by breaking the system into units so that each receiver can be tested on its own. For diversity systems, any faults which cannot be identified as being in a particular receiver are probably due to the diversity switching/combining section. Where data detection is being used, be rigorous about checking specifications, the data-detection process may rely quite heavily on a function of the receiver output which has little or no effect on the quality of the audio signal.

Finally, as a point of basic philosophy about the fault-finding/servicing process, if you receive a service call and after checking you find no evidence of a fault, do not assume the operator raised a false alarm and forget about it. In all the fault-location work I have done I have never yet found a case where a user of equipment reported a fault where absolutely nothing was wrong. It may be that the fault is intermittent, in which case it will do your reputation no good at all to have declared the system to be working perfectly and then have the same symptoms reappear a few days or weeks later. It may be that the operator does not fully understand the operation of the equipment; that does not mean there is no fault, just that the fault is not a technical matter and it is cured by you spending a few minutes instructing.

CHAPTER

3

Coupling to the Environment for Frequencies below 30 MegaHertz

3.1 Why Only for below 30 MegaHertz?

For very low frequencies with long wavelengths, the coupling between the electronic boxes and the radiation field is via a combination of elevated conductors (the aerial) and a good connection to the mass of the Earth. In most low-frequency receiving stations the length of transmission line between the electronics and the base of the aerial is electrically short, much less than one wavelength. On the other hand, for operating frequencies in the VHF and UHF bands and higher, the aerial is mounted on a high tower many wavelengths above ground; the connection to it must be formed so that connection to the bulk of the Earth is not needed and the transmission line is normally many wavelengths long. The frequency of transition between these two quite different methods of coupling is in the 30-MHz range, but that figure is subject to quite wide variation in individual installations. For instance, mobile stations on the low-VHF band (70 to 80 MHz), in which a quarter-wave whip is mounted in the center of the roof of a small passenger car, can be treated exactly as a scaled-down version of a quarter-wave vertical aerial for use on the MF broadcasting band. In the other direction, defense and government stations sometimes use separate transmitter and receiver sites for long-range HF communication services with aerials mounted on very high towers for frequencies in the 10- to 30-MHz band; VHF techniques are required for these installations.

This chapter deals with those cases where a connection to the Earth forms part of the radiating system. The specification of frequencies below 30 MHz is not intended to imply a sharp cutoff in any legalistic sense, and there will be references to this chapter in later ones dealing with some VHF services.

For most aerials operating in the VLF and LF bands and for many in the MF band, the aerial falls in the class of "electrically short antennas." Sections 8.3, 8.4, and 8.5 in *How Radio Signals Work* give relevant information on quarter-wave vertical and electrically short antennas, and this chapter is written on the assumption that you have read and understood the relevant sections of that book.

3.2 Feedlines and Shielding

Once you have selected a suitable receiver and arranged a workable power supply and those parts of the system are able to pass the test suggested in the end of Section 2.2 you need next to attend to collection of the off-air signal. On the MF broadcasting band a sensitive receiver will hear all the signals that are strong enough to overcome atmospheric noise with only a few meters of wire as an aerial and a very rudimentary earth connection. If you do an initial test with exactly that by hammering a metal fence post (these are called "star droppers" in Australia) into the ground for an earth connection and taking a short length of insulated wire and just throwing the end of it over a tree or some bushes for an aerial, arranged as shown in Figure 3.1, you may find that the amount of noise your receiver hears may be changed by switching nearby electrical equipment on and off. If that happens it indicates you need a shielded feeder to connect your receiver to the aerial.

To do that test when your receiver has a coaxial socket as the aerial connection make the following connections:

- connect the bare end of the earth lead to the outer conductor of the socket (held with a hose clamp or bound with thin string, dial cord, or fine wire)
- poke the bare end of the test aerial lead into the center conductor of the coaxial socket

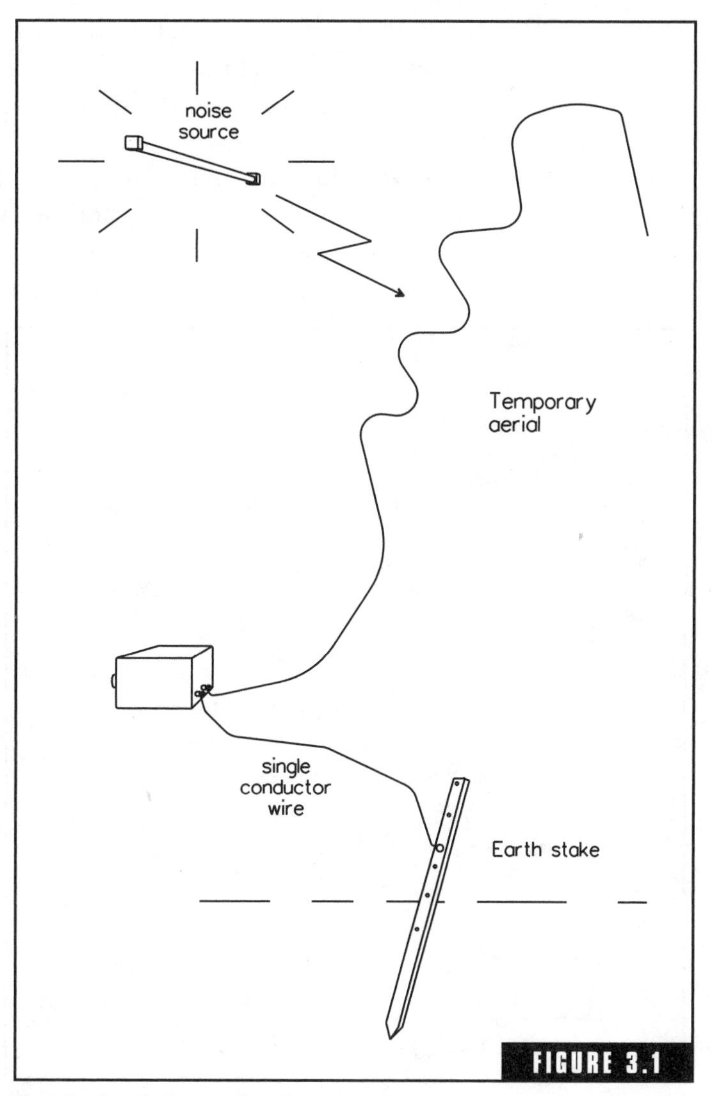

Test for local electrical noise.

For other frequency bands that test may need the following modifications:

Toward the high-frequency end of the HF band there may be difficulty in making the earth connection short enough to give a definitive result. In that case leave off the earth connection and make up a test aerial using a small tuned loop after the pattern shown in Figure 3.12 (suggest about half-meter diameter) connected to a few meters of coaxial cable and then use that as a search coil to identify particular objects as sources of electrical noise.

For the LF and VLF bands a longer length of wire may be needed for the test aerial. The length needed for a sure result may be as long as 5% of a wavelength.

For the ELF band the lead length of the earth connection will be less critical, but induction of power mains voltages may be a problem if multiple earth points are used. To avoid that, make the connections in coaxial cable so that the earth is carried on the outer of the cable then connect the test aerial to the inner of the coax at the point where the cable is actually connected to the earthed object. Figure 3.2 illustrates the required arrangement.

If you find your site has low electrical noise you can plan the permanent aerial and earth system to use an unshielded feeder. The requirements for an effective earth are detailed in the next section. Connection for the aerial will probably depend a bit on what connections are provided for on the receiver; if the receiver has screw or spring terminals and you intend to use it with a random length wire you can in a low-noise location simply lead the aerial wire straight to the set, bare the end of it, and plug it or screw it in.

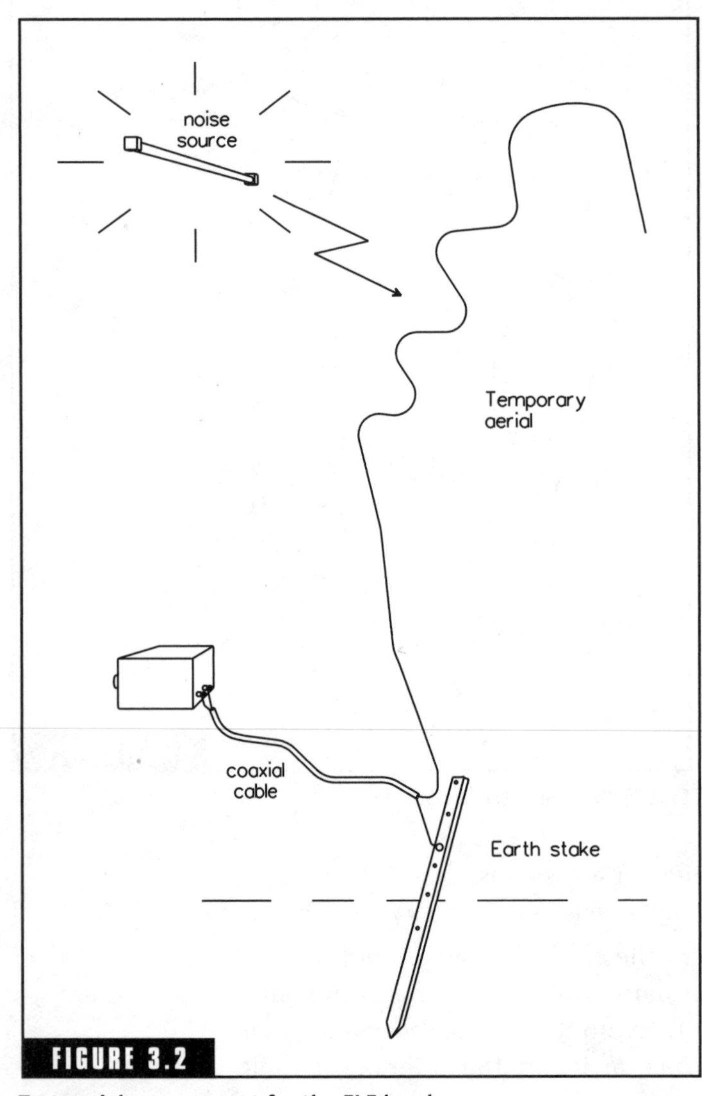

FIGURE 3.2

Test aerial arrangement for the ELF band.

For receivers which have a coaxial socket for aerial connection you may find it is tidy and cost-effective to use coax to run through the building to the eave, whether that is needed for shielding or not. If a shielded feeder is required, coaxial cable is usually the best choice in modern times. For receiving installations on frequencies below 30 MHz the cheapest type of coax that will match to the socket on the back of the receiver will be quite adequate. When coaxial cable is used as a shielded feeder, the following are important points in its use:

The cable has two separate conductors; the outer braid

must be connected to earth, the center conductor is used for the aerial lead.

Shielding is achieved by having the local interfering signal induce RF currents on the outside surface of the outer conductor of the coaxial cable, and these must be led straight to earth with minimum coupling into the receiver input circuits.

If the coaxial cable is more than one-twentieth of a wavelength long (ranging from 6 m for 1600 kHz up to 18.2 m for 535 kHz) it may need to be treated as a transmission line and impedance matching may be important.

If your receiver has a coaxial connector as its aerial terminal it will almost certainly be designed as an impedance-matched input for either 50 or 75 Ω, so matching at the receiver end will be correct if coaxial cable of the appropriate characteristic impedance is used. Section 7.12 in *How Radio Signals Work* gives some information on characteristic impedance and matching of transmission lines. The only place where impedance matching is required in that case is at the aerial end of the feedline.

If your receiver has two connections (screw terminals, for instance) labeled something like A1 and A2 or A+ and A− and checks with a multimeter show low resistance between the two terminals and isolation between the pair of them and the rest of the set, then those connections are probably the two ends of an RF transformer coil which is coupled to the first preselector tuned circuit, possibly arranged as in Figure 2.6. They will probably be designed to match to a somewhat higher impedance than a coaxial cable, although for an initial test they could be tried with the outer conductor of the coax connected to one terminal and the inner to the other. If matching is drastically wrong the effect may be to damp the selectivity of the input-tuned circuit, which will cause objectionable interference from signals at the image frequency. See Section 9.4 in *How Radio Signals Work* for an explanation of image frequencies. If damping is not a problem, go ahead and use the receiver connected in this way. To reduce the effect of damping you may be able to use a small series capacitor (which may need to be tunable) or use the circuit of a preselector as an input-matching transformer.

For receivers that have only one terminal for aerial connection the design is usually intended for connection of a short, unshielded random length wire as the aerial. If that is the type of aerial you intend to use they can work well if the aerial wire is not made too long (suggest no more than one-eighth the wavelength for the highest intended operating frequency). For almost any other type of aerial connection with that type of receiver a preselector will be needed, and if a shielded feedline is to be used the lead between the preselector and aerial terminal must be as short as possible.

There are a number of other arrangements possible for aerial connection and in most cases you will need to at least have access to a service manual to make a sensible guess at what is needed for matching to a shielded feedline. There may be times when a matching transformer is needed; these are usually based on a tuned circuit, set up as shown in Figure 3.7, with provision for selection of either tap position on the coil or adjustment of capacitance ratio. Because of the tuned circuit these must be readjusted each time a new signal is selected, so tuning of the receiver becomes a two-stage process. The receiver is first tuned to the channel then sensitivity is peaked with the matching transformer.

Those receivers in which the first tuned circuit is arranged so that its coil directly picks up the signal by being wound either on a ferrite rod or in the form of a large coil in the back of the case will not give good results in installations where a shielded feeder is necessary. Ideally the receiver should be constructed in a metal box or with sufficient internal shielding so that when it is operated at maximum sensitivity with no aerial connected there will be no signals heard anywhere across the band, even in strong-signal areas.

Matching required for reception of the MF broadcast band and lower frequencies is less critical than is required for a transmitter. The signal-to-noise ratio is mainly determined by atmospheric noise, so quite gross loss of sensitivity can be tolerated before much degradation of the final output is noticed. If the coupling is too tight the damping of the input tuned circuit mentioned above may be significant, so if you cannot match exactly it is usually best to err on the side of reduced coupling.

3.3 Earthing for Lower Frequencies

Several aspects of the earth connection are important. If the feedline is coaxial cable its outer conductor must be earthed and so must the case of the receiver, but if possible, each must be earthed at one point only. Radio-frequency earthing is not the same as is required for power line safety. For radio frequencies, the distance to the earth reference point is more important than the lowest possible figure for DC ohms. If the length of conductor from the receiver to the earth is an electrical quarter-wave the connection is a complete open circuit for radio frequencies even though it may be read with a multimeter as a complete short circuit for DC. On the other hand, it is possible to lay out quarter—wavelengths of wire in a radial pattern and have them totally DC-isolated from the physical ground, and yet they will behave as a perfect earth for radio frequencies.

The arrangement you use for connecting to the physical earth will depend on the location of the receiver and the type of soil around the building. In some cases achieving a connection which is both low loss (resistance component) and at the same time offers a low impedance to radio frequencies can be quite difficult. The following examples are some that should be workable in practice and be adaptable to other related situations.

If the receiver is in a metal-framed building connect its earth terminal to a near point of the frame using a short direct wire; then hammer a metal stake as far as you can drive it into the ground at each corner of the building with a connecting wire (also short and direct) to the nearest point of the frame as shown in Figure 3.3.

If the building is metal-clad so that it forms a shielded box for radio signals run the aerial feedline in coaxial cable with the outer conductor bonded to the building cladding as shown in Figure 3.4. Use earthing stakes at the corners as in Figure 3.3 and treat the building cladding as the reference for the aerial system. Inside the building treat the inner surface of the cladding as an earth reference for the receiver electronics. For radio frequencies, the two surfaces can be treated as quite separate conductors due to skin effect, and there can be separate patterns of currents and voltages on the inner and

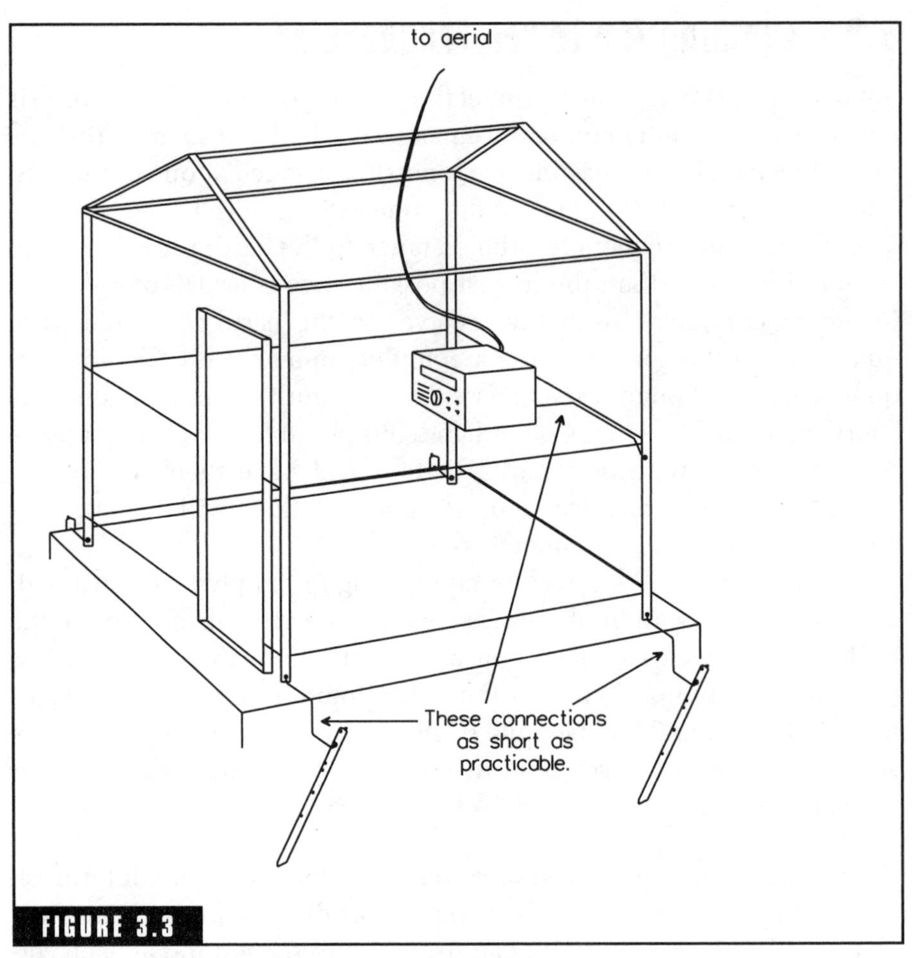

FIGURE 3.3

Earthing for a metal-framed building.

outer surfaces of a box of this type with no interaction between them.

A building constructed of reinforced concrete may behave the same as a metal-clad building if the amount of reinforcing mesh used was enough to make a shield for radio signals. An ideal earth if you can get the builder's permission to do it would be to make a permanent electrical connection to the reinforcing mesh close to the point where the receiver will be installed. Note that shortness of the connection is the important factor; it would be no good, for instance, in a multistory building to lead the earth wire down a couple of floors to an official electrical connection in the basement. If you are not permitted to make connection to the reinforcing mesh at a suitably close point, lay a thin sheet of metal on the floor under the receiver and use that for an earth connection. With regard to the area covered, the bigger the better; if you can cover the whole floor of the room, that would be ideal. With regard to thickness of sheet, you can make it as thin as you can make reliable electrical connection to and it does not even need to be a solid sheet—brass flywire will do as well as a solid membrane. The metal itself can be almost anything that is electrically conductive; for instance, the metallized "Kraft" paper that is widely used as a reflective surface for building thermal insulation would be good enough but the metal is so thin that making a reliable electrical connection to it would be a problem. If you can solve that practicality then that type of paper spread under the carpet of the whole room would make a very good radio frequency earth.

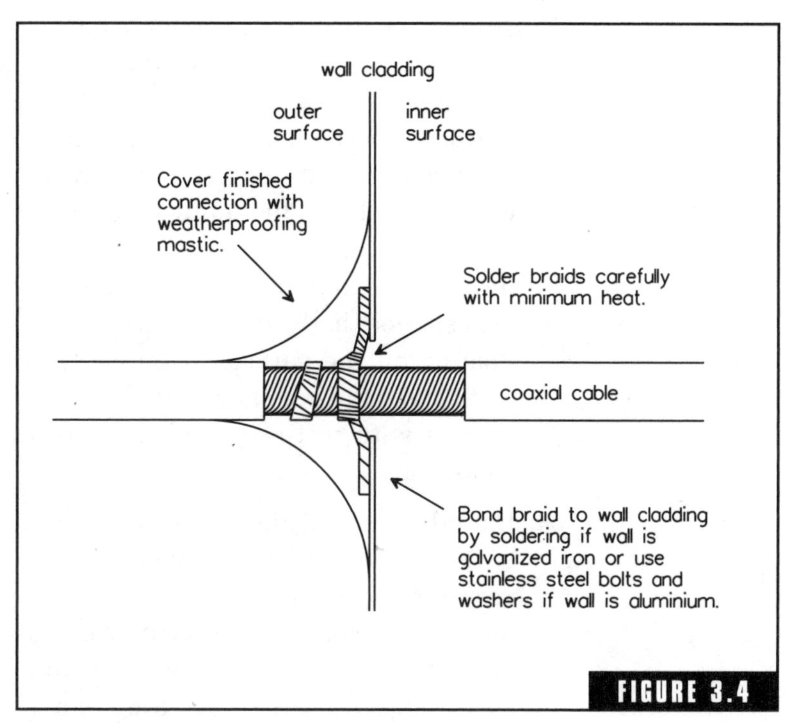

Aerial entry to a metal-clad building.

The same arrangement of a sheet of conductive membrane can be used in any multistory brick- or timber-framed building, but in this case you may need to add a DC connection in the form of a wire to an earth stake at ground level or clamped to a metal water pipe if one is available. (Note that in this case the conductive sheet takes care of the need for short connection; the DC lead can go a couple of floors down to an official connection point if necessary.)

If the receiver site is close to the sea or a salt lake that never goes dry a short direct connection to the salty water (shown in Figure 3.5) will give the best possible earth. There will be a practical problem due to corrosion; however, a highly efficient connection is not needed, so something that is cheap enough to be replaced regularly will probably do the job best. If you are near a salt lake that goes dry you will usually find at least some damp mushy salt if you dig through the dry surface. On a sandy island there will usually be a lens-shaped body of rainwater (poor conductivity) in the sand, floating on the surface of the salty stuff. In both these cases the earth stake must be hammered far enough into the ground to be in contact with wet salty soil, even in the most adverse of weather conditions.

At the other end of the conductivity scale, a desert sand dune may be almost completely nonconductive to electricity and thus a very poor connection for a receiver earth; however, if the sand dune is windblown and the original ground surface was a clay pan or salt lake you may be able to make a very good earth connection by making the earth stake long enough to go through the sand to the more conductive soil below, arranged as shown in Figure 3.6. There are rods specially made for this purpose which can be connected together end to end so that when one is driven to its full depth another can be added to the top and driven further down. Similarly, ice and permafrost are both electrically insulating and will need a connection that goes straight through to the more conductive layer below. This arrangement works when the sand or ice are pure enough to behave as electri-

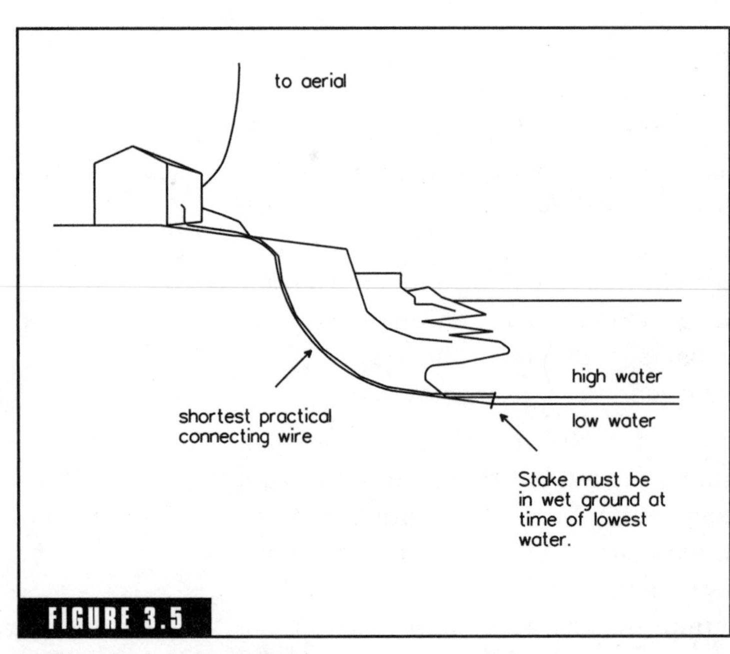

FIGURE 3.5

Earthing for locations near the sea.

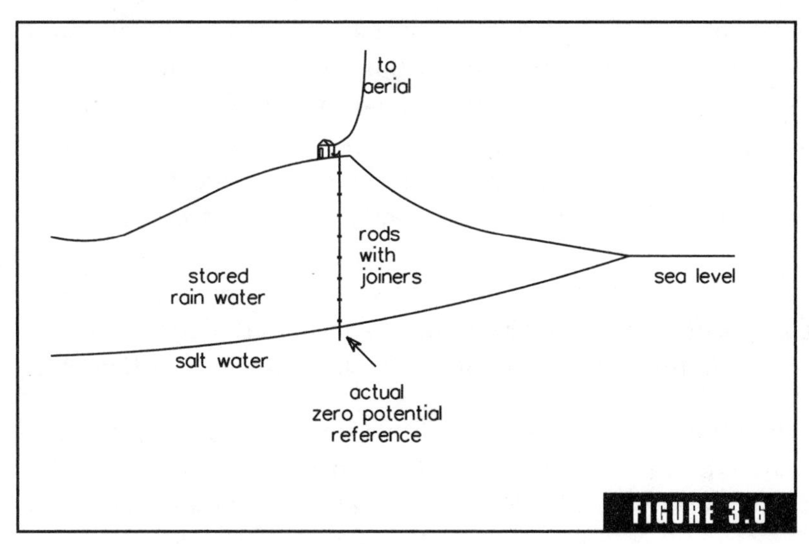

Earthing through an insulating layer.

cal insulators and are therefore almost transparent to the radiowaves. Note that in all these cases the earthing rod should be as near vertical as possible and with the shortest possible length of wire connecting to the point where the reference is required; the length of conductor between the surface of the conductive layer and the reference point will behave as part of the aerial element and will have the unavoidable effect that the receiver is partially tapped along the active element of the aerial.

The worst soils for radio-frequency earthing are the ones that are partially conductive because they absorb the energy of the radio signal but do not ever make a really low-resistance connection. There are some clay soils which have little nodules of ironstone ("gibbers") mixed through them which are notable for their poor performance in relation to radio signals. In these locations you will get best results by running several wires away from the earth reference point in a radial pattern. In noncritical applications this may only need to be 4 wires at right angles to each other going out a few meters to earth stakes. For really high performance there may need to be up to 120 wires at 3-degree intervals around the compass, each a full quarter-wave long, arranged as shown in Figure 3.20. At that level of complexity it does not really matter much whether the wires have earth stakes at the end or not; in many installations of this type the wires are bare copper buried in the ground, but in a few cases the wires are suspended a half-meter or so above the ground and purposely kept insulated. These need to be cut to an exact length so that the open circuit at the end is reflected to the central point as a short circuit. See Section 7.13 in *How Radio Signals Work* for an explanation of the quarter-wave transformer. This form of "earth" is called a "counterpoise" and is sometimes used as the base for the antifading radiator of a transmitter station. For receiving, the simpler structure of Figure 3.7 is usually sufficient.

When the aerial is a vertical with a low-impedance coupling point at its base there must be a very good earth reference at the base. It is best if the receiver can be located very close to that feed point; the one

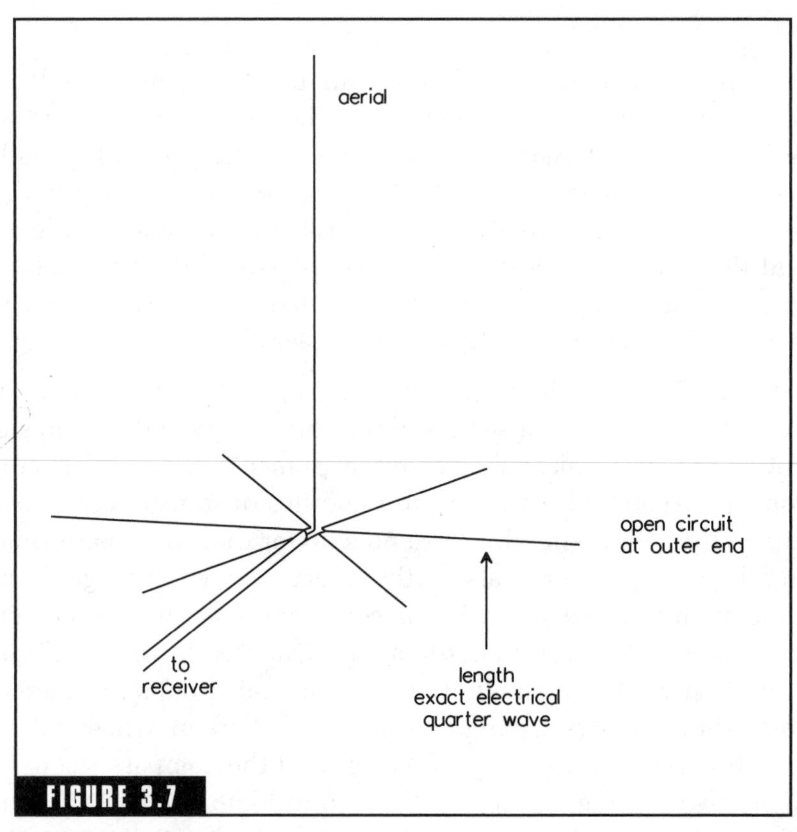

FIGURE 3.7

A simple counterpoise.

reference can be used for both aerial and receiver. In cases where the receiver must be located away from an aerial that requires its own earth there is a risk that the two separate points of earthing could cause electrical interference to be induced into the receiver. Receivers in which the aerial connection is a coaxial socket will usually be designed to minimize the possibility of this problem. Those in which the aerial terminals connect to the primary coil of an RF transformer can be used with a shielded feeder by earthing the shield at the aerial end and maintaining DC isolation at the receiver. When the aerial terminal is a single connection to a tap on the coil of the first tuned circuit, that is an indication the receiver is really designed for use with a random-length wire aerial with no shielded feeder and is probably not the best choice for this station; however, if it must be used then it should be with a matching transformer with inductive coupling of the signal and DC isolation between primary and secondary coils of the transformer, as in the schematic circuit of Figure 3.8.

3.4 Aerials for MF Band Domestic Reception

At the start of planning for a receiver aerial for the broadcast band there is a need to establish whether the expected signal will come via skywave or groundwave and from which point of the compass it will come. Other stations on the same frequency will also need to be considered as well as the directions from which interfering signals are most likely to arrive. Skywave signals are those that come at night and are available from all stations whenever the whole path from transmitter to receiver is in darkness. When the sun rises the ionospheric D layer forms within about the first 20 minutes of daylight and skywaves are completely absorbed. Groundwave signals are available at about the same strength all the time; they are strongest for local signals and get weaker with increasing distance at a fairly predictable rate. If you find that what

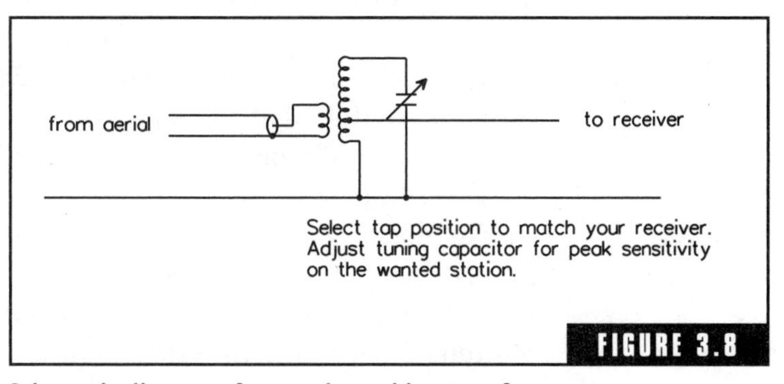

FIGURE 3.8

Schematic diagram of a tuned matching transformer.

you think is a groundwave signal is being affected by the time of day, what is probably happening is that the receiving site is in a zone where at nighttime ground- and skywaves are about the same strength but are interfering with each other because they are out of phase. See Section 6.4 in *How Radio Signals Work* for an explanation of the cause of this phenomenon.

In the remote areas of the continents (and they don't really have to be very remote at all) and on small oceanic islands it is quite common during the daytime for most of the MF broadcast band to be quite empty of strong signals. At night, however, signals come from all of the dark hemisphere; the band is a bit over 1 MHz of spectrum space, each channel occupies either 9 or 10 kHz of it, and there are many thousands of broadcasting stations worldwide. If a sensitive nondirectional aerial were used, every channel would be heard as a jumble of many programs. There is almost a need for different aerials for daytime or nighttime because the performance requirements are so different.

For groundwave signals the aerial must be vertically polarized and the aim is to make it as efficient as possible at collecting all available signals. At night strength of signals is less of a limitation, but tailored directional response is the key to success. On the MF broadcasting band groundwaves are most effective at the 600-kHz end, so all the stations that seek to offer a daytime long-range service will have negotiated for allocation of highest possible power with a channel allocation as near as possible to the low-frequency end of the band. For receiving groundwave stations, a quarter-wave vertical aerial sitting on an efficient earth radial system as in Figure 3.20 would theoretically be ideal; however, the dimensions of such a structure, for instance, for 600 kHz, include using a height of 118.6 m with an earth mat of 120 pieces of wire each about 120 m long. Even so, after all that was put together it would only be ideal for one particular channel. For receiving, it would give strong signals over all the band because atmospheric noise is so much in excess of that produced internally in the receiver.

The design of an aerial which will efficiently collect and deliver power to a coaxial cable over all of the approximately 3:1 range of frequencies represented by the MF broadcasting band represents the resolution of several conflicting requirements. The next three sections give examples of three different principles which can each be used as the basis of a practical MF broadcast reception aerial. In addition to all of these suggestions, all the designs mentioned in the section on rebroadcasting systems are available for use in the domestic scene; the only limitation is cost. A factor that can be used at very little cost but which may make as much difference as spending several thousand extra dollars on the aerial is the selection of a favorable site in relation to the standing wave microclimate. This is the subject of Section 3.8.

3.5 Short Loaded Verticals

Shorter than quarter-wavelength vertical aerials can give good results when used with LF and MF band broadcasting receivers. If shielding of the feeder is not needed, 20 m or so of wire connected to the aerial terminal directly at the back of the set that leads out a window to an insulator on the eave with the remainder of the length supported as near vertical as possible will give good reception of groundwave signals if used in association with a good earth connection. The endpoint of such a structure has a very high impedance with respect to earth so a good-quality insulator or maybe two insulators in series is required to avoid loss of signal. If the receiver is physically close to the base of the vertical element then there would be sufficient signal to overcome background noise over most of the band with a tower height of only 6 m, with the 20 m of wire running at an angle to the vertical. The length of wire needed varies with the environment; in tropical areas and in cities and industrial zones the length required to collect all signals strong enough to overcome noise may only be 10 to 12 m. On the other hand, in a polar region the atmospheric noise is so low that 25 to 30 m of wire could be useful.

For a short vertical to work when the feeder must be shielded, however, there must be no more than a couple of meters of shielded cable from the aerial base to the receiver input connection. When a vertical aerial is shortened below a quarter-wavelength its output impedance rises and includes a capacitive reactance component; a vertical as short as that presents a very high source impedance and the capacitance of a length of coaxial cable will divert most of the signal straight to earth.

The high source impedance means that economies can also be made to the earth system; a high-impedance aerial can work effectively with a more rudimentary earth connection if reductions in both aerial and earth are roughly kept in proportion.

Short vertical aerials can be used with coaxial feeder cables if the aerial is loaded sufficiently to make it appear as a full quarter-wavelength. An inductance coil added to the base of the vertical element will add the required electrical length but the implication of a loading coil is that it is a tuning component, so tuning is only correct at one particular frequency. Provision for adjustment is required to operate over the whole width of the MF band. A loading coil with either an adjustable capacitor in series with it or with switched taps (illustrated by Figure 3.9) can be used at the base insulator. In general, center loading and top loading (other than the "inverted-L" construction, described later in this section, or a capacity hat) are not practical when adjustable tuning is needed. Note that the loading coil must be placed physically very close to the base of the tower and between the aerial element and the coaxial cable, so if there is any distance between there and the receiver either a motor drive system for the tuning capacitor or relay switching of the coil taps will be needed.

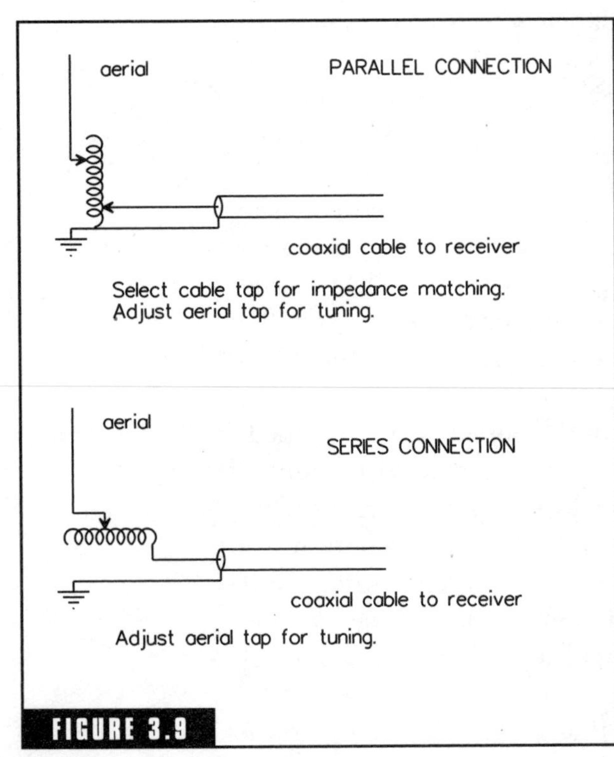

FIGURE 3.9

Using a loading coil.

In both of the schematic circuits shown in Figure 3.9 the tuning adjustment can also be achieved by using a fixed connection from the aerial to the coil and placing a variable capacitor from the aerial end of the coil to earth. If you choose to use a capacitor, however, be aware that you may need actual capacitance values several times larger than the range of commonly available tuning capacitors and the relationship between capacitance and frequency may not be a simple function.

Loading with an inverted-L section, T section, or a capacity hat will give wider bandwidth; however, the length of the horizontal section is not adjustable and there is a limitation on how long it can be. For the inverted L, the total length of both vertical and horizontal sections (refer to Figure 3.10) must be no more than about 0.4 wavelengths at the highest frequency required for operation. If that length was extended to 0.5 wavelength it would present to the coaxial cable the very high impedance of a parallel tuned circuit. For the T section and capacity hat versions the same limitation on electrical length will apply, but the physical length of conductors that is required to give that electrical effect may be slightly different in some cases. For those countries where the band edge is 1600 kHz that length is 75 m; for instance, with 20 m vertical, the horizontal section should be no more than 55 m.

A combination of inverted L and loading coil is possible. In that case the total length of wire should be no more than a quarter-wavelength at the highest operating frequency; the loading coil needs to be the series version of Figure 3.9 with an adjustment range from zero inductance to sufficient to tune the aerial to the lowest required frequency.

In all these examples of impedance adjustment, if the aerial is adjusted for proper tuning its source impedance will be close enough to the characteristic impedance of the cable so that there is very little

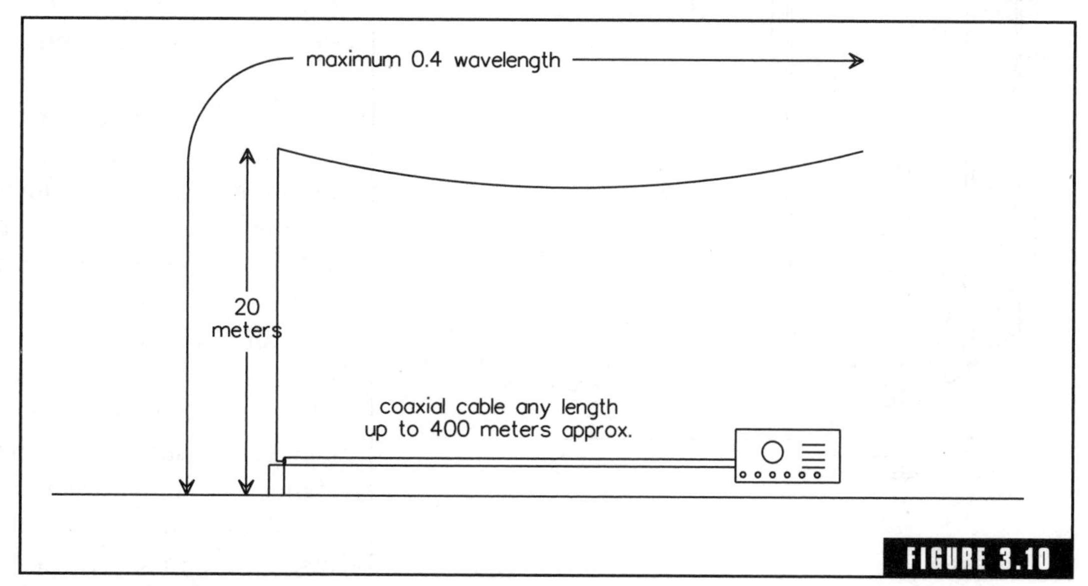

Inverted-L aerial dimensions.

limit on the length of cable used. If the cable was long enough losses would accumulate and weaken the signal, but on the MF band several hundred meters would be required for that to be important, and if it was a factor it could easily be cured by adding a few meters to the vertical section of the aerial.

3.6 Broadbanding

Another way of achieving lower source impedance with a short vertical is to make its effective diameter very large (several meters across). In practice that can be done by making the vertical element with many strands of wire spread out from a central point at the base with a structure as in Figure 3.11 that looks a bit like an upside-down birdcage. The effect of making the diameter of an aerial element bigger is to make it less sharply tuned, which gives closer to a resonant response. You can also get an idea of what it is doing by thinking of each of the wires as a separate vertical element, each of which forms a high-impedance source but when all are connected in parallel the impedance of the combination is lowered.

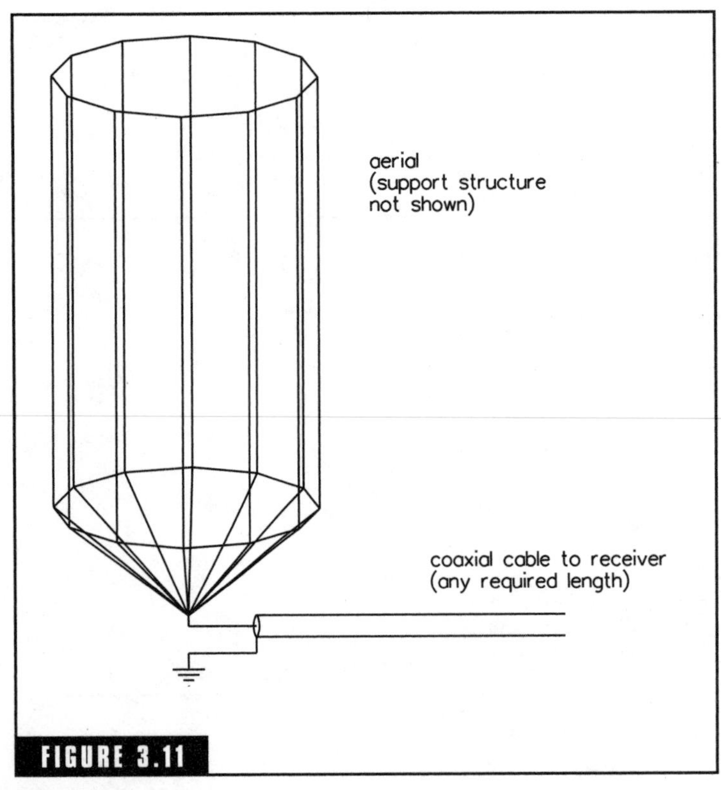

FIGURE 3.11

Broadbanding.

Several different forms of mechanical mounting are possible with a cage element of this type; it can be slung between two towers in the manner of a vertical with a T section top, or it could be hung around a central pole. The space inside the cage is electrically shielded from the signal, so an earthed solid metal pole or guyed lattice mast can be used if the conductors are electrically insulated from the support. If guys are used outside the cage they must be broken with insulators into electrically short sections.

A similar electrical effect can be obtained in a mechanically different form if there are trees or poles conveniently placed near the required aerial feedpoint and if you attach several wires in parallel at the feedpoint and take each to a separate pole or tree in such a way that each wire makes an angle of about 45° to 60° to the vertical; the completed structure has roughly the shape of an inverted cone with its apex at the feedpoint.

3.7 Tuned Loops

Atmospheric noise comes from all points of the compass fairly evenly spread, so if your aim is to receive a particular station (or a group of stations all from the same direction) a directional array may be worth considering. The simplest directional aerial that will work on the broadcast band is a tuned loop arranged as in Figure 3.12.

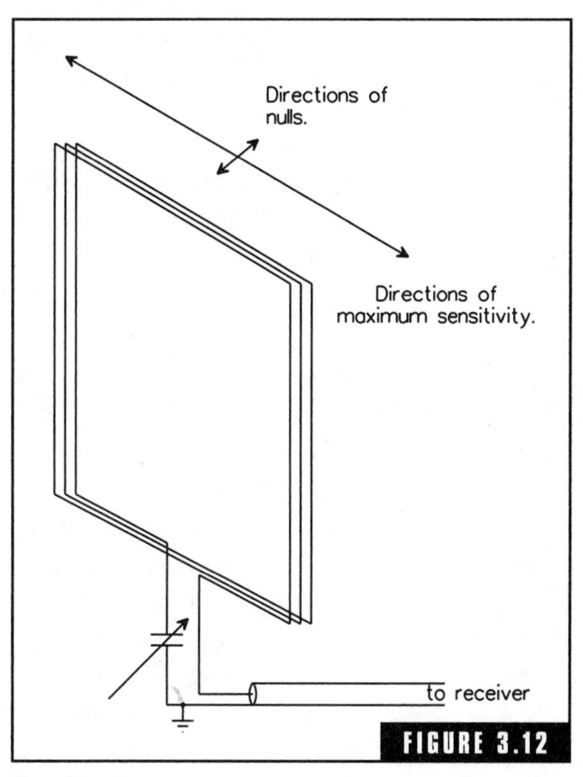

FIGURE 3.12 Tuned loop.

Practical open-air loop aerials for the broadcast band will usually be in the size range from 1 to about 5 or 6 m on each side and the shape of the loop does not matter—any regular polygon from a triangle to a circle (infinite number of sides) will work almost equally as well as a square. The number of turns can be adjusted to suit individual requirements, but the total length of wire used should be kept to less that about 0.45 the wavelength of the highest operating frequency to avoid stray resonances. A loop made like that does not have quite such a high sensitivity as a quarter-wave vertical, but it does have the directional properties in the horizontal plane of a dipole aerial (doughnut pattern; see Sections 7.4 and 7.5 in *How Radio Signals Work* for an explanation) so will give the possibility of a $1\frac{1}{2}$- to 2-dB better signal-to-noise ratio. As with the short verticals the tuning components must be at the base of the aerial, so remote control by motors or relays may be needed. The tuning component for a loop is a series capacitor which because of its series connection will not necessarily have a linear relationship to frequency, so switching of capacitors may be needed to achieve the required range. Because of their relatively low sensitivity, loops are most useful when you wish to receive a broadcasting station from just outside its primary service area.

The ferrite rod aerial in a portable radio is a tuned loop. The effective size of the pickup coil is the physical size multiplied by the magnetic factors of the rod. If your aim is to increase the sensitivity of one of these sets a bigger lump of magnetic material will do that. You can either use a bigger rod if one is available or a bundle of several rods side by side. The tuning coil will have to be redesigned with a bigger former and fewer turns, and if the original coil was tapped there will have to be taps at the same proportions of the reduced number of turns. There are practical limitations on how much you can gain by doing this; it will only be an advantage when the weakest usable signal is determined by internal noise or limited maximum gain of the receiver itself. If the signal-to-noise ratio of the received signal is set by factors external to the receiver then building in a bigger magnetic circuit will be no advantage.

3.8 The Standing-Wave Microclimate

The exact signal strength at a particular site is affected by a host of minor factors, many of which change signal strength by a total of less than 1 dB. Individually, any of these factors can be ignored and most of them can hardly be measured accurately, but the total of all their effects combined means that for each signal there is a microclimate of locations of stronger and weaker signals. Across the face of any reasonably large area there is a semiregular series of zones of strong and weak signal locations related to a number of overlapping standing-wave patterns. One of the most pronounced is a factor related to the vertical angle of arrival of the incoming signal.

In the horizontal plane there are a multitude of potentially reflective objects so there are liable to be a large number of less pronounced patterns overlapping. Standing-wave patterns are due to reflections from fences or a variety of other conductive objects. Local variations in topography and variations in ground conductivity also cause local variations in signal strength. The combined effect of all of them means that if you were to do a detailed check of field strengths over a square area of a couple of wavelengths or so you would probably find some spots that have signal up to 6 to 8 dB stronger than the signal from the same transmitter in the weakest spots.

If you intend to use skywaves and your location includes a substantial hill overlooking a stretch of flat ground or open sea in the direction you expect signals to come from you will be able to site your receiver at a particular height to favor a particular vertical angle. To use this factor you will need to know the wavelength of the wanted signal and the vertical angle at which it is expected to arrive at your location.

Wavelength can be calculated from this equation:

$$\lambda = C/f$$

{wavelength (in meters) = 299.7/frequency (in megaHertz)}

The vertical angle is potentially a more involved calculation but in those countries which have a service of solar astronomers who use their observations to predict conditions in the ionosphere (in Australia it is the IPS Radio and Space Services) they can for a very reasonable price give you a prediction which includes the expected vertical angle. The angle will vary slightly depending on weather conditions on the Sun, but if you notify them that you are interested in the range of possible values the predictors can give you figures for all expected conditions during the 11-year sunspot cycle. Several modes of propagation are possible for skywaves in the MF and HF bands; you may be given a prediction that includes several which may be designated as 1E, 2E, 1F, and 2F. If so, for the MF band select the lowest numbered E-layer prediction whose vertical angle is greater than 8°. (The F-layer predictions are more relevant to frequencies in the HF band.)

The combination of direct and reflected signals will give a pattern of nulls and lobes of reinforcement, as shown in Figure 3.13, and the

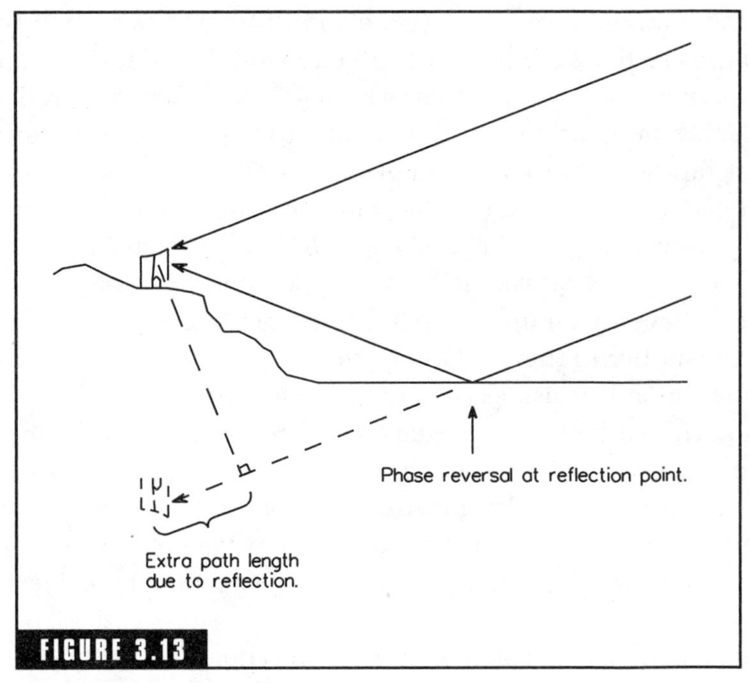

FIGURE 3.13

Effect of a reflective surface.

vertical angles will be determined by the height in wavelengths above the reflective surface. Some examples are:

	Vertical angles in degrees		
Height in wavelengths	**Lowest lobe**	**Null**	**Second lobe**
$\frac{1}{2}$	30.0	90	—
1	14.5	30.0	48.4
$1\frac{1}{2}$	9.6	19.5	30.0
2	7.2	14.5	22.0

Those figures are theoretical and would be correct if a half-wave dipole were suspended over a perfectly flat surface with perfect electrical conductivity. With real conditions the figures will not be exact but should usually be within 5 to 10%. For adjustment to an exact angle they may be interpolated with a straight-line relationship with accuracy good enough for most practical cases. Above two wavelengths in height the angles for this first lobe and null can be calculated with fair accuracy by taking simple proportions, i.e., if you double the height you halve each of the angles. There is a practical limitation on transmitter aerial performance, which means that in most cases signals launched at angles less than about 8° above the horizon will not have a useful amount of power available to propagate as a skywave. Therefore the tabulation shown above should give a workable basis for calculation of vertical angle for all cases that are useful in practice.

These angles may be used to select for either enhancement of the wanted signal (about 3 to 6 dB of improvement is possible) or for rejection of interference. The depth of null may be up to 15 to 20 dB but will depend very much on the reduction of power at the reflection point due to losses in the soil. Seawater is a very low loss reflector of radio signals, so the effect of height will be most pronounced at seaside locations.

This factor of height of the receiving aerial is a particular aspect of the general theory of Fresnel zones (see Appendix section A1), but in this case the transmitter aerial location is the ionospheric reflection point, which is a rather indeterminate point high in the sky. The one thing that can be said for sure about that point is that it is far enough away for the ray lines of the incident wave to be effectively parallel so the calculation is based on angles rather than distances related to Fresnel zones.

If you wish to use the height factor in practice there is a book published by The American Radio Relay League called *The ARRL Antenna Book,* which explains the effect in great detail and shows graphs of radiation patterns in a variety of circumstances. It is specifically directed for use by amateur radio operators in the United States, but the references to the 160-m amateur band can be adapted for use worldwide on the broadcast band.

The signals from potentially interfering stations and distant discrete sources of electrical noise will also show the same microclimate effect, but because they come from different directions the actual spots of strong and weak signal will show a different pattern for each transmission. Interference from local electrical noise sources will be even more affected by location. On the other hand, site location will have less effect on the strength of atmospheric noise. Atmospherics are themselves caused by a host of separate events at a host of different locations at all compass bearings from your site which each have coupling to different spots in the area under test, so the combined result of all of them is smeared into a more even concentration of energy at all locations.

The practical implication of all this is that if you have a large-enough area available to select from and time to fiddle around with a field-strength meter you could vary the final signal-to-noise ratio of your received signal by up to 12 dB or so just by selection of the correct site. The microclimate effects are almost impossible to predict by theoretical calculations, but they do show standing-wave patterns so what is sometimes possible is that if you find a group of high readings in a line you can expect there may be a similar line of high readings about half a wavelength either side of it, and if you find a line of particularly low readings you could expect to find high readings by moving about a quarter-wavelength or a bit more perpendicular to that line. Figure 3.14 shows a typical pattern of standing waves that may be found on a receiving site of a several-hectare area; it is probably simpler than most in practice because it only shows the effects of reflections from two sources and nothing from other possible factors.

As an added complication remember that what you are really looking for is not the location of lines of high signal strength at ground level but the equivalent lines at 10 to 20 m or more elevation. The locus of points of high signal is actually a three-dimensional surface which may be curved and may be intersecting the Earth's surface at any angle. Figure 3.15 is an attempt to show the effect in three dimensions of a standing-wave pattern similar to one of those shown in the surface plan of Figure 3.14. The lines of Figure 3.14 become three-dimensional surfaces in Figure 3.15, and they are drawn as if they were solid like the blades of a turbine to show some of the effects of their three-dimensional nature. The straight lines at the front and rear illustrate the Earth's surface, and the small symbol in the center shows a typical receiving aerial with towers about 20 m or so high to give an indication of scale.

Microclimate effects are only significant over distances comparable with half a wavelength or so, and for the MF broadcasting band you need several acres or hectares of space to move around in to map them. If the only space available to you is a suburban building block or smaller the most important factor will be locally generated electrical noise so you should select a site based purely on distance from noise sources and the practicality of day-to-day living with a large metal structure in that spot.

At the high-frequency end of the HF band the scale of the standing-wave patterns is comparable with the size of a suburban building block, so for omnidirectional aerials operating in the 27-MHz CB band and the 15- and 10-m amateur bands the exact siting of the aerial is potentially a significant factor. When directional aerials are used the fine-grained patterns from reflecting objects behind the aerial are much reduced and only those reflections from objects close to the beam of the aerial are noticed, for reflections from objects close to the line of arrival of the signal standing-wave patterns are on a larger scale. (See Section A1 in the Appendix for explanation.)

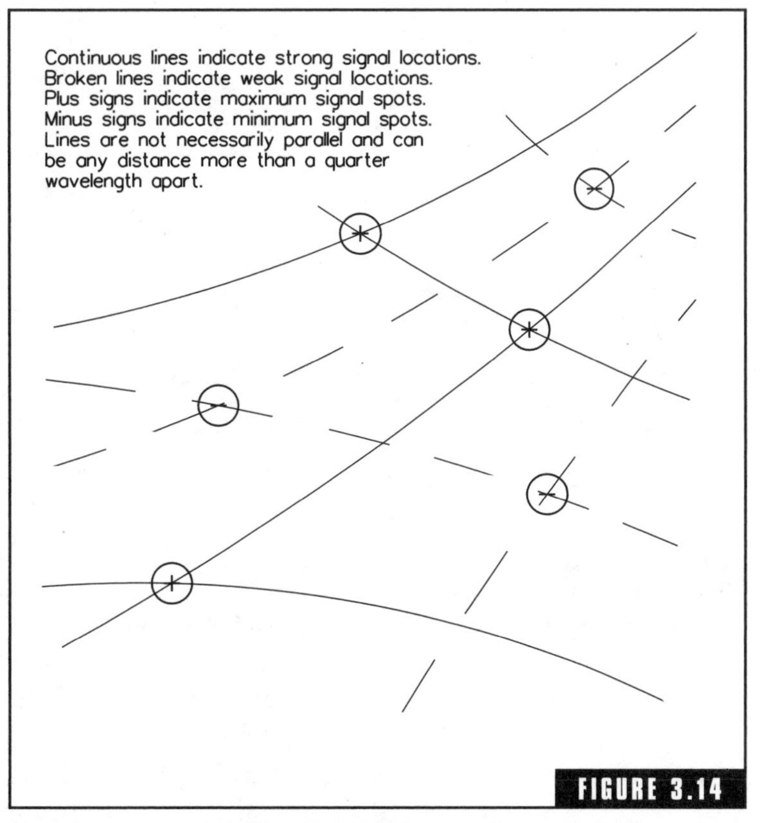

FIGURE 3.14

Plan view of a receiving location showing standing-wave patterns.

3.9 Aerials for the Lowest Frequencies

For signals in the frequency range of 30 to 500 kHz all the techniques used for groundwave reception on the MF broadcasting band will work if appropriately scaled up; however, the 20-m vertical with loading coil that gives good results at 1600 kHz would need to be slightly over 1 km tall to give equivalent performance on 30 kHz. The scale of the microclimate effects described in the previous section is also proportionately larger, which probably means that in most cases it is not

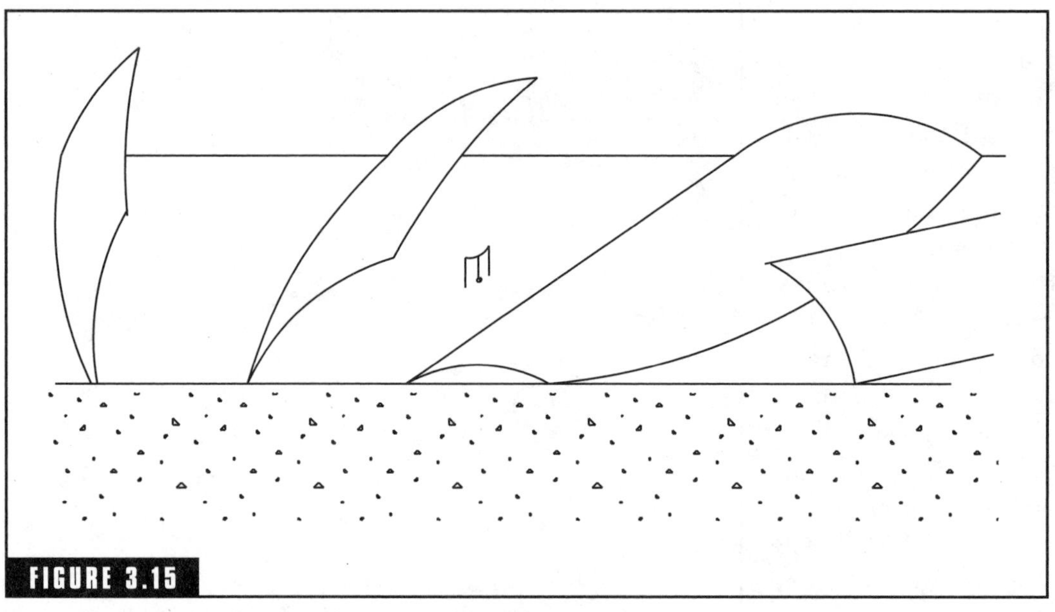

FIGURE 3.15

Isometric sketch showing the effect of height on Figure 3.14.

practical to explore for high-signal spots on an individual site. For the VLF and ELF range the scale of designs using the same principle as aerials for the broadcasting band and higher frequencies leads to such enormous structures that they are beyond the practical range of use, and different basic principles are required.

Active aerial designs give good results for the LF band and for some of the higher frequencies in the VLF range. They can also be used on the MF broadcasting band in cases where a shielded feeder is needed and there is such a severe restriction on space available that even 20 m of wire is too much. The active aerial relies for its ability on the very high input impedance of an insulated gate field effect transistor (FET) used in the "source follower" circuit arrangement. The FET can work with signals from a very short length of conductor, which may present an output impedance of several megaohms. Because the FET is out in the environment in a relatively naked form it must be protected from lightning induction and build-up of static charges.

Figure 3.16 shows a circuit diagram of an active aerial that could be used on any channel from 20 to about 1600 kHz with the appropriate choice of components and wire lengths. The primary control of sensitivity is the length of vertical conductor used with it; that can be ad-

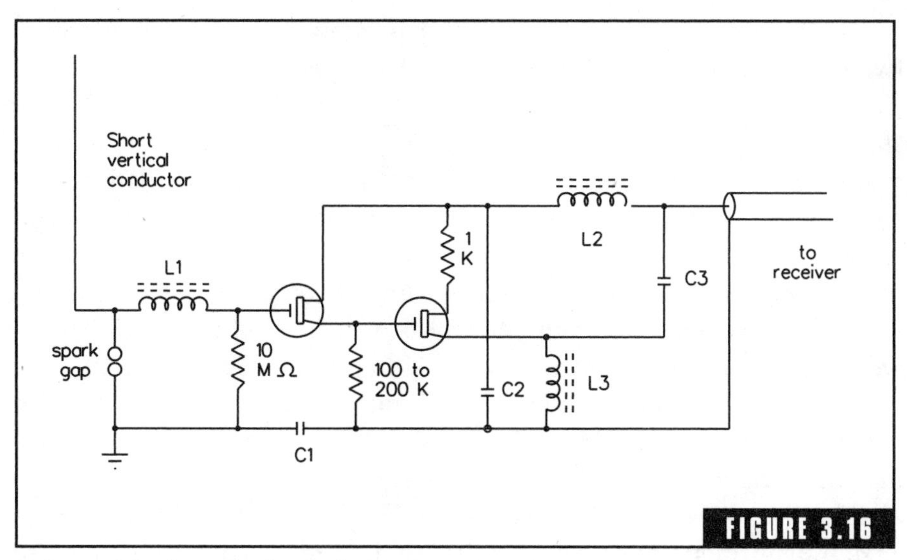

Active aerial schematic circuit diagram.

justed longer if there is no interference and the wanted signal is marginally weak. On the other hand, if the receiver initially suffers interference from a strong local station the vertical conductor can be trimmed shorter. The impedance conditions that apply generally preclude the use of traps or tuned circuits on the input side of the FET.

In the circuit of Figure 3.16 the spark gap, the 10-MΩ resistor and inductance L1 are for lightning protection and static bleeding. In addition, in a practical use of this circuit the first FET should be mounted in a location where it is easily replaceable. The situation of a vertical conductor connecting straight to the input of an insulated gate FET means that this circuit will be affected by lightning; the protection components will help in the case of induction from nearby strikes, but in the case of a direct hit some damage is likely and the first FET is the fuse that will blow.

The three capacitors can all be made the same value. The figure will depend slightly on the expected range of operating frequencies; they must be big enough to provide a low impedance (i.e., less than about 50 to 100 Ω) for the lowest required frequency but not so large that there are stray resonances in the range of the highest frequency used. The value will probably end up somewhere in the range 0.05 to 1 μF.

The maximum usable value for L1 is about 100 μH. For the other two coils they need to present as high an impedance as possible to all signals in the operating range but within the limitation that there should be no self-resonances in the significant operating range. If either of the amplifying stages is possibly going to be overloaded by a strong local signal the highest frequency of that range could be as high as the third harmonic of the desired frequency. The coils can be wound on a suitable ferrite as either toroids or pot-cores. If the mechanical construction precludes the use of magnetic material there would be a problem of stray coupling between air-cored coils. Very careful attention to alignment for minimum coupling would need to be attended to during the design. Possibly building the box with internal shielding would make the device workable.

Amplifiers arranged in this source follower mode work as impedance transformers; however, this device does not completely match to the input of a coaxial cable. Output impedance of the second FET is liable to be in the range of a few hundred up to about 4000 Ω, so the length of coaxial cable used should be kept to no more than about 30 to 40 m. One cunning possibility is to make the length of coax as near as possible to a quarter-wave (even if that is more than 40 m) and then use an impedance transformer to match to the active aerial by making adjustments at the receiver. The amplifiers need a DC power-feed of perhaps 12 to 20 V and the components L2 and C2 are working as a power-separation filter so that both the DC and RF signals can be fed along the same coaxial cable. If a cable with three conductors is available then components L3 and C3 can be left out and the output of the second amplifier taken directly to the cable. That arrangement will require a DC connection between the signal lead and the outer of the coax, which will be taken care of if there is any form of transformer or tapped coil coupling at the receiver end.

For reception of electromagnetic waves in the audio-frequency range and the ELF band the factor that is really most important is length of conductor. Conductor height above ground does not appear to make much difference; ELF aerials are generally too short to have any directional properties so the factor of importance that is left is length of conductor. In the early days of the 20th century there were some very effective ELF aerials constructed quite unintentionally when telephone lines using wires at the top of poles were extended long distances into country areas. The pioneering Overland Telegraph Line from Adelaide to Darwin had all the features required of a very good ELF aerial; it used one single wire at the top of a series of tall wooden poles. Each section was a continuous length of conductor for about 250 km and it ran through country that had very little noise and interference from man-made machinery. If any of the 19th-century network of telegraph lines still exist they would make very good aerials for research into signals on ELF and audio frequencies.

3.10 Aerials for 2 to 30 MegaHertz

The first point to make about this subject is that all the information given in this section will only be a very sketchy outline; the subject of aerials for receiving on the HF band is so vast and detailed that many books have been written on it and in most cases even the most detailed of books only cover a portion of the subject each. The factors that make the HF band unique in regard to aerials are:

- it is the frequency range in which skywave propagation is useful. This implies that the vertical angle of arrival of the signal is more relevant here than on other bands
- a regular sequence of frequency changes is required to follow cyclic changes to propagation conditions
- the band is a transition region between the "aerial-and-earth combination for groundwave reception" of lower frequencies and the "isolated aerial with two-conductor feeder and no relevance to an earth," which is characteristic of the VHF and UHF bands

The cyclic variations of propagation (Section 6.5 in *How Radio Signals Work* describes the major sources of variability) require that aerials for the 2- to 30-MHz range have multifrequency capability. This does not necessarily require continuous broadband operation; it can be achieved by making the aerial system sensitive to a number of particular spot frequencies. All of the design principles described in Chapter 8 of *How Radio Signals Work* can be used on the HF band, but those involving a screen (screen-backed arrays, corner reflectors, and parabolics, for instance) are not common because they involve an uneconomically large structure for the screen.

Resonant structures must be adapted for multifrequency operation; there are two principle ways that can be achieved, illustrated by Figures 3.17 and 3.18. In Figure 3.17 the length of wire between the tuned circuits is made resonant for the higher frequency of operation and the tuned circuits are adjusted to present an open circuit at that frequency by being tuned for exact parallel resonance. At the lower operating frequency the tuned circuits are reactive and the length of the outer section of wire on each side is cut to have the effect of a loaded quarter-wave section with the tuned circuit acting as a loading coil. Three-frequency operation can be provided by adding a second tuned circuit on each side of the dipole. There is no theoretical reason why a fourth or more operating frequency could not be provided, but the practicality would be a structure that would be very unwieldy to adjust.

The multiple dipole assembly shown in Figure 3.18 uses a separate set of conductors for each operating frequency. The intrinsic limit on how many separate channels could be provided for by this means is in theory very high, but in practice if an attempt is made to include more than about three or four elements or to confine the elements too close together there will be quite severe interactions between them which will make difficulties for tuning and matching. To keep these interactions reasonably tame there needs to be at least about a 30° angle between each element and the next nearest one to it. Matching of

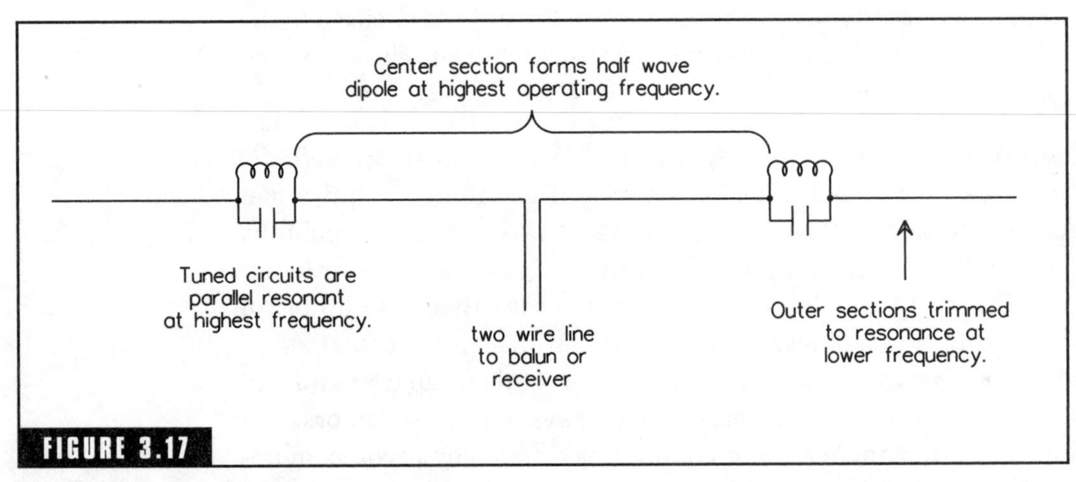

FIGURE 3.17

Two-frequency trap dipole.

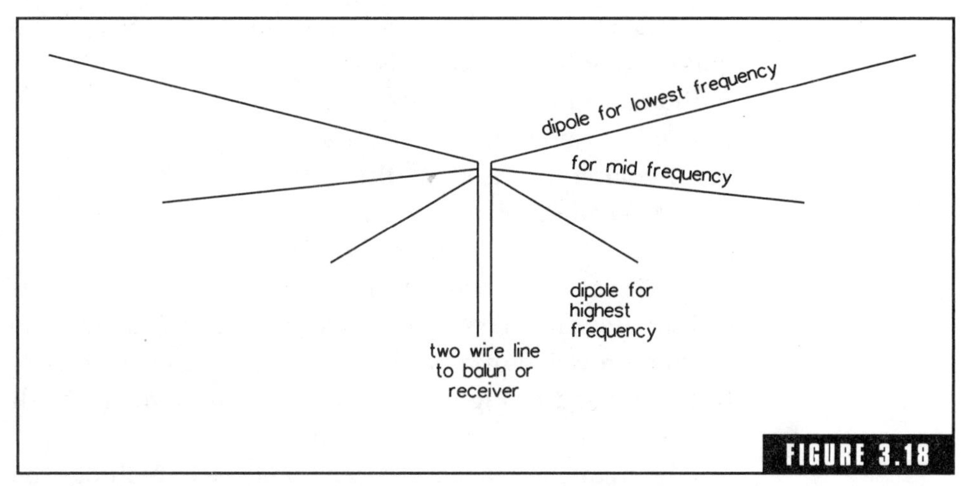

Multiple dipole arrangement.

impedance to the feedline is more of concern to a transmitter than a receiver because for most cases in the range below 30 MHz the limit of receiver sensitivity will be atmospheric noise, as in the case of reception of MF broadcasting signals, so a significant degree of inefficiency can be allowed for in omnidirectional systems before the aerial would be judged as unacceptable. If only reception is ever to be considered it may be possible to build a multiple dipole structure for up to about six frequency bands.

The electrical performance of trap and multiple dipole aerials can be made very similar; the only factors deciding between them are practicalities such as ease of installation and/or appearance in relation to a suburban garden. The trap dipole will usually require a certain amount of on-site adjustment, which may not be a great problem if it is an amateur or recreational project but could be prohibitive if you are paying wages to someone else to do the job. The trap dipole is only a single wire with a few small lumps along it and will always be much less visually intrusive than a spiderweb of multiple dipoles hung over the garden.

As higher frequencies are considered the economics of directional and rotatable aerials becomes more attractive. Simple aerials such as dipoles, long wires, and loops have some directivity, so directional effects using fixed aerials can be provided for all frequencies in the 2- to

30-MHz range. For domestic and amateur purposes the lowest frequency at which a rotatable array could be considered is somewhere in the range of 7 to 14 MHz. In almost all cases aerials that are used as rotatable beams are based on scaled-up versions of aerials that are more commonly used on the VHF and UHF bands. For commercial and government services physical rotation of beams for frequencies below 10 MHz is usually only provided in the form of an adjustment which is set and permanently clamped as part of the installation procedure. Rotation as part of day-to-day operation is rare and if required is often provided by electrical slewing of fixed arrays using the principle illustrated in Figure 3.22. There is more detail on directional and rotatable aerials in the next section.

The frequency range from 1.8 to 30 MHz includes the international amateur bands. One of the major items in the amateur's charter is experimentation and dissemination of the information collected. Because there have been so many amateurs experimenting all over the world there is a good chance that something similar to whatever structure you propose to use as an aerial has been tried by somebody. If you are an amateur operator or short-wave listener you should be able to find the design that fits or can be adapted to your purposes either from books and magazines published by the various relevant societies and associations or from personal contact with active amateur operators in your area. In Australia the body that represents the interests of radio amateurs is The Wireless Institute of Australia. There is an equivalent organization in most of the countries in which amateurs are active. If there is no body such as that local to your area then the American Radio Relay League and the Radio Society of Great Britain are both organizations that have worldwide interests in promoting amateur radio and will happily correspond with you from anywhere in the world. If you have access to the Internet both have sites with relevant contact addresses.

On the short-wave broadcasting bands an aerial designed for use on the nearest amateur band in the spectrum will work well if all the dimensions are scaled to suit. Note, however, the following points:

The scaling factor should be calculated from the center of the amateur band to the center of the short-wave broadcasting band.

The amateur bands are harmonic-related and some multiband designs trade on that fact. Single-band designs can always be adapted by scaling, but that will not necessarily work for more than one band. The trap and multiple-dipole designs illustrated in Figures 3.17 and 3.18 will work when bands are not harmonic-related.

Designs for amateur use will handle transmitter power which may be up to 1 kW. For use in the "receiving only" mode the specified insulators may be more expensive than necessary. However, if you change to smaller, cheaper insulators check that the size of the original was not a critical dimension for the radiation performance of the aerial.

In other uses of the HF band, such as for point-to-point communication links, single channels are allocated and channels are not often harmonic-related. All of the designs for amateur use will work at least as well for point-to-point services and may be able to be made even better if the amateur design includes any components especially to provide a bandspreading function. In general aerials that give good performance on a single channel only are not of great interest to the amateur even though they may give very good performance for point-to-point communication. Some differences will be required in the assessment criteria. So long as you keep that factor in mind you will always find the amateur fraternity, either in the form of personal contact or though books published for use by amateurs, is a useful and helpful source of practical technical information.

3.11 Rebroadcasting

This section deals with aerials for all those cases where a received signal is needed to be somewhat clearer and with lower distortion than would be considered the minimum acceptable for a listener in a domestic situation. It may be a signal to be used as a program source for retransmission, it may be used in a public-gathering broadcast over a PA system, or it may be recorded for another use at a later date. The assumption of this section is that the signal must be received with as little degradation as possible and that performance is important enough to justify a considerable sum of money being paid for a high-performance aerial system.

In the case where the output of the receiver is fed to a transmitter and rebroadcast, the performance of the system must have high sensitivity to give a clear incoming signal but also must be absolutely impeccable with regard to overloading by the signal from the strong local transmitter, and that situation must hold when there is no signal being received and the receiver's AGC has advanced its gain to maximum. If that aspect is less than perfect there is a danger that the receiver-and-transmitter combination could become a giant and very embarrassingly public signal generator. Figure 3.19 shows the feedback path that could cause that effect.

The engineering calculation that determines whether a circuit such as in Figure 3.19 is stable consists of adding all the gains and losses around the complete loop. With working in decibels, which is usually the most convenient way of doing it, the total loop gain must be lower than 0 dB.

In the case of Figure 3.19 there is a coupling loss between the two aerials, which in the case of a station with transmitter and receiver in the same building may amount to 60 to 80 dB; the trap could be relied on for a further 25 dB or so of loss, and there is a conversion loss at the receiver input which will depend on the actual signal level presented to the receiver and the separation in frequency between the transmit and receive signals. The actual figure for conversion loss could vary

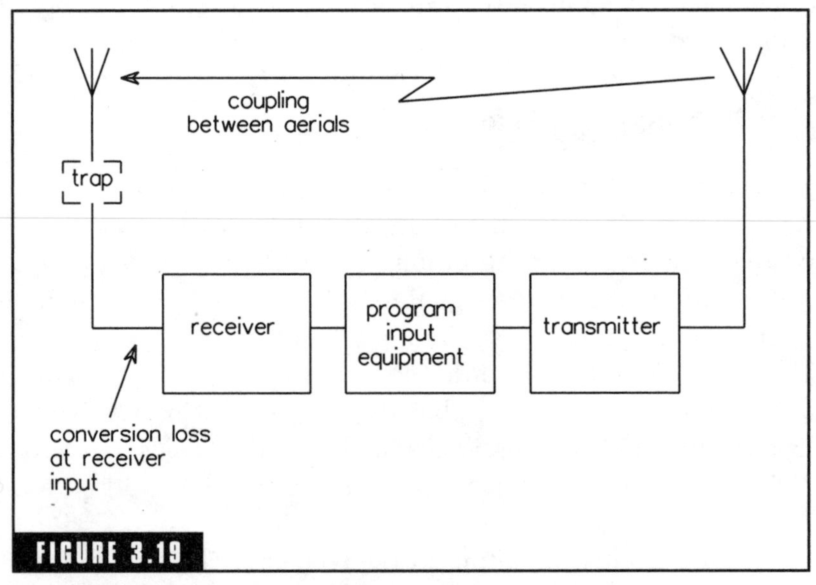

FIGURE 3.19

How feedback can occur during rebroadcasting.

from about 20 to 70 dB. Thus the total loss between transmitter output and receiver input in this instance will be a figure between 105 and 175 dB. If the receiver requires 10 μV to give full audio output with the AGC adjusted to maximum gain, that corresponds to 2 pW in a 50-Ω cable, which is 117 dB below 1 W. If the transmitter output is to be 1 kW then the total gain between receiver input and transmitter output is 147 dB, which would be satisfactory provided that the conversion loss figure for the receiver (due to front-end selectivity) was at least moderately good. If the transmitter is a 250-kW short-wave relay station then total gain is 171 dB, which in most cases would mean that the receiver could not be operated in the same building. Finally, if the receiver is made more sensitive so that full audio output is available with a signal of 0.5 μV, that represents another 26 dB of gain, increasing the total from receiver input to transmitter output to 197 dB.

To increase stability of a rebroadcasting system any or all of the following measures can be taken:

- increase separation between transmit and receive aerials (remove the receiver to another site, for instance)
- make the receiver aerial directional with a null in the direction of the transmitter
- arrange for cross-polarization between transmitter and receiver aerials
- choose the site for the transmitter so that it is on the near edge of its service area, which will allow for a directional aerial at the transmitter with the main beam pointing away from the receiver
- add selectivity (note passive components only) in the line between receiver aerial and receiver—possibly in the form of a preselector in addition to the trap
- select a receiver which has a very high dynamic range to minimize the effect of overloading
- limit the range of control of the receiver AGC so that maximum gain is just adequate for the weakest signal normally received

The points on the above list that deal with receiver specifications have been dealt with in Chapter 2. The rest of this section will concentrate on aerial systems with particular directional properties. Building directivity into structures designed for groundwave reception is different from what is needed for receiving skywaves so the two cases are treated separately.

If the signal to be received is arriving by groundwave, aerials based on quarter-wave verticals or tuned loops may be used. The tuned loop was described in Section 3.7. Figure 3.20 shows the general layout of a full quarter-wave vertical.

Quarter-wave verticals are potentially useful in the frequency range from 1 to about 2.5 MHz. Below 1 MHz the structure is so big that it would be too expensive to build and the sensitivity it offers could not be used because of atmospheric noise; a loaded vertical would be more economic. (Refer to Section 3.5 for details.) For frequencies above 2.5 MHz, the daytime skywave is liable to be predominant at all receiving sites which are more than 20 to 30 km from the transmitter. Dimensions for a quarter-wave vertical tuned to 1 MHz require a height of 71.2 m with the length of radials about 75 m. When tuned to 2.5 MHz these figures are 28.5 and 30.0 m respectively.

The "matching section" shown in Figure 3.20 can be an electrical quarter-wavelength of a coaxial cable of the correct characteristic im-

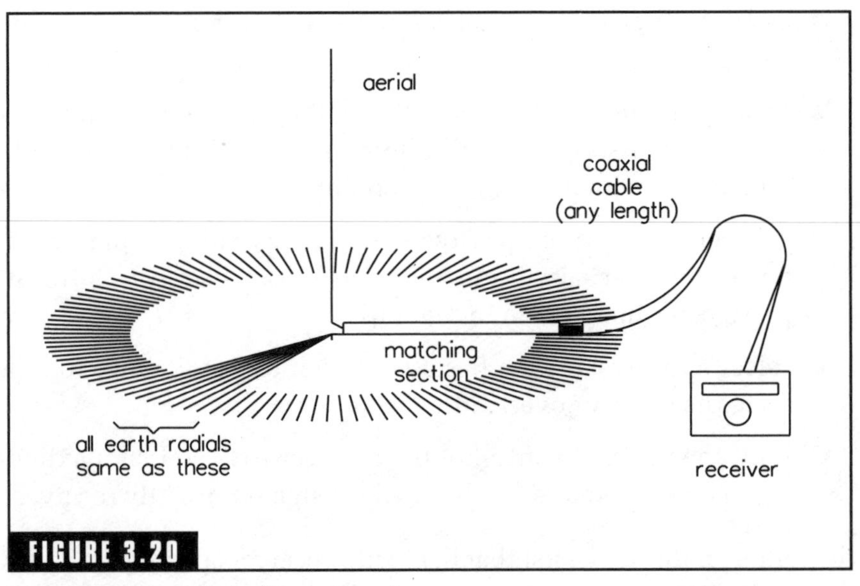

FIGURE 3.20

A quarter-wave vertical with earth mat.

pedance. The output impedance of a quarter-wave vertical aerial will usually be about 35-Ω resistive and for perfect matching the impedance of the matching section must be the geometric mean between whatever that figure for your aerial is and the characteristic impedance of the rest of the feeder. For example, the geometric mean between 35 and 75 is 51.2, so in practice an electrical quarter-wavelength of a 50-Ω cable will often match between a quarter-wave vertical and a 75-Ω cable with a better than 1.05:1 SWR. An electrical quarter-wavelength of coaxial cable is less than the free-space wavelength; it can be calculated by multiplying the free-space quarter-wave by the velocity factor of the cable, which is usually in the range 0.65 to 0.68. The complexity of a matching section would normally only be needed for a receiver if this was one section of a complex directional array. They are commonly used in aerials for transmitting stations where exact power matching is more important. For most other receiving sites the location of the receiver would be close enough to the aerial base so that the small losses due to standing waves could be ignored.

An increase in signal-to-noise ratio of about $1\frac{1}{2}$ to 2 dB may be gained by using two vertical elements spaced more than a quarter-wavelength apart and with their feedlines brought to a common point and combined with the correct phase. Phase can be adjusted by changing the length of the feedline and combining must be done in such a way that there are no impedance mismatches to cause reflected signals. (Section 7.12 in *How Radio Signals Work* refers to this.) The two elements do not both need to be quarter-waves on high-efficiency earth mats; the simpler arrangement shown in Figure 3.7 can usually be used, or they can be loaded by lumped tuning components or be made as inverted L aerials, but they must have identical electrical characteristics. Whatever is done to one must be exactly duplicated in the other.

If the two elements are between a quarter- and a half-wavelength apart and fed in phase with equal power, the horizontal polar diagram of the combination will be exactly that of a half-wave dipole. If the elements are placed further apart there will be more than one null and maximum combination on each side of the central axis, but the signal-to-noise ratio will stay the same. Directions of the nulls can be adjusted by making an unequal feed to the two elements; relative power can be varied by adjustment of impedances at the combining point, and phase relationship can be adjusted by changing the total electrical length of feeder between the combining point and the aerial base. The angle between the nulls can be varied from 180° to almost 0° in either direction within the limitation that the pattern is symmetrical about the line joining the two elements. The *ARRL Antenna Book,* published by The American Radio Relay League, includes a full chapter on this subject under the title "Multi-element Arrays," and if you have a need for more detailed information on the subject that chapter would be a good place to start looking.

To increase the signal-to-noise ratio more elements are needed, each of which must be brought to the combining point and added in the correct phase and in a way that does not generate reflected signals. There is no technical limit to how large and complex an array could be built up by this means, but the result would be physically huge, very expensive, and fiddly to adjust for maximum performance. During World War II arrays of four vertical elements were used for radio direction finding as a navigation aid for aircraft. In these stations the four feedlines were brought to a central station, as in Figure 3.21, and could be switched into various combinations and adjusted in phase using delay lines to steer the nulls of the polar diagram to required points of the compass. These stations probably represent about the maximum complication that is economic for aerials of this type.

The directional pattern of a multielement array does not need to be fixed. The major beam direction and the directions of some nulls can be made adjustable as an operator's control by electrical slewing. In some cases the range of adjustment may be limited; if slewing is achieved by variation between two structures which each have directivity in themselves, the two individual directivity patterns must match and the range of slewing available is only from one side of those patterns to the other. The principle is shown in Figure 3.22.

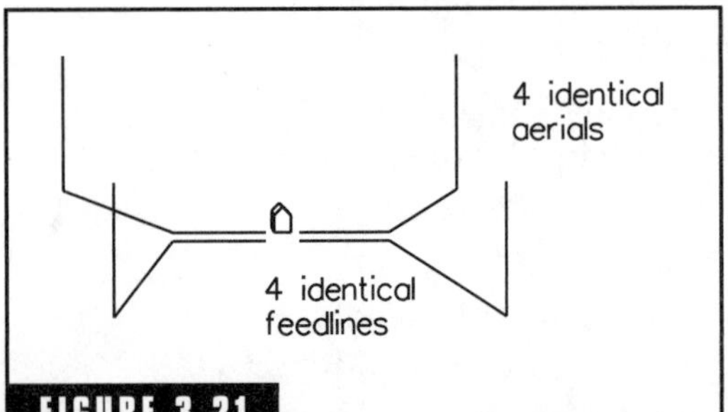

All the multielement arrays that can be built up with quarter-wave verticals can also be built with tuned loops. The loop, however, embodies the effect of two vertical sections built in, which means it has by itself a horizontal polar diagram that includes fixed nulls at right angles to the line of the loop.

4 identical aerials

4 identical feedlines

FIGURE 3.21

Direction finding station—general layout.

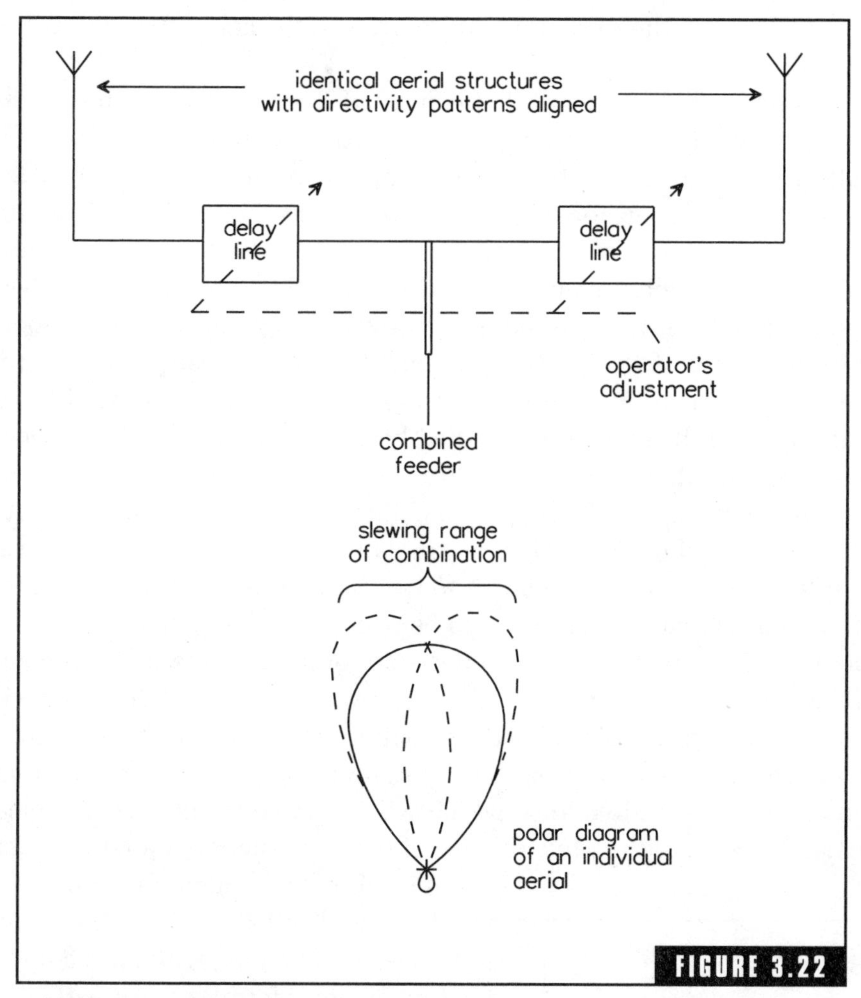

How electrical slewing works.

These nulls cannot be modified by slewing or variation of power division. Arrays made with a combination of loops should always have all the loops aligned in the same direction.

For skywave signals vertical polarization is not needed; the polarization of the signal is changed by reflection in the ionosphere and the change is usually not constant or predictable. The returned signal is sometimes described as "elliptically polarized" and will give at least some signal in a conductor placed at almost any angle. On the MF broadcast band at nighttime the wanted signal is usually strong enough to be received with a quite small (perhaps 10 m or so) length of wire, but it is the directional properties of the aerial that will determine its performance.

A tuned loop will do all that is needed if you happen to be in a location where the desired signal is only suffering interference from one station which happens to be at a compass point approximately 90° away. To use a loop for this purpose set it so that the null is giving maximum rejection of the interfering station. The lobe of maximum sensitivity is much broader and will show little difference in received signal strength of the wanted station over a range of directions about 15° either side of the perfect alignment. Two verticals can have the same effect but in a more flexible form in that by adjusting the distance apart of the elements and the phasing of the feeders the null can be steered to other than exactly 90°.

If you have the available space a horizontal long wire is probably the simplest type of directional aerial to plan and get working; it can be as simple as a wire strung from tree to tree in open land. However, on the MF broadcast band it will have to be over 0.5-km long before much in the way of tailored directional effects are possible. Note that if the end of the wire is left open so that signal currents are reflected the aerial will have four directions of maximum sensitivity; if the far end of the wire is terminated in a resistor to earth the pattern can be reduced to maximum sensitivity in two directions, which will improve the chance of a clear signal. Because of the physical size of the construction a long wire aerial cannot be rotated for control of directional effects. The directional properties of a long wire have a three-dimensional form; the direction of maximum radiation, which looks like two lobes in the horizontal polar diagram, is in fact a cone of maximum radiation as depicted in the isometric sketch of Figure 3.23.

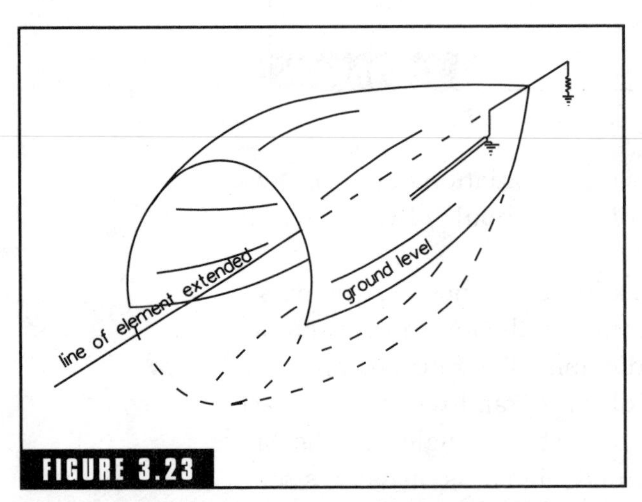

FIGURE 3.23

Long-wire radiation depicted in three-dimensional form.

If you plan for a long wire to have a vertical section for groundwave reception and then turn horizontally for receiving skywaves with some directional effect; remember that it is the section nearest the feedpoint that is most effective for groundwaves, so that section should go nearly vertical for as far as possible before it turns horizontal. If the wire must be allowed to droop closer to the ground try to ensure that the low section is well separated from the feedpoint.

If four long wire sections are used and aligned so that all their effects combine in one direction they can be used as a "rhombic" aerial. These can give high gain (up to 25 dB has been reported) and are electrically simple but physically extremely large. Depending on the gain required a rhombic for the broadcast band could be about 3 km on the long axis and almost 1 km wide. They are not steerable except that it is possible to make them reversible, and the electrical slewing described above would not give good results for these. Rhombics are physically simple; they do not need to be much more than four straight very tall fence lines with only one wire at the top of each, but surveying of the correct angles is the key to successful operation and must be accurate.

A rhombic could be used on an oceanic island where there is interest in receiving all the broadcasting stations local to a particular city. It could be made broadly tuned enough to cover the whole MF broadcasting band but could only be used in one particular direction (and the opposite direction if the rhombic is made reversible; see Figure 3.24). For higher frequencies, such as the international shortwave bands, the dimensions of a rhombic aerial are more economic; the longest dimension may only be a few hundred meters. Rhombics are not favored for use as transmitting aerials be-

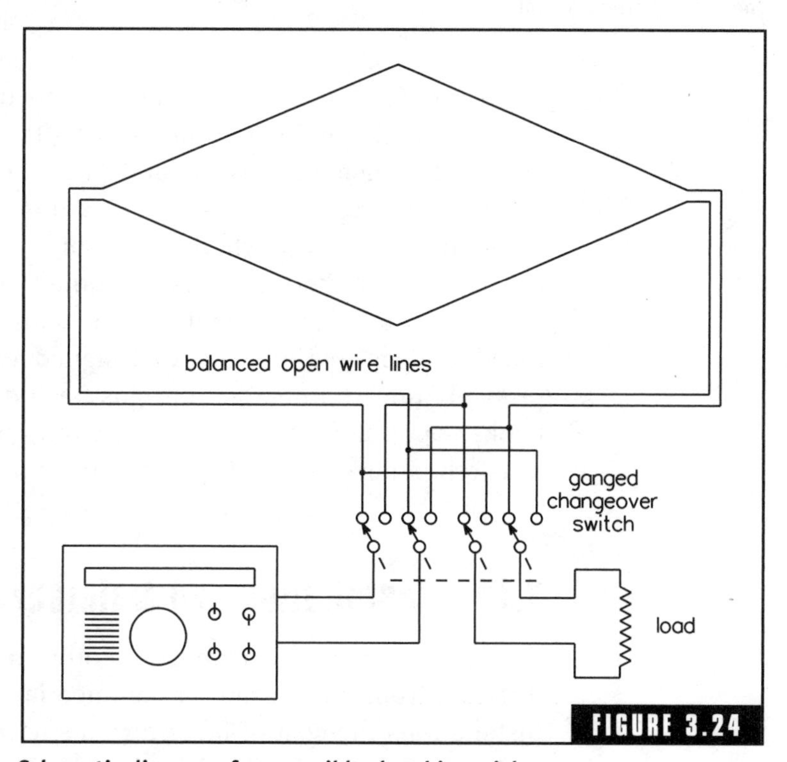

Schematic diagram of a reversible rhombic aerial.

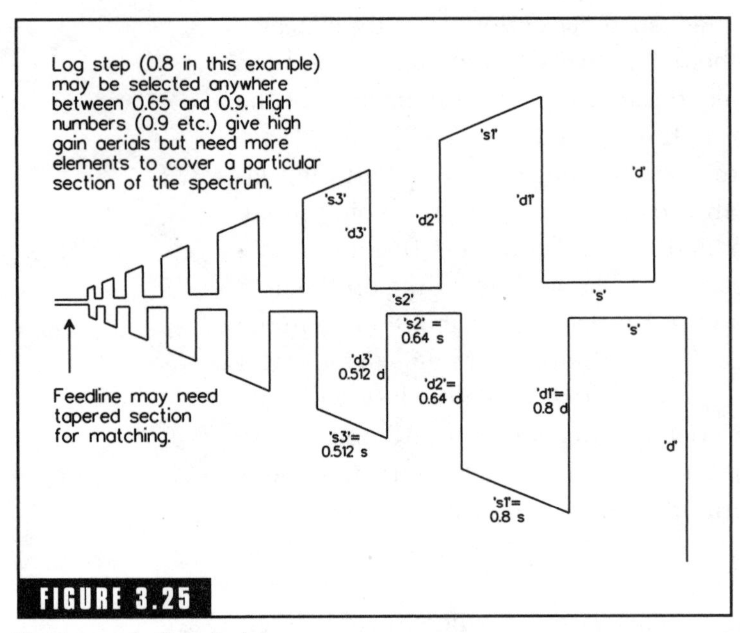

FIGURE 3.25

The log-periodic principle.

cause half of the transmitter power must be dissipated in a terminating resistor, but their use is indicated for receiving when the required signals are scattered over a broad band of frequencies but all come from close to the same direction.

When the receiving aerial must be rotatable to receive signals from several different directions the design will usually be based on either a Yagi (the principle is described in Section 8.7 in *How Radio Signals Work*) or the log-periodic principle, as outlined in Figure 3.25. Log-periodics are inherently broadband devices. Yagis can be made for multifrequency operation by using the "trap dipole" principle on each of the elements. Multiband Yagis designed for the HF amateur bands using traps on the elements are mass-produced and sold commercially. Other uses where directivity, multiband operation, and the ability to be rotated must be combined are so specialized that commercially made equipment is not commonly available; aerials for those services are individually designed and craftsman-built. You will probably not find detailed designs for them in any volume-produced books or popular magazines. They would be the subject of engineering research papers, some of which may be published in specialist scientific or technical journals.

3.12 Installation and Maintenance Practicalities

Aerials for domestic, amateur, and other recreational uses can be quite different from those used for commercial purposes. With commercial installations the cost of labor for on-site installation is a major factor and if the installer is working on a contract any unpredicability in the job can mean the difference between profit and loss. The type of installation that is most satisfactory for commercial use is one where components simply bolt or fasten together and work reliably as soon as they are assembled. On the other hand, if the installation is being done by the owner, who does not pay out money for labor, the major cost becomes the cost of components, so a certain amount of trial-and-error adjustment and gradual refinement of a basically working system is quite acceptable. Of all the parts of the installation the aerial structure is the one where you are most likely to have a chance to make great savings of money for only a small sacrifice in performance. You may pay as much money for the receiver as you would for the aerial, but unless you happen to be able to buy the facilities you need in a good secondhand item you will probably be stuck paying about that amount of money for any model of receiver that will give you the required performance.

As a point of practicality, before you start actual construction work on an aerial system, think about electrical safety. A broadcast band aerial is a large piece of conductor placed high in the air. If there are power lines close try to keep the aerial well away from them, try to avoid passing the aerial under a power line, and ***never ever*** put one over the top. If there are not any power lines closeby then it is likely your installation will be the highest metal object for some distance around and therefore subject to lightning strikes. Protection of a sensitive receiver from pulses induced by lightning is possible (see Section A3 in the Appendix) but takes careful design of the system and if needed should be designed in right from the early planning stages, not tacked on as an afterthought.

Think also about the height and falling. Structures that can be hinged down so that you can work at ground level are usually safer to you than those that require you to climb to the top. If work at height is unavoidable then the structure must be a great deal more robust to take the weight of a person than is needed simply to support a piece of wire. Get into the habit of checking that all attachments to the tower really are attached; bolts must be properly tightened and have a locking device (locknut, spring washer, etc.)—there should be nothing left loose on platforms or just hanging somewhere. A question to be asked is "If a particular guy or aerial element broke would the tensions on the remaining stays cause the tower to fall into the property of a neighbor or some place the public has access to?" An incident of that type could involve you in a public liability lawsuit that would be best avoided if possible.

It often may seem attractive to use trees as supports, and that can be done, but the upper branches are being constantly tossed around by the wind, the maximum movement may be several meters in any direction, and the vibration pattern may cause an attached wire to whip violently. If you can solve these mechanical problems a tree can be a very economic form of tower. One measure that will help at the end point of the wire is to attach the aerial near the top of a major branch then cut off all vegetation above that attachment point; that, of course, will need to be planned with aesthetics of the finished tree kept in mind. Remember also that for most trees the removal of the end of a branch which has an established sap flow will result in a tuft of very thick foliage sprouting from close to the cut. For supports along the length of the wire the forces can be decoupled to a certain extent by hanging the aerial a meter or so below the attachment point with a vertical wire which only supports the weight of the aerial without placing any tension on it as shown in Figure 3.26.

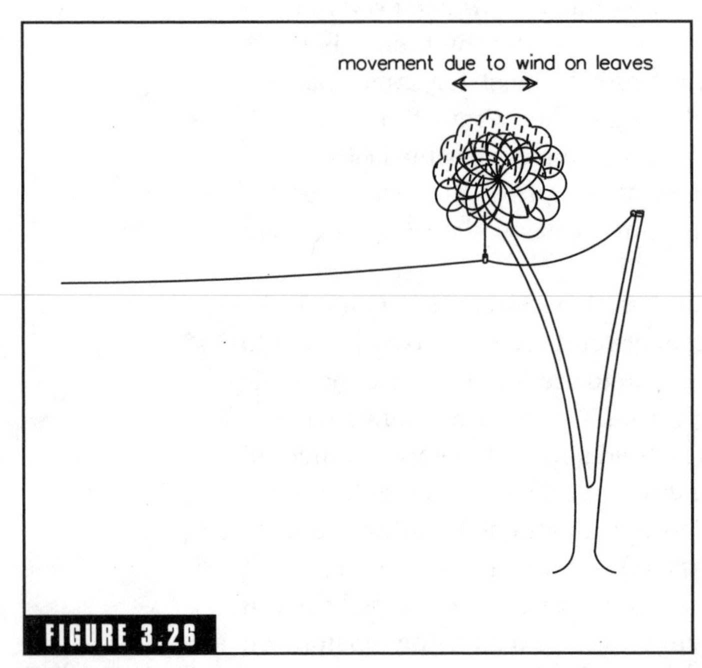

FIGURE 3.26

Using a tree.

The attachment to a tree branch must allow for sap flow. A wire wrapped tightly several turns around the branch and twitched tight or a collar bolted tightly around the whole branch will ringbark that branch and whatever is above that point will die. A bolt hole drilled right through the center of the branch is much less damaging to the tree.

With the aerial, tower height is one of the most negotiable factors; you may find the most economical arrangement is to use one or two lengths of whatever is the standard length of water pipe in your area with an insulator screwed to its top to carry the wire and then plan for some modification of the wire length to compensate for the low position. The most economic height for an aerial depends partly on the type of soil and vegetation at the site. Clay soils, which are highly lossy to radio signals, and dense sappy vegetation are indicators for setting the aerial wire as high as possible; on the other hand, on a leached sandhill in a high-rainfall area the effective earth surface for radio signals is liable to be many meters below the physical surface, so a wire just layed out on the clear sand would give fair results as an aerial. Overall there are so many factors contributing to the equation of maximum microvolts at the receiver per dollar expended that you will probably have to be content with a best guess that will get you close but not perfect. If in doubt try the simplest arrangement first but keep an eye on the possible need to upgrade particular parts of the system later.

The time to start thinking about maintenance is at the planning stage, before the start of construction. Do you intend that the whole aerial should be a throwaway item that is scrapped the first time it fails? If you intend to repair faults at all then you need to think at the planning stage about how you will get access to each joint once the aerial is up in the air. For instance, if you have used a tilt-over tower then welded it in position you will have to cut the weld each time you lower the tower. If you have strained-up the aerial element with a winch or agricultural-type wire strainer, tied it off, cut the excess wire or cable short, and taken the strainer away, as is commonly done for agricultural fencing, you will have great difficulty when you next wish to release the tension and lower the aerial.

On the subject of maintenance the most important aspect of it is periodical inspection. The actual signs you look for will depend a bit on the type of construction; for a galvanized steel tower held in place with guy cables the items of major interest are guy tension, rust, and any visible movement in the tower base or guy anchors. If the tower is wooden then rot, insect attack, and splitting are the important factors. Having a look at each tower every few months will prevent all of these from becoming catastrophic failures, and if each is repaired in good time the life of the structure will be as long as is needed.

CHAPTER 4

Fringe-Area Television in the Computer Age

4.1 The Special Characteristics of Television Signals

Signals radiated by analog television transmitters have a complex mixture of components to carry several different types of information. Amplitude, frequency, and phase modulation are all used as components of the complete signal. Worldwide there are a number of different standards (formats or algorithms) used for the transmission of signals, but there is a basic agreement on the principles to be used. The following list details the major components of a color television signal; actual numbers used are related to the PAL 625 line standard, which is the one used for terrestrial transmissions in Australia, but the types of signals transmitted are the same for several standards; only the numbers are different:

The picture to be transmitted is scanned into lines and the lines are transmitted in time sequence with information on the brightness of each spot in the line being transmitted as an analog modulation of the vision carrier over a restricted range of modulation percentages. A spot of maximum white causes an instant of transmission of the vision carrier at 20% of its maximum power and a spot of blackest black gives an instant of radiation of the vision carrier at

76% of the maximum level. Scanning of the lines is very fast by comparison with mechanical processes; time is measured in millionths of a second. The complete scanning of one line takes 64 μs and the smallest detail is transmitted in about 0.2 μs. The technical name for this smallest possible detail is "pixel." The complete analog signal is called "video luminance modulation."

To signal the end of each line there is a short pulse of transmission of the vision carrier at 100% of the maximum level with no other modulation; that pulse is called the "line synchronizing pulse." The video luminance signal ceases transmission slightly before the start of the sync pulse and does not recommence until several microseconds after the end of it.

When the scanning process has reached the bottom of the picture a signal called the "frame synchronizing pulse" is transmitted. In this pulse, video luminance modulation is suspended for several lines and replaced with transmission of the vision carrier at 100% of the power level. Line synchronizing pulses continue during this period but are transmitted as reduction of the vision carrier power from 100 to 76% for the time of the synch pulse.

The combination of these three signals is described as the "luminance signal" and the carrier wave used to transmit it may either be described as the "vision carrier" or "luminance carrier." A diagram of the way these components fit together is shown in Figure 4.1.

Exactly 5.5 MHz higher in frequency than the vision carrier is another signal, which is a frequency-modulated transmission of the audio information which accompanies the moving picture. This signal is called the "sound carrier," and during reception it is intermodulated with the vision carrier to produce a frequency-modulated signal with an accurate carrier frequency which is always 5.5 MHz no matter which channel the receiver is tuned to or where the fine tuning of the receiver is set. The sound carrier is transmitted at one-tenth the power level (for monophonic sound) of that of the vision carrier at synch pulse tips, and the 20% of vision carrier power which is present even when the picture is transmitting peak white is needed for this intermodulation process. If stereophonic sound transmission is required the power of this carrier is reduced

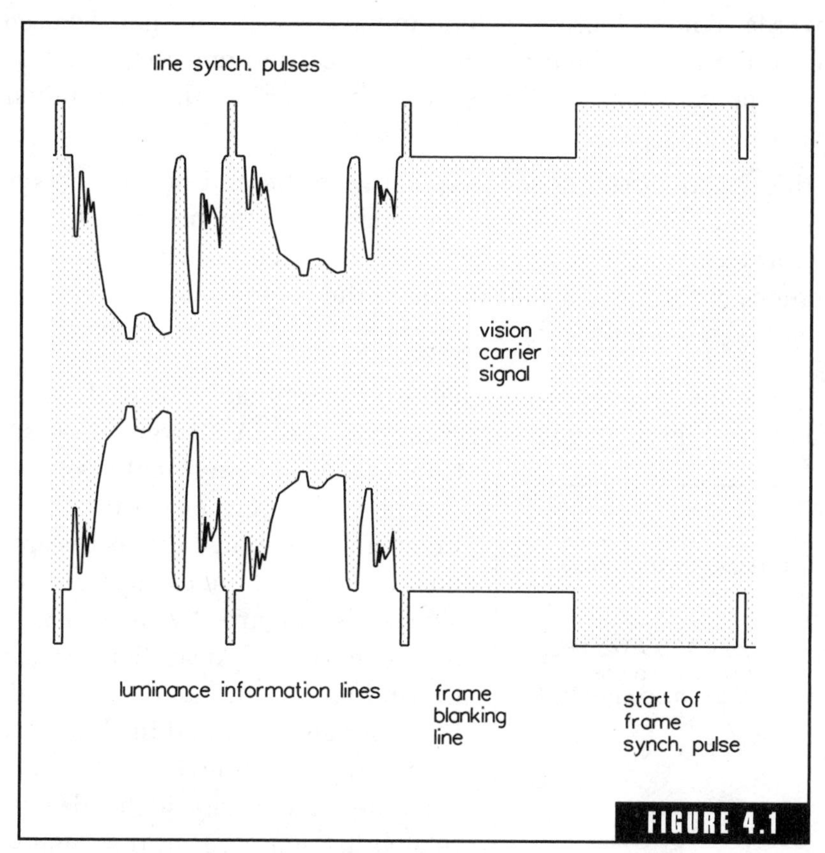

Modulation envelope of the vision carrier.

to 5% of the peak power of the vision carrier and a subcarrier is added 242 kHz higher in frequency at 1% the power level of the vision carrier. The main sound carrier transmits a signal which can be resolved as a mono sound signal, and the subcarrier carries multiplexing components which a suitably equipped receiver can use to resolve the mono signal into a pair of stereo channels.

At a very exact frequency (approximately 4.43 MHz) above the vision carrier there is a subcarrier which carries information on the color components of the signal. This subcarrier is modulated in both amplitude and phase and the combination of both forms of modulation is used to transmit information on the three original color signals (red, green, and blue) generated by the video camera. The words "color," "chroma," "chrominance," or "hue" may be used to describe the modulation on this subcarrier, and the signal itself may be called either the "color subcarrier" or the "chrominance subcarrier."

On the vision carrier at the time between the end of the line synch pulse and the start of video information for the new line there is a short burst of phase and amplitude modulation which gives a reference signal for the modulation of the color subcarrier. This reference signal is called the "color burst."

The complete television signal is the combination of all these components and there may be some others as well. A diagram of the spectrum of a television signal is shown in Figure 4.2; its purpose is to show the frequency relationships between all the carriers and subcarriers and their sidebands. Figure 4.2 is specific to the signal of an Australian terrestrial transmitter operating on channel 2. For other channels and in all countries where the 625-line PAL standard is used the carriers and sidebands will be related to each other in the same way, but actual frequencies may be different. In countries where other standards are used the general relationship between the components will be similar, but the actual frequencies of the sidebands will be different even if the vision carrier happened to be on the same frequency; for example, in the 525-line NTSC standard commonly used in the United States the frame rate is 60 Hz and the line frequency is 15,750 Hz, and these figures will determine the actual frequencies of the sidebands.

Sound carrier, 5 500 MHz. above vision carrier; frequency modulated.

Colour sub-carrier, 4.43361875 MHz. above vision carrier; quadrature (phase and amplitude) modulated.

Vision carrier frequency, 1.25 MHz. above low limit. Vestigial sideband, amplitude modulated.

channel low frequency limit

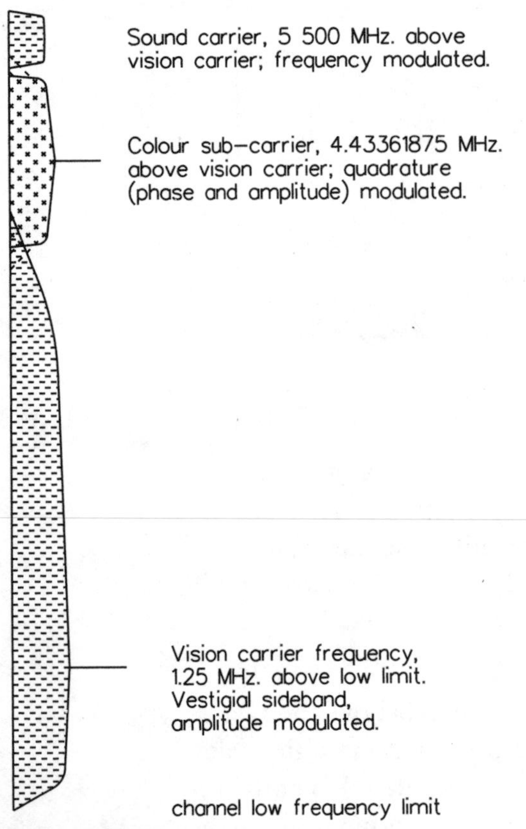

FIGURE 4.2

Spectrum of a television signal.

Correct operation of the receiver

uses comparison between the components for many of its functions so it depends on all components being received at the correct time with the correct relative level and free enough from noise and interference for the comparison to be made accurately. One important result is that a television signal must be strong enough to be relatively noise-free, but another is that the sensitive receiver is easily overloaded so the dynamic range available for television reception is much less than for most other classes of radio service. For an individual transmission there will be some noise noticed on the picture when the strength of the received signal is less than about a millivolt and for some receivers overloading becomes noticeable in its effect if the input signal is greater than about 30 mV. These two figures represent slightly less than 30 dB of dynamic range (compare this with a figure of 80 to 100 dB for an AM broadcasting receiver and up to 120 dB for a voice communication system).

When the aim is to receive a group of channels over a broad frequency range the situation becomes more critical. If all were exactly the same strength the overloading point would be reduced by 3 dB each time the number of stations available is doubled so if a total of eight equal transmissions are present the dynamic range is reduced to 21 dB, and if the signal you want is not the strongest in the group you can easily find that the dynamic range available in practice is reduced to almost zero. Often the requirement to have the wanted signal strong enough to be noise-free without causing the front end of the receiver to be overloaded is one of the most critical factors in the design of a television receiving site.

4.2 Defining a Fringe Area

Television is transmitted in the VHF and UHF bands. Each channel occupies several megaHertz of spectrum space (in Australia it is 7 MHz and for some other standards even more per channel) so carrier frequencies must be high enough to accommodate the modulation bandwidth; in Australia the low-frequency edge of the lowest channel is 54 MHz. Channel-frequency allocations are above the range in which skywave propagation is normally expected so coverage of each transmitter is line-of-sight-plus-a-little-bit. The "plus-a-little-bit" varies with the frequency allocation and to a certain extent the transmitter power; it is substantially more at the low-frequency end of the VHF band (50 MHz range) than at the high-frequency end of the UHF TV allocation (800+MHz). Transmitters are usually sited on the most prominent mountain in the area and the aerials are often placed on towers over 100 m tall with multielement arrays in each direction. The aerial array does give some gain, but its most critically adjusted feature is the careful tailoring of the beam to give as nearly as possible an even field strength over the whole of the primary service area.

A typical major-city television station may be sited on a location which is 500 to 1500 m above the surrounding countryside with actual transmitter power of between 2 and 10 kW for the synch pulse tips of the vision carrier, and the aerial gain may be quoted as 10 dB. This will give a primary service area which in most directions will extend out to 60 to 100 km from the transmitter site. Figure 4.3 shows what a typical map of such an area could look like; there is a detailed description of what the lines on such a map actually mean in "Measuring Field Strengths" (see section A5 in the Appendix). The aim will be that receivers close to the transmitter (perhaps 1 to 10 km away) can be used with a rooftop aerial with zero gain and have a signal not greater than 30 mV at the input to the receiver and that those at the far edge of the primary service area can get a signal of at least 1 mV at the receiver by using higher-gain (8 to 10 dB) aerials in chosen clear locations.

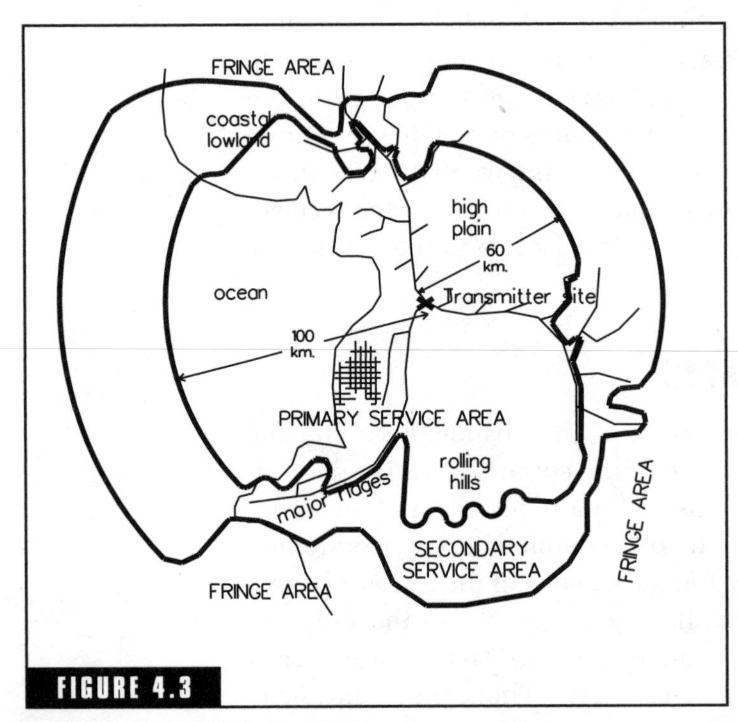

FIGURE 4.3

Example of a map of field strengths.

Outside of that primary service area there will be a region called the "secondary service area"; perhaps another 30- to 40-km radius (for VHF transmissions) in which good signals can generally be expected but will require that the receiving aerial is placed on a substantial tower in a clear location. There will be some places where signals will show some noise and some times when propagation disturbances will disrupt the program.

Beyond that region is the fringe area. Very high towers and high-gain aerials are needed to receive anything at all and reception must rely on some assistance from a favorable propagation mechanism, and even then it will be normal for the received picture to show some noise and be easily affected by weather conditions and local sources of interference. Refraction and diffraction (explained in Sections 5.4 and 5.5 in *How Radio Signals Work*) are the mechanisms with most effect in open country and rolling hills.

Diffraction is the result of application of some of the basic laws of optics to the radio signal in the particular geometric configuration of that path. Its effect is to cause the "shadow" to have a fuzzy edge; if at a particular time you were to take measurements of field strength at various heights over a particular location you would measure a gradually rising level of signal with increasing height rather than a sudden jump to strong signal as you climbed over the line of the geometric horizon. Diffraction effects can be observed over the whole electromagnetic spectrum from VLF radio to the gamma ray/cosmic ray end of the range. The portion of the spectrum most affected depends on the size of the object causing interruption to the wavefront. For shadows caused by obstructions, the size of the curve of the Earth the effect of diffraction is most noticeable for radio signals at the low-frequency (long-wavelength) end of the range. For television transmissions, objects the size of hills and mountain ranges are relevant. Given a particular effective radiated power and assuming the receiver requires a particular strength of signal in terms of millivolts at the receiver input, the effect of diffraction on practical signals is that for whatever distance past the radio horizon a UHF TV signal is workable, a signal at the low end of the VHF band will be effective for several times that distance. Figure 4.4 illustrates this principle.

Refraction occurs in the lower layers of the atmosphere and is greatly affected by local weather conditions at the time; its actual effect is to cause the signal path to curve, but the subjective effect to an observer is that the apparent radius of the Earth and therefore the distance to the horizon appears to vary with changes in weather conditions. In exactly the same way that red light passing through a prism is

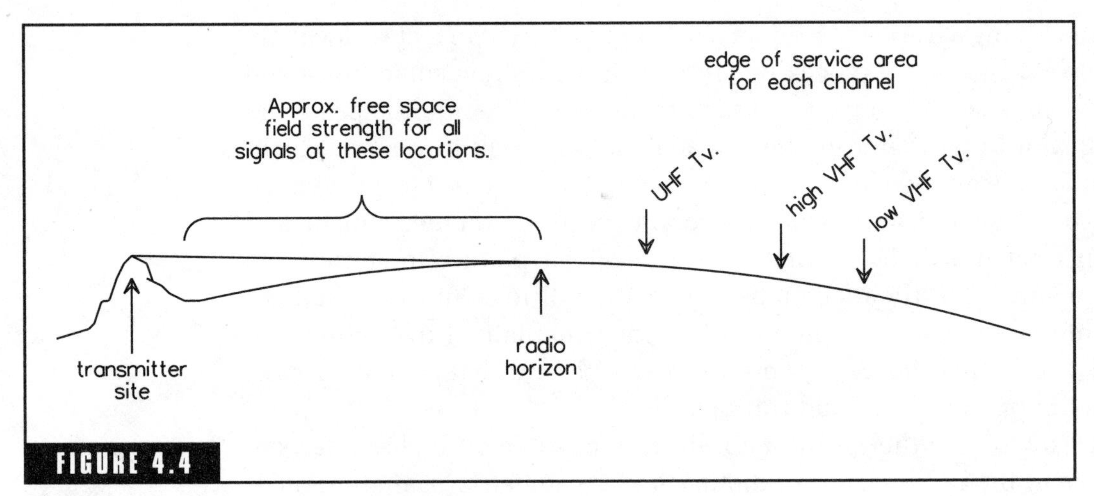

FIGURE 4.4

Edge of service area for different channels.

bent more than blue light that enters on the same path, low-frequency radio signals are bent more than higher frequency signals in their travel through the atmosphere, with the result that for low-channel television signals the radio horizon itself is at a greater radial distance from the transmitter. Also, once the signal gets past the radio horizon the angle of the refracted ray line is closer to following the curve of the Earth than is the case for a higher-numbered channel. This factor works in the same direction as the effect illustrated in Figure 4.4 and amplifies its effect. When the receiver aerial is slightly below the radio line of sight the variation due to weather changes the observed position of the receiver aerial with respect to the fuzzy edge of the shadow (diffraction), which results in the received signal strength being varied by changes in the weather (see Figures 5.4 and 5.6 in *How Radio Signals Work* for an explanation of the term "radio line of sight").

In mountainous regions reflections are possible, and also there may be a particular form of propagation described as "knife-edge refraction," which occurs when a signal travels over a sharply peaked ridge and a portion of it is deflected downward by interaction with the partially conductive surface of the ridge. Knife-edge refraction is a significant factor in those cases where the signal path to a receiver crosses the ridge at close to a right angle and the top of the range is high enough to be within the radio line of sight of the transmitter aerial. Its effect is hard to theoretically predict because one of the important factors in its magnitude is the radius of curvature of the ridge at the top of the mountain range. Knife-edge refraction, however, is not varied by weather conditions; if it is present at a particular location it will be present all the time.

Reflections (apart from those of the type shown in Figure 4.5) are generally not welcomed for television signals because the reflected signal arrives by a different path from the main signal. The total length of the path is different for the two signals and this different length of path means that one is slightly time-delayed in relation to the other. Unless the time delay is very short (less than 100 ns equivalent to a path difference of about 30 m) the delayed signal does not reinforce the other but causes a second ghost image which reduces the definition of the original picture.

4.3 Weak Signals in Low-Noise Locations

The information in this section typically applies to rural locations which are between 100 and 500 km from a major city (that is, large enough to have their own television transmitters) and far enough

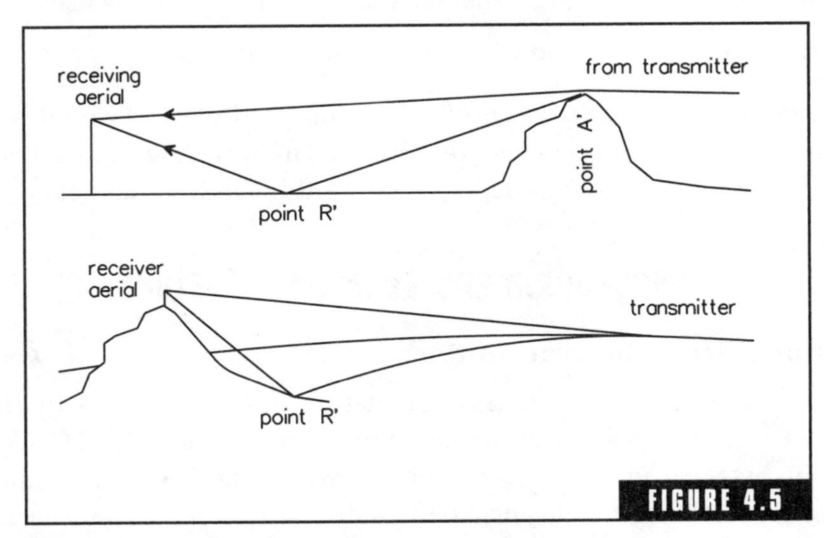

FIGURE 4.5

Ray lines for a path with one reflection.

away from the nearest neighbors to ensure that any local sources of interference, if they occur, are due to facilities owned by the person or people who are receiving the television signal. Typical locations are farm or pastoral homesteads or mining camps. In most of those locations the limit of sensitivity of the receiver is set by thermal noise generated internally in the receiver by its first amplifying stage. There is some value in making a careful selection of which receiver to use; a set designed for good performance when signals are strong may require a millivolt or so of signal for proper operation. On the other hand, a set designed for high sensitivity may be able to produce a quite intelligible (although perhaps noisy) picture with a signal of only 100 μV. When you have selected the most sensitive receiver available the clarity of the picture on the screen will be determined by the strength of signal that can be applied to its input, which will depend on:

- aerial height in relation to distance from the transmitter
- features of the terrain between transmitter and receiver
- aerial forward gain
- the number of channels being designed for
- length of feeder and whether a masthead amplifier is used
- density of vegetation close to the receiving site

In addition, the quality of the final picture will depend on whether sources of local interference are present. The next few sections contain some short notes on each of these points as individual items.

4.4 The Geographic and Geometric Factors

Distance from the Transmitter

Based on the 4/3 Earth radius calculation (see Section 5.5 in *How Radio Signals Work* for an explanation of what the term '4/3 Earth Radius' means) in those cases where there are no hills or mountain ranges in the way the distance to the radio horizon for various heights of transmitter aerial over flat ground with only light vegetation are shown below:

Height of transmitter aerial (in meters)	Distance to radio horizon (in kilometers)
100	41
200	58
300	71
500	91
1000	129
1500	158
2000	183

Note that these figures refer to the difference in height between the transmitting aerial and the surrounding flat ground; altitude above sea level may have little relevance to that figure. All receiving locations which are closer than whichever of those distances applies to your particular transmitter are within radio-line-of-sight, and it should only be necessary for the receiving aerial to be clear of local obstructions in order to receive a signal at very close to the free-space field-strength figure which applies for that distance, transmitter power, and aerial gain.

Aerial Height

If your receiving location is beyond the radio horizon then the grazing ray (as shown in Figure 5.6 in *How Radio Signals Work*) will pass above you by the following amounts:

Your distance past radio horizon (in kilometers)	Height of grazing ray above smooth Earth (in meters)
10	6
20	24
30	54
40	96

These figures can be related to sandhill or slightly undulating country (hills up to about 10 m high) if you take the tops of the hills as the flat surface; in heavily forested country the tree tops correspond to the flat surface. In theory, the signal strength at the height of the grazing ray should be about 6 dB less than the free-space field strength for that distance from the transmitter.

For all receiving aerials which are below the grazing ray line, the signal strength will be determined by the interaction of diffraction, refraction, and weather conditions as explained in Section 4.2. In practice high-gain aerials can be used to counteract at least part of the loss of signal strength and the final combination of tower height, aerial gain, and how much noise can be tolerated is determined by an economic calculation similar to that described in Chapter 1.

When the receiver site is close to more substantial hills or is located at the top of a small rise the smooth relationship between aerial height and signal strength may not apply. In those locations where the signal grazes the top of a distant hill and then travels over an area of relatively flat ground there will be quite rapid variations of received signal strength with relation to height. The signal reflected from the flat foreground will generate interference fringes which can make the signal at the aerial up to 6 dB above the strength of the incident signal or almost completely nulled out. Complete nulling only occurs if both signals are exactly the same strength so it doesn't happen very often, but total variations of 6 to 8 dB are common. A similar effect occurs if the receiving aerial is on the top of a small rise with a gentle slope in the direction of the transmitter. Figure 4.5 shows the mechanism at work, and Figure 4.6 shows in graphical form the effect of height on signal strength.

There is a phase reversal when the signal is reflected so the "in phase" condition of maximum signal applies when the path of the reflected wave is an odd number of half-wavelengths (1, 3, or 5, etc.) longer than the direct path.

In the simple case of one obstruction it is possible to make a reasonably close calculation of the height of the signal maxima if you know the wavelength of the signal, the distance to the obstruction, and the height difference between the reflection point and the obstruction (see Section A1, "Fresnel Zone Theory," in the Appendix).

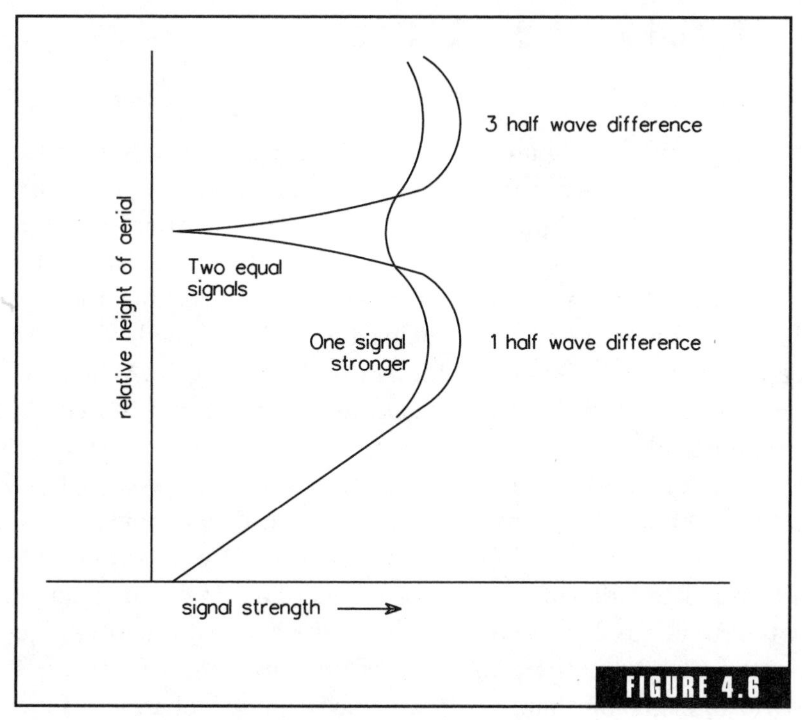

FIGURE 4.6

Signal strength versus height for a path with one reflection.

Terrain along the Path

In those cases where the surface of the Earth between transmitter and receiver is not what a radio signal would see as flat and smooth there are possibilities of obstruction, reflection, or knife-edge refraction and the prediction of a figure for expected signal strength will require a "path profile" calculation. This calculation is routinely done by engineers designing radio links for broadband telecommunications bearers but is not usually considered worthwhile in the case of an individual television receiver; the cost of the engineer's time is more than the cost of buying a receiver and aerial and just employing trial and error. A path profile may be used, however, if all the people in a particular community are suffering from weak signals and they are considering installation of a repeater or CATV system.

There is more detail on path profiles in Chapter 9 ("Radio Communications). Repeaters and CATV are dealt with in Sections 4.8 to 4.10.

4.5 Aerial Gains and Losses

Aerial Forward Gain

In theory, when the signal is weak it should be possible to make the forward gain and/or capture area of the aerial as high as is required to deliver a noise-free signal, but there are certain practical and economic limitations which modify that theory. In the UHF band aerial gain of 15 to 20 dB is quite readily usable, but as the frequency is made lower the practical limit on aerial gain is reduced.

Bandwidth is one factor that limits usable aerial gain. When antenna assemblies are combined to make a higher gain aerial the combining process normally causes restriction of bandwidth. It is possible to start with very wide-band antenna units to counteract that process, but these usually are made with many elements so the finished product becomes an enormous structure which cannot economically be supported on a high tower. The sidebands and subcarriers of a television signal all occupy the same number of megaHertz so for signals at the low-frequency end of the VHF band the proportion of the spectrum used (i.e., relative bandwidth) is greater than it is for higher-frequency channels. In principle, antenna sections can be stacked as shown in Figure 4.7 without limit using quarter-wave transformer sections to adjust impedances for minimum SWR, but each section is frequency-dependant so each additional stage of stacking and matching reduces the overall bandwidth. Design engineers assess the performance of particular types of structure in terms of a factor that they call "gain-bandwidth product."

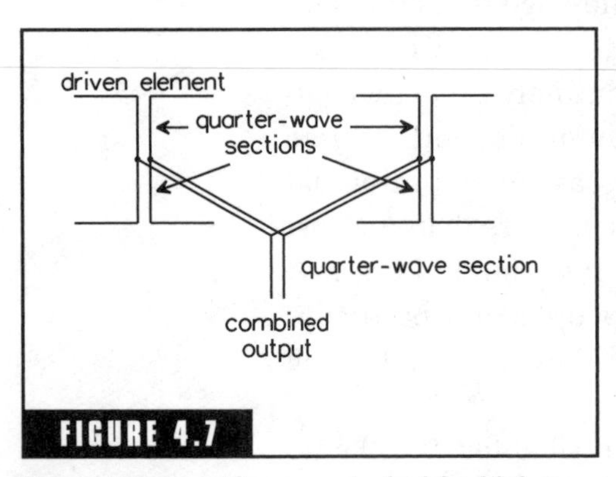

FIGURE 4.7

How antenna sections are stacked for higher gain.

The size of individual elements is also a factor. In most cases, length of the element must be close to a half-wavelength so when the wavelength is longer the elements are longer, must be made of heavier-gauge metal for mechanical support, and are spaced further apart on the boom so overall the structure required for a particular figure of forward gain is bigger, heavier, and more costly.

It is unusual for multichannel aerials for the low end of the VHF television band to be made with gain higher than about 3 to 4 dB, and in the high-frequency VHF range 8 to

10 dB is about the economic maximum. Single-channel VHF beams (usually Yagis) can be made with gain perhaps up to 6 dB above these figures. There is not usually any need for single-channel aerials for UHF; the channels are small enough in relative bandwidth so that a Yagi, even with no intended broadband features, will cover several channels. The spectrum managers usually try to arrange stations so that all UHF channels transmitting from a particular location are within the passband range of a single Yagi.

Designing for a Certain Number of Channels

As explained previously, there is a tradeoff between forward gain and bandwidth. In some cases, however, multichannel performance can be achieved by simple structures with stations which are harmonically related. When television systems were initially being set up the spectrum managers tried to arrange channel usage so that the populations of the largest cities had the most opportunities for economic aerial performance; later stations at the provincial cities were fitted in so that stations which were close to each other were protected as much as possible from both adjacent channel interference and interference due to intermodulation products.

Your receiving location may be in the fringe area for several cities, and in that case you may have to make choices about which stations to favor. If you attempt to give the overall aerial sensitivity for all available stations and combine them into one multichannel input to the receiver, the range of signals picked up may be so great that the dynamic range problem described in Section 4.1 means that none of the desired signals is able to produce a clear picture on the screen. There are two broad classes of ways to reduce the effect of that problem: you can either mount a broadband aerial on a rotator using the arrangement shown in Figure 4.8 or set up a particular beam for

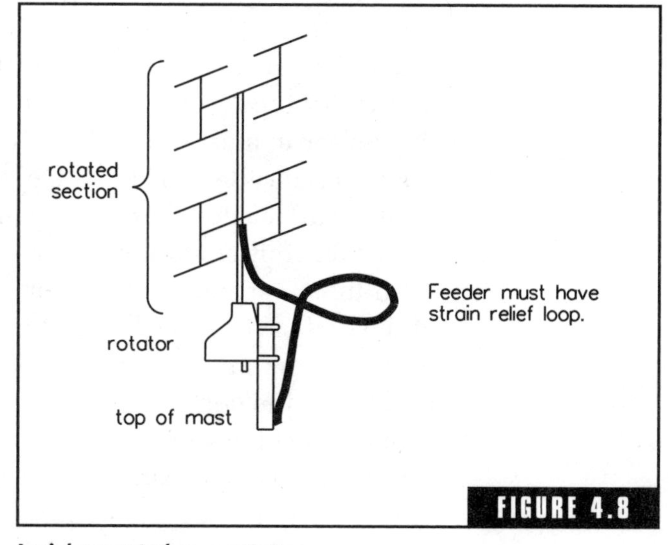

FIGURE 4.8

Aerial mounted on a rotator.

each desired station and then switch between them. There will be more detail on this in a later section.

There is an advantage to only providing for reception of those stations which you know the viewers will require; in general, limiting the number of stations provided for will give opportunity for better quality of reception of those that are desired.

Length of Feeder

The downside of having a high tower is that it also needs a long feeder, and there comes a point where the gain in signal strength due to making the tower higher is all lost in the extra length of feeder cable. The effect of cable losses is more severe for high frequencies than for the low end of the band so will be mainly noticed on UHF channels, which usually means that a higher operating frequency requires a more expensive feeder.

There are relatively cheap small-diameter coaxial cables whose attenuation can be in the range of 30 to 50 dB per 100 m at a frequency of 400 MHz; these are totally unsuitable for use on UHF, but they can sometimes be used for short runs on the low-frequency end of the VHF band. Their attenuation at 50 MHz may only be 5 to 10 dB per 100 m. Cable diameter affects attenuation. For cables made with the same materials and construction methods doubling the diameter will approximately halve the loss per 100 m. The type of material and the construction method also have an effect. The lowest loss-insulating material of all is a vacuum, but that is usually impractical; the next best is dry inert gas, and almost as good is dry air or dry nitrogen. If the dielectric must be solid, one of the best (and also most expensive) is a material called PTFE, which is a refined grade of the Teflon that is used to surface nonstick frying pans and cooking pots.

At transmitter stations the coaxial feeder is often made with two solid tubes separated by spacers, as shown in Figure 4.9, with most of the dielectric space being filled with dry inert gas (argon), dry nitrogen, or dry air. In installations of this type the outer sheath of the feeder and the connectors are made airtight, and the whole assembly is pressurized by being connected to either a cylinder of the required gas (through a pressure reducer) or a small air compressor/dryer unit.

For receiving stations coaxial cables are made in which the dielectric is either made to foam or shaped into cells so that most of the volume between the two conductors is filled with dry air. A practical point concerning installation of these cables is that compared to a solid dielectric cable, the minimum bending radius is much greater; solid dielectric cable can be bent with a radius of curve comparable with its own diameter, but if that was done with one of these, the conductors would be kinked and press against each other and the cable would be permanently damaged at that point. Unless you have a manufacturer's specification to work to, it would pay to regard the drum the cable is supplied on as a measure of the minimum radius to use. However, when properly installed, good-quality foam or cellular dielectric coaxial cables with a diameter in the range of 10 to 12 mm may show an attenuation figure at 500 MHz, which is as low as 3 to 4 dB per 100 m.

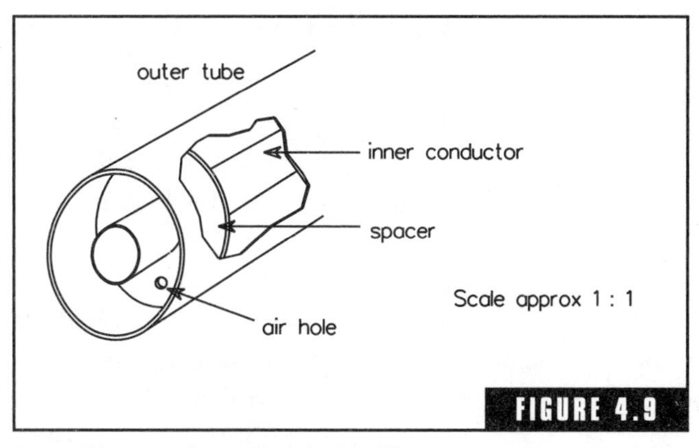

FIGURE 4.9

Transmitter station transmission line.

There are many designs and construction methods possible to achieve the basic aim of mechanical support with maximum possible air space between the conductors. Figure 4.10 shows one type that is relatively easy to manufacture because the dielectric shape can be extruded around the center conductor with very little more difficulty than applies to normal single-insulated electrical wire. Other designs can give a higher proportion of free air space but most of the more advanced designs require more complicated processes than simple extrusion and wrapping.

Air-spaced twin open wire lines such as the 300-Ω ribbon or ladder feeders made especially for television use give lower loss figures than even the best of coaxial cables, but they are so much more susceptible to pickup of off-air signals and interference that they are hardly ever used in critical situations.

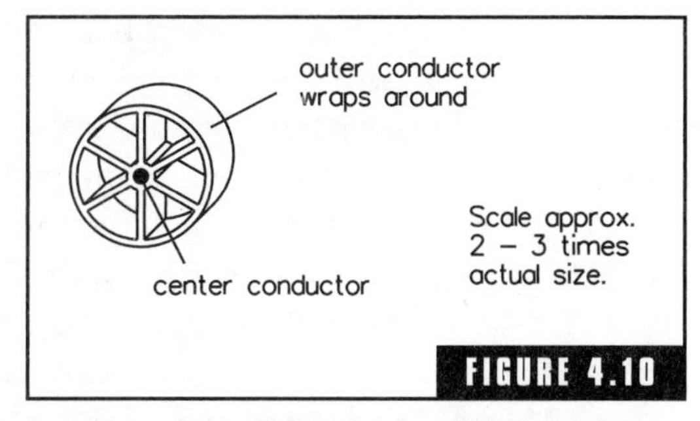

FIGURE 4.10

A section of the dielectric of a UHF TV receiver feeder.

Masthead Amplifiers

An amplifier can be placed at the top of the tower within a meter or so of the aerial output connector to overcome the losses due to the feeder and in ideal conditions can be so effective that the limitation on tower height due to feeder loss is almost totally removed. With one of these masthead amplifiers, the total length of feeder can be up to about 300 m, so in some cases there is opportunity to remove the aerial and tower to the top of a local hill where stronger, clearer signals are available. Care is needed in their use; the sensitivity of the whole system is controlled by noise and intermodulation generated in the first amplifying stage, which in this case is the input stage of the masthead amp. That amplifier must be made as sensitive as possible, but it must also be broadband so if you are trying to receive a weak signal in the presence of stronger ones you will have to deal with the dynamic range problem in its worst form. Masthead amplifiers can be fitted with traps to control overloading and part of the installation process may include setting of the trap frequency. There is more information about the effect of interfering signals on masthead amplifiers in Section 5.6 of the next chapter.

4.6 Density of Vegetation Close to the Receiving Site

In terms of radio signals vegetation acts as a resistive (lossy) conductor. In theory, all vegetation has an absorbing effect on all radio signals, but in practice for low frequencies the effect is usually not significant. On the VHF and UHF bands, however, it must be taken into account. For a signal at 1000 MHz any screen of trees thick enough to obstruct vision will also be thick enough to obstruct a radio signal. At lower frequencies the effect is less severe; at 100 to 150 MHz a moderately thick forest can increase the path loss of the signal by about 15 dB.

Thick vegetation very close to the tips of antenna elements can upset the operation of the aerial by detuning those elements. As a general rule of thumb, the aerial should be kept clear of vegetation by a distance equal to the longest dimension of the assembly (i.e., whichever is longer of the boom or the elements) and if possible a bit more than that in the direction of the forward beam. There is also an advantage in siting the aerial so that the line on which the signal will arrive is at least a wavelength or so above local treetops.

Vegetation which is more than a quarter-wavelength in height will absorb signals at reflection points and also at points where knife-edge refraction would otherwise be occurring. These are the points shown in Figure 4.5 as "point A" and "point R." This effect, of course, is related to frequency; for instance, an herb garden in which plants are thick and lush but only about half-a-meter high at the point of reflection would have no effect on a low-VHF channel but would completely remove the reflected signal on UHF; on the other hand, a line of taller trees such as poplars at that point may have more effect on the lower-frequency signal.

Occasionally, the absorbing effect of vegetation can be put to good use; it is a broadband effect so it will absorb interference as well as desired signals. If there is an unavoidable source of interference such as a welder or arc lamp in a place where it can be hidden from the receiving aerial by a group of trees then the disturbance will be reduced. Note, however, that trees are in a state of continual slow change; they grow taller and wider, branches drop off, they grow old and die. If you plan to use vegetation as a screen in this way you will need to continually manage it for the proper effect.

4.7 Ghosts

Ghosting is often classified as a form of interference, so it is dealt with in the next chapter, but all ghost images must be due to a signal which originates from the same transmitter as the direct signal but arrives at the receiver by another (time-delayed) path; that is, by a reflection of some sort. So in terms of the technicality of fixing the problem it must be treated as a propagation defect. The distance apart on the screen between the main image and the ghost can give you a rough indication of the extra path length traveled by the reflected signal, but be warned—it is only a rough indication! There has been a lot of discussion over the years in the technical media over whether measurements of the ghost image on the screen can be made accurate enough to be used to suggest a particular object as the reflection point. One difficulty is due to the need to make measurements on a moving picture with small-enough errors to make a sensible estimate of time difference from which path-length difference can be calculated. Another is that once a path-length difference is calculated it specifies a line which is an ellipse, with the transmitter aerial at one focus and the receiver aerial at the other, and the reflection point can be anywhere on that line; a wheat silo 200 m behind the receiving aerial can give an effect indistinguishable from that of a rock face halfway between the transmitter and receiver and 20 km off the line of the direct signal.

Figure 4.11 shows the conditions that might apply if you estimate that the extra path length for the reflected signal is 10 μs ±15% (which is about the closest you would be able to estimate with a moving picture on the screen). The point of reflection can be anywhere between the two lines. If you read through the section in the Appendix on Fresnel zones (Section A1) you will recognize the diagram of Figure 4.11 as related to them. When Fresnel zone theory is used to calculate the path loss on a radio-communication link it is zone numbers 1 or 2 and sometimes number 3 that are considered, but in this case the reflector may be in zone number 10 to a couple of hundred or more, and you will never be able to put an exact figure to the zone number. With those provisos all the results of Fresnel zone calculations can be applied to TV ghosts.

There is one particular case where the appearance of a ghost image can be used to identify its source. If you are using a very long feeder (more than about 70 m for coaxial cable) and the power transfer between the cable and the receiver is poorly matched, the standing waves produced will be time-delayed enough to show as a ghost image. A

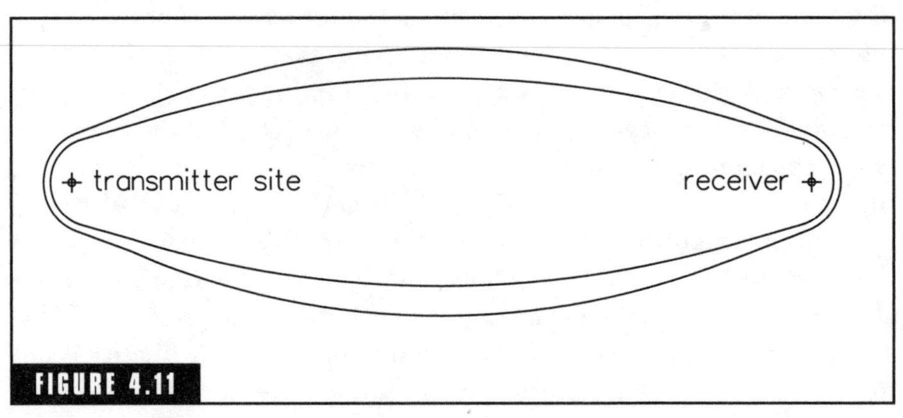

FIGURE 4.11

Possible locations of a reflection point.

ghost from that cause will usually show as a number of closely spaced images all exactly the same distance apart, and the separation distance can be correlated with the length of feeder (remember that the velocity factor for most coaxial cables is in the range of 0.65 to 0.68).

The practicality of ghosts is that in most locations where they will be a problem there will be several possible sources of them so rigorously identifying and removing the effect of any one will not do a lot of good; in those sort of locations the selection of aerial should be for maximum front-to-back ratio and minimum side lobes. This is a brute-force-and-ignorance method of reducing the effect, but in many cases it will be as much as you can do at the receiver.

4.8 CATV and Repeaters

The letters CATV stand for "community antenna television."

One of the significant characteristics of signals in fringe areas is their variability in strength at different locations. In small communities where many people have difficulty in reception there may be a location which has good signal strength and is free from interference and a single aerial at that site could be used to give a signal for the whole community. In multistory buildings (hotels, motels, etc.) it is often impractical for each television receiver to have its own individual aerial on the roof so one aerial is provided with multiple outputs.

The received signal can either be distributed to the receivers by coaxial cable or it can be used as the input to a low-power transmitter operating on a different channel from that of the main transmission. The choice of which system is used depends on factors such as the area to be covered, the density of population, and how many channels must be provided for. In a multistory building, for instance, a cable system is usually chosen because cable runs are short and usually channels over the whole band must be provided for. At the other end of the scale, in the case of an isolated farming community spread over the area of a river valley that has social and commercial ties with only one other town and that town has only one or two television stations, then a repeater or pair of repeaters would be preferred. If a repeater is to be used each channel needs its own equipment so the need for multiple channels shifts the economic balance in favor of cable.

4.9 Cable Distribution

Small cable distribution systems can be much simpler than the most basic of repeaters, but it is not acceptable to simply attach several coaxial cables to the one aerial; a splitter must be used to keep impedances matched in order to prevent reflected signals. Splitters can be simple passive devices consisting of resistors arranged as in Figure 4.12 or, if isolation between outputs is needed, the splitting can be done by a pair or group of amplifiers arranged so that their inputs are in parallel but each has a separate output.

For minimum reflection of signal in a passive splitter the following set of conditions must apply:

- Each of the three resistors must match the characteristic impedance of the cable and be noninductive.
- The input of each receiver must appear to the cable as an accurate nonreactive termination.
- The splitter must be mounted in a box or on a ground-plane assembly that preserves the characteristic impedance and shielding conditions of the cables.
- If either of the receivers is removed, it must be replaced with a nonreactive termination to prevent reflection on that cable.

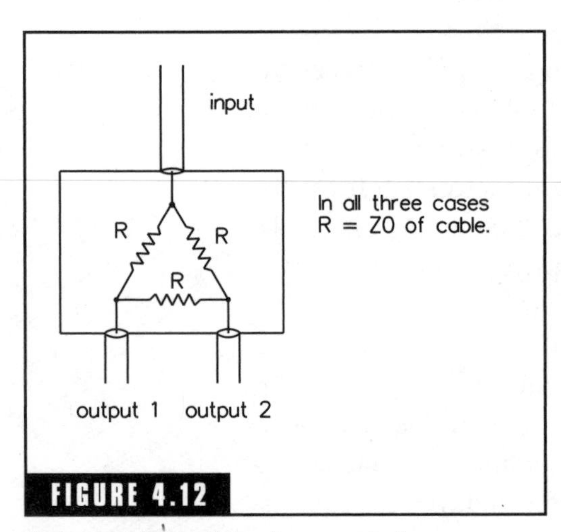

FIGURE 4.12

How a passive splitter is connected.

Each time a passive splitter is used there is a 6-dB loss in signal strength. The output of one passive splitter can feed the input of another, and that process can be repeated as often as is needed providing that there is enough signal level to accommodate the 6-dB loss at each stage. If many outputs are needed, however, less overall loss and usually better isolation between outputs can be achieved with unequal splitters arranged as shown in Figure 4.13. In an unequal splitter the through connection is intended to be connected to other splitters, and the split line may show a loss of between 12 and

70 dB compared to the incoming signal. In most cases, the signal on the through line is unaffected by lack of termination on the split line.

All the examples shown in Figure 4.13 will work with any impedance of cable; the actual resistor values shown are the nearest in the 5% preferred value series to the exact values required to give 20 dB isolation of the split line using 50-Ω cable. To use this design with cable of other impedances multiply the exactly calculated values (3.8, 247.8, and 61.3) by the ratio between the Z_0 of the cable being used and 50 Ω. In the capacitive splitter the isolation is determined by the size of the plate and its spacing from the center conductor of the through line, and in the inductive splitter the length and spacing of the pickup line controls the coupling. In general, capacitive splitters are most suitable when the required isolation of the split line is greater than 40 dB, and if the required isolation is in the range between 10 and 20 dB the resistive version works best. An inductive splitter with about 30 dB isolation can be configured as a directional coupler, which may reduce some of the effects of ghosting if the source of the ghost is mismatch on the downstream side of the through line.

In a large community-antenna television system the output of the aerial will go straight to an amplifier, which will boost the received signal sufficiently to overcome losses in all the following cable and splitters. Power output required of this amplifier may be up to 5 W in the largest of systems. The most critical of the specifications required of these amplifiers is that they must have absolutely impeccable

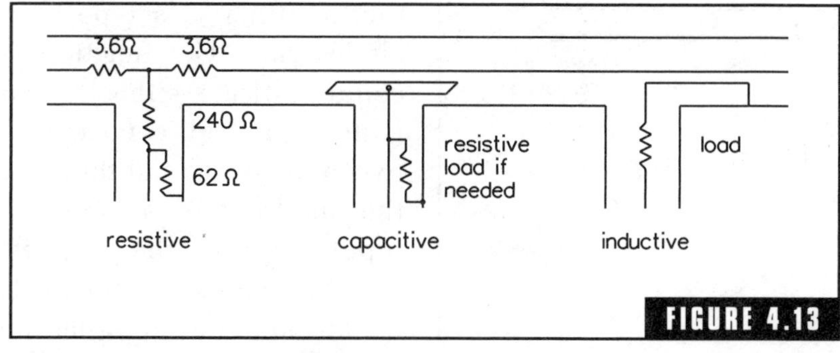

Construction of unequal splitters.

performance in regard to intermodulation and cross-modulation. Amplifiers designed for this service can be made so that the output is split into several separate feeds for use where the building or community can be divided into several sectors.

The most common fault mechanisms on CATV systems are wrong signal level (either too high or too low) and ghosting due to a faulty termination on a cable. If you are called to fault-find a system of this type you will need to know the layout of the splitters; your first check should be to see which receivers are affected. It is also appropriate to ask at an early stage whether the system has ever worked properly in its present configuration; if it has, you must be able to find and fix a faulty component. If it has never worked properly, then you may be able to produce good results simply by juggling splitters, attenuators, and terminations. To localize the fault do a careful census of which receivers are affected then consult the chart to find the splitter at the head of the common stream; the fault will probably be associated with that unit. In the case of ghosting problems, be aware of the situation shown in Figure 4.14 where the fault indication is to a particular splitter but the physical location of the faulty component is many meters away and not obviously associated with any of the affected receivers.

Electronic faults on the amplifier will give equal indication on all receivers and are most likely to be observed as either intermodulation or a reduction of output power. If a fault is observed as a noisy picture on all receivers and the amplifier is a type that includes an AGC function, that could indicate either an aerial fault or a bad connection between the aerial and the input of the amplifier or a fault in the input stage of the amp itself. Ghosting faults are not likely to be related to the amplifier unless there is at least 70 m of cable between the aerial and the amp.

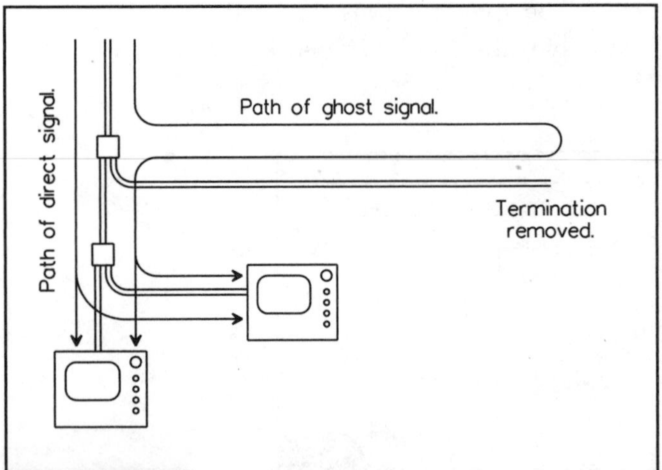

FIGURE 4.14

How ghosts can be produced in cable distribution.

4.10 Repeater or Translator Types

If a repeater is the preferred method of collecting and distributing the signal in most cases it will need to be licensed, and in most cases the administrative procedures required for the license will be the slowest part of the setting-up process and so should be started first. In a few cases, as described later in this section, a passive repeater can be used and these do not need to be licensed (apart from possibly needing a building permit from a local council), but in all cases where the signal is to be received, translated to another channel, and retransmitted, then licensing will be required. The initial approach to the licensing authority should be made even before land acquisition is finalized; the first procedure is a check on the effect of the new service on all existing radio services in the area, and this may bring to light a very good technical reason why the site that looks most desirable to you may not be usable and another site must be selected.

When the site is licensed the document will authorize operation on a particular channel with a specified output power and directional properties, and polarization of the aerial will also be specified. There are companies that operate a "turn-key" service, and for those you can simply provide them with a copy of the license and a check for the appropriate fee, and they will provide the equipment and install it so that all you have to do is turn the key and start using it (Figure 4.15).

If you wish to be more closely involved than that, the project breaks down into the following portions:

- building the tower and hut
- power supply (and storage if needed)
- lightning protection system (extremely important for hilltop sites)
- receiving aerial and its feedline
- transmitting aerial and feedline
- repeater electronics and power amplifier
- remote alarm and fault reporting if needed

Local conditions at the proposed site will have some effect on all of these. In modern times the electronic equipment is almost universally

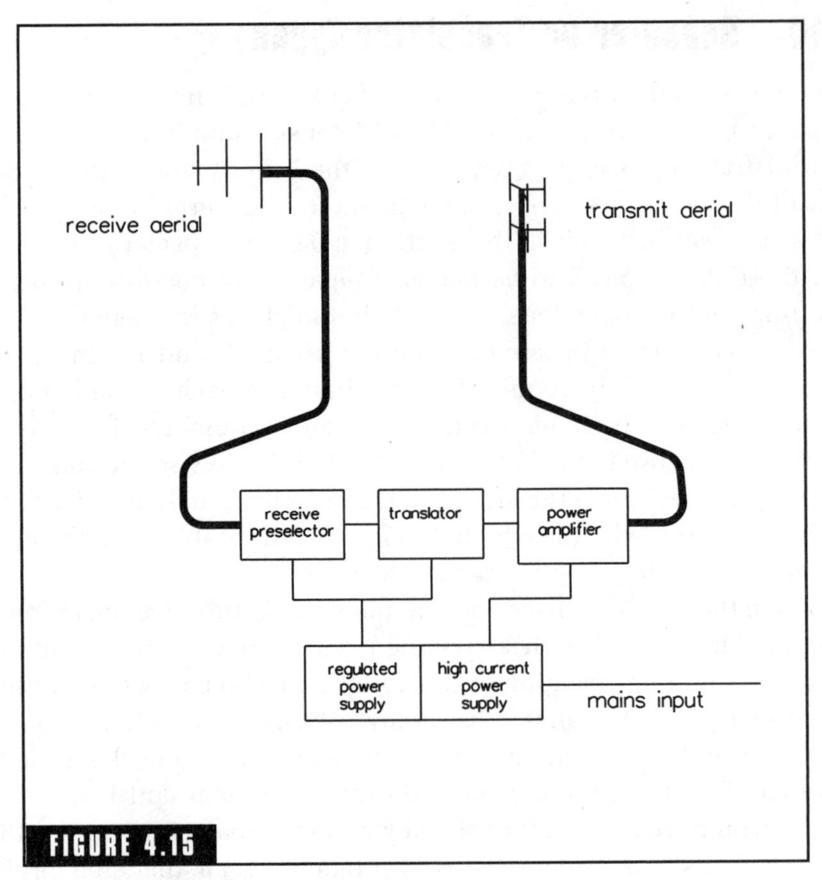

FIGURE 4.15

How a typical repeater is arranged.

solid-state so this means the power must be in the form of a DC supply usually nominally of either 12 or 24 V. The equipment can be built into a rack mounting panel usually between 200 and 300 mm high (four to six rack units; in Australia a standard for rack mounting panels is a horizontal measurement of 19 inches or 482.6 mm and vertical rack units of 1¾ inches or 44.5 mm). The received signal may simply be translated from the incoming channel to the output channel and amplified and retransmitted or it may be fully demodulated to a composite video baseband and then used as the input to the modulator of a small transmitter. Demodulation to the video baseband allows opportunity for some processing of the signal, such as stripping and renewing of synch pulses, or correction of frequency response and group delay, but it does add some distortion to the signal so if no processing is required then operation as a translator is preferable. The output power of the repeated signal at the input to the aerial feeder will usually be in the range between 100 mW and 5 W.

If the site happened to be alongside of a farm homestead which used direct current from storage batteries for household light and power then no additional power plant would be needed for the television repeater and the equipment could be housed in a weatherproof box mounted directly on the tower. On the other hand, a site that was isolated from the outside world by a long road that may be subject to being closed by adverse weather conditions would require a substantial building big enough to contain a power converter and a set of batteries and may possibly be equipped with a work bench and some basic repair tools; it may even need to provide emergency accommodation for a serviceperson trapped by a sudden change in the weather.

If the repeater site is on a hilltop, lightning protection may well be the most important factor in long-term reliability. The essential features are as follows:

- a copper rod mounted vertically at the top of the tower with its top being a sharp point at least a meter or more higher than any other part of the tower
- a heavy-duty conductor to carry current from that rod directly down to an earth stake
- some radial conductors (at least four) connected to that stake and leading to other earth stakes at the perimeter of the site

Lightning protection of radio sites can be a complicated subject; more detail is provided in Section A4 of the Appendix.

When the repeater is initially switched on it is wise to advise intended viewers that the first 3 months of operation are a test period and that unannounced breaks in transmission will be possible. With solid-state electronic equipment there is a definite "burn-in" period which lasts for about the first 3 months of normal operation; once that period has passed faults on the electronic equipment itself will be very rare. If you are a serviceperson called to a repeater that has had at least 3 months of fault-free operation, the most likely causes of fault are listed below:

- mechanical damage or corrosion on the aerials and feedlines
- corroded battery terminals
- power supply faults
- vandalism!

All these are things that can be checked by relatively basic inspection and testing, and if a fault is indicated that suggests returning some parts to a manufacturer for bench repair then that should not be done until integrity of aerials and power supplies has been proven.

The proportion of fault-finding work that is performed on-site compared to what is referred back to an equipment supplier will depend on the technical capabilities of people living in the local area. There is certainly great economic benefit in constructing the equipment in a modular fashion so that a definitive test can be done to isolate the fault to a particular section and then have someone who lives locally and is familiar enough with the system do those tests as a "first-in maintenance" activity.

An ordinary receiver can give much useful information when used as a monitor. If it is equipped with an input socket of the same type as the input of the repeater, the receiving aerial can be checked by simply disconnecting its feeder cable from the repeater and looking for a strong clear signal on the monitor. On the output side of the repeater an unequal splitter adjusted to give about 70 to 90 dB of reduction in signal can be used with the monitor receiver to give a useful quick check on power output and picture quality at the input to the transmitter aerial. Figure 4.16 shows where opportunities exist for definitive tests to isolate faults to a particular section of the equipment.

The attenuators shown in Figure 4.16 as 15 and 70 dB respectively may not be exactly these figures. In the initial setting up of the repeater these and the off-air pickup aerial should be adjusted so that with all the equipment working correctly there is a clear but slightly noisy picture on the screen of the monitor receiver. In operation any loss of signal strength at the output of the attenuator will immediately show as an increase in noise on the picture. Normal signal on the channel of the main transmitter at the output of the 15-dB attenuator is a check for correct operation of the receiver aerial and its feeder ca-

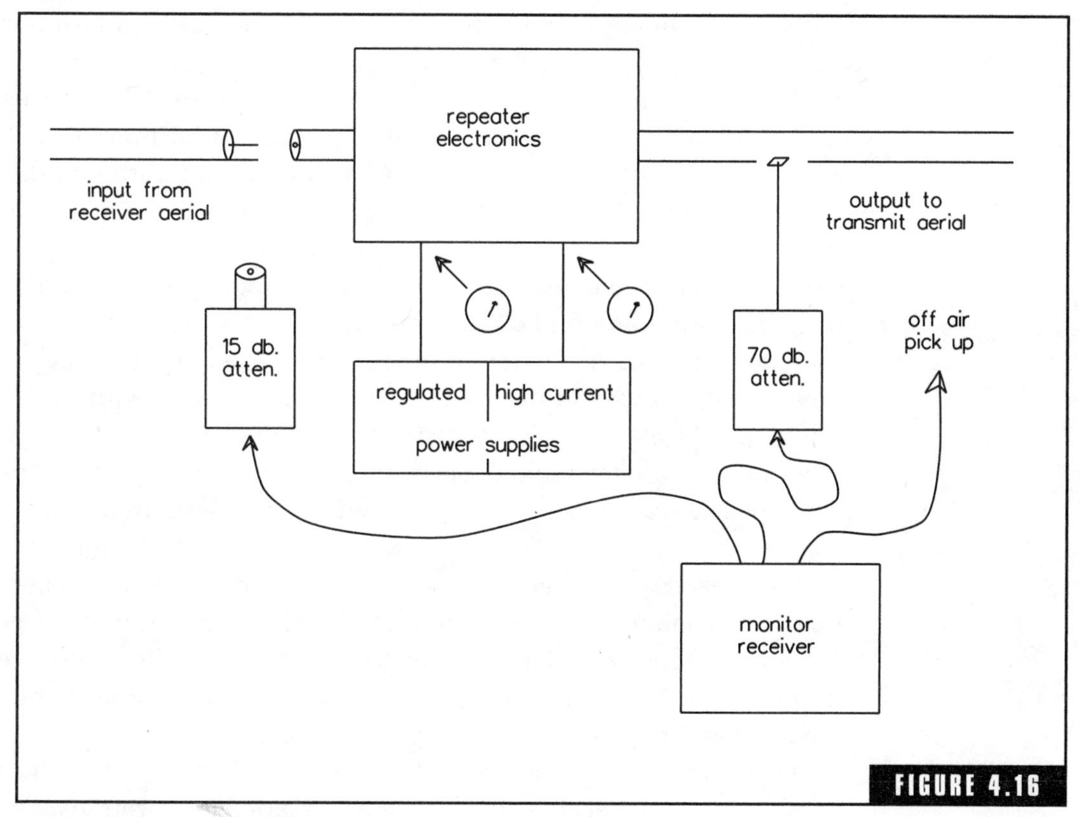

FIGURE 4.16

Isolating defective sections.

ble. When that is proven correct, check next at the output of the 70-dB attenuator; a normal signal there indicates that the repeater electronics are working correctly. Finally, check the transmit aerial and its feeder by observing the off-air signal. Transmitter output and transmit aerial operation can also be checked with a directional power meter and these meter readings will be more definitive and accurate than the indications of a monitor receiver, but they do require that the sensor head of the power meter be permanently installed in the line close to the transmitter output connection.

If the repeater uses two boxes to produce the transmitted signal they will probably be a translator whose output is the final signal at very low power (typically 1 to 100 mW) and a power amplifier. In that case there is opportunity for another monitor point in the form of an unequal splitter in the line that carries signal from the output of the translator to the input of the power amp. The required split proportion will probably be in the range of 50 to 60 dB.

The power supply lines are points for metering of DC voltages. The regulated voltage should stay absolutely constant for months or years at a time, and if it is seen to vary, check first the calibration of the meter. In most cases, a change in this voltage will indicate a fault in the power supply except in the case where a reduction of voltage coincides with excessive heating of either the power supply or the repeater box; that combination of circumstances indicates a short circuit in the receiver or translator section of the repeater. The output voltage of the high-current power supply may vary slightly in step with variations due to power mains fluctuations or state of charge of the battery, and there may be a detectable drop in voltage when the power output of the repeater transmit section is turned on from a muted or standby condition, but this drop should be less than 0.1 V. In many cases, faults on aerials and feeders can be identified by physical inspection. Use of these monitoring and metering facilities should make it possible for a person with basic technical skills to definitely prove electronic faults into a particular box which can then confidently be removed and sent to a service depot for repair.

There are a few cases where change of direction of a signal is needed but not necessarily with amplification and retransmission. One instance is in mountain valleys close to a major transmitter but over the edge of a watershed line so that there is no part of the horizon that is directly illuminated by the signal from the transmitter. In these locations, the signal may be directed into the valley by reflection from a "passive repeater."

A passive repeater can be as simple as a large flat sheet of conductive material sited on a ridge which is in direct sight of the main transmitter and aligned so that it reflects like a mirror to the center of the service area. Solid metal is not needed; mesh, fence netting, or a grid of rods is equally effective. In cases where the signal must only be turned a few degrees, the sheet would have to be aligned almost end-on to the signal and would need to be impossibly large to be effective; in those cases, a pair of sheets can be used if they are arranged so that each turns the signal though about 45° in the form shown in Figure 4.17. This can give the same effect on the signal with a smaller total structure. Another form of passive repeater that has been used

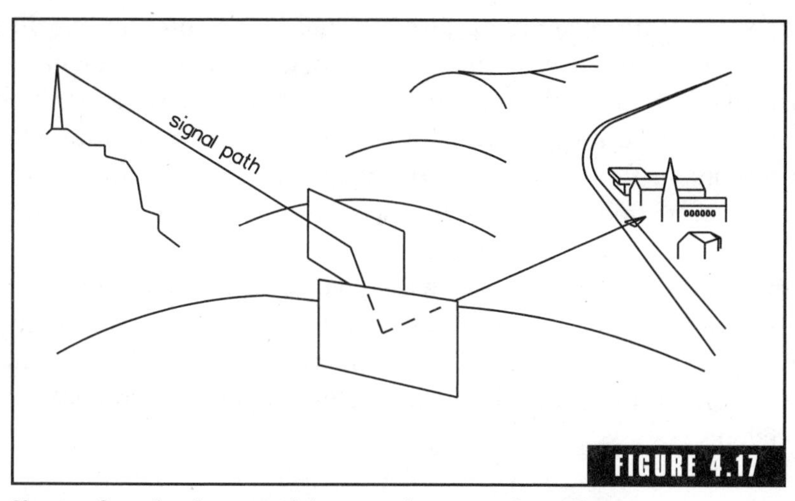

FIGURE 4.17

Sheets of conductive material arranged as a passive splitter.

occasionally is to simply connect two high-gain aerials back-to-back with one receiving the signal from the transmitter and the other pointing to the receiving position. However, systems of that type will usually only pick up enough signal to work over a couple of hundred meters.

Passive repeaters depend for their effectiveness on being close enough to the transmitter and big enough to collect a useful amount of power and redirect it to the receiver. If a screen 10 m square is placed 1 km from a transmitter which gives effective radiated power of 100 kW the power collected by it is 796 mW, and if the distance is increased to 10 km that figure drops to 7.96 mW. For a repeater of the type shown in Figure 4.17 to present 100 m^2 of collecting area to the signal, the collecting sheet would actually have to be 10 m high and 14.2 m long. There will be about a 50% loss of power at each reflection, which means the output of a device consisting of two sheets 10 × 14.2 m sited 10 km from the transmitter would be about 2 mW.

The output sheet of effectively 10 m^2 will form a directional beam with an aerial gain of somewhere between about 12 to 30 dB depending on the wavelength of the signal so the effective radiated power at the output of that passive repeater will be somewhere in the range from 8 mW up to possibly a bit over a watt. In cases where an isolated valley is located within a few kilometers of a powerful transmitter but is prevented from receiving a direct signal by an intervening ridge, a device such as shown in Figure 4.17 may give sufficient signal without the need for an active repeater installation. If the ridge has a sharp peak there will in most cases be a usable signal due to knife-edge refraction, but in those cases where the ridge has a rounded top there will be no refracted signal and a passive repeater could be used.

If you plan to install a passive repeater you will not normally need an equipment license but you will have to satisfy local government building regulations and environmental protection requirements. In many cases, the people who control these regulations do not know a lot about electronics or radiophysics so much patient explanation will be needed.

4.11 TV Dx-ing

This is a hobby activity that can be occasionally thrilling and often frustrating and can bring out the competitive spirit that we all have in a way that pits the human intelligence against the mindless elements and that occasionally produces a piece of information that has value to the transmitter operator or the general scientific community. For most television watchers the program is only of interest if the signal is strong and steady enough so that the variations due to propagation changes and interference do not intrude too much on the enjoyment of the program; for that to apply the signal strength must be close to that which is suggested in Section 4.2 and so requires that the distance between transmitter and receiver be close to those suggested in Figure 4.3.

At much greater distances the signal is not usually watchable as a television program, but on some very rare occasions it can be detected for long enough to be identified. The people who seek out these rare times are not usually interested in the program content of the transmission; their interest is in identifying the station and proving that the signal really did cover the particular number of miles or kilometers that lies between them and the transmitter. Television Dx enthusiasts rely for their success on freaks of propagation such as:

- skywaves by sporadic E-layer or transequatorial reflections
- "super-refraction," which occurs when weather conditions favor inversion layers

- ducting
- reflections from meteor trails

There are also people who regularly receive television programs from thousands of kilometers away by artificial satellite. This is a slightly different class of service; they are interested in the program content of the signal and once the equipment is properly set up they expect the program to be present every time they switch to that channel. There is a separate chapter later in the text dealing specifically with reception of satellites and other extraterrestrial signals.

The frequency range of television signals is higher than that normally returned to Earth by skywaves but not so high that it is out of the range of chance. There are some times mostly close to the peak of the sunspot cycle when the F-layer MUF for the very longest of paths is higher than the frequency of the lowest one or two television channels in a particular country. There are also times when local spots of very intense ionization occur in the E layer for a few minutes to a few hours at a time, and those spots are concentrated enough to reflect signals at frequencies up to 80 to 100 MHz for the short time they exist. A mechanism called "transequatorial propagation" exists where if the transmitter and receiver are located roughly on the same meridian, about equal distances either side of the thermal equator and far enough apart to be out of the skip zone range, signals in the low-VHF range of frequencies can be returned to Earth. In all cases of television via skywave the signal is not reliable enough to interest the normal viewers of programs although there can be some seasonal conditions when a reasonable expectation of a signal for an hour or two each day is possible; signals are most likely for the lowest frequency channels. There is always a substantial skip zone, but signals, when they do appear, are strong and clear except for some times when the MUF actually falls within the channel, at which times a range of funny effects are possible depending on the exact relationship between the transmitted frequencies and the MUF for that path. For more detailed information on how skywaves are propagated, refer to Section 5.8 in *How Radio Signals Work.*

Ducting and superrefraction are closely related propagation mechanisms, and both are closely related to the process which forms visible mirages. There is a normal range of relationships between height above local ground, air temperature, pressure, and humidity. In a well-mixed atmosphere a range of measurements from just above tree-top height up to perhaps a couple of kilometers would show that the higher air is cooler, less dense, and dryer than that below it. When an inversion layer is formed that normal relationship is disturbed; instead of gradually becoming cooler at higher altitudes there may be a time when the higher air is hotter than that lower down, and at those times the refractive effect of the air on radiowaves is altered possibly enough to cause the signal to be returned to Earth by the process of "total internal reflection." When there is one inversion layer the signal is returned to Earth, and that process is called "superrefraction"; at those times quite strong signals may be available from stations several hundred kilometers away and there may be some evidence of the same sort of skip zones that are seen with skywaves.

When there are two inversion layers the signal can be trapped between them and propagate in a manner similar to that of a signal in a waveguide, and that process is called "ducting." For regular television viewers and commercial users of the VHF and UHF bands, ducting is usually a nuisance because unless it happens to connect between transmitter and receiver it deflects the desired signal away from its normal destination. For TV Dx enthusiasts, however, the mechanism can carry signals from one side of a weather cell to the other, which can make signals occasionally available at distances up to a couple of thousand kilometers. The inversion layer must be thick enough in the vertical sense for the signal to be turned gradually and the required thickness is related to wavelength so superrefraction and ducting effects are more common at higher frequencies although are occasionally seen for frequencies as low as 100 MHz (wavelength equals 3 m). For microwaves, ducting is a daily occurrence on some commercial broadband bearers and is particularly common when the path of the signal is partly over water and partly over land.

Meteor scatter propagation is almost totally unpredictable except for a few times each year when the Earth passes through known showers. The trail is composed of minute solid particles and separate atoms which have been ionized by the very high speed passage of the meteor through the upper atmosphere. Trails which are too faint to be seen visibly glowing can sometimes give a short time in which weak but workable signals can be detected by reflection. Trails which are bright enough to be seen glowing can be used for reflection of workable signals for several times longer than the time the glow is visible. The size of the reflective surface is small as seen from transmitter and receiver so signals are always weak but can be detected with high-gain aerials pointed at an appropriate angle up into the sky. The propagation does not necessarily obey the normal laws of reflection from a surface; the ionized trail is a new point-source of radiation so the maximum signal is detected if the trail happens to be aligned with the maximum sensitivity of the receiving aerial, which means that the ray line from transmitter to receiver can follow a sawtooth shape or arrive from a direction other than the direct line of the great circle path between transmitter and receiver. Trails are usually located between 40 and 100 km above the Earth's surface; the exact height of maximum ionization depends on the size, speed, and direction of travel of the original particle which formed the meteor. Signals from meteor trails are almost totally unpredictable, and reliability in terms of proportion of time a signal is available is very low, but the process can give workable signals for a few minutes at a time over distances of 500 to 800 km.

There are a number of propagation mechanisms which can carry very weak signals over several hundred kilometers with fair to good reliability. The combination of these is used commercially for broadband bearers (telephony, data, or television point-to-point links) in "forward scatter" radio-communication systems. They require enormous, very high gain aerials at both ends of the link to overcome the high path loss. The same mechanism applied to television broadcasts would require an aerial beyond the economic reach of most private citizens for a program to be made strong enough to be comfortably watchable, but with smaller (but still very large) aerials a picture can be made strong enough to identify. The signal only needs to be a few microvolts for the synch circuits of the receiver to lock onto; at that signal level even a test pattern cannot be identified. There is a trick that can be used with a camera at these times. Set the camera on a tripod and focus it to show the screen of the television receiver. Take some photographs with the shutter speed set for as long as you can guarantee the picture will be stable and with the aperture set very small to avoid overexposing the film. The genuine picture on the test pattern is repetitive and gradually builds an image on the film; the noise is random and each time the raster is written, the image of the noise from one frame tends to cancel the effect of others. A photograph of a test pattern good enough to identify the callsign of the station can be produced from almost any signal strong enough to be locked by the synchronizing circuit.

If you are a person who is specifically interested in record breaking, then you will still need to spend a lot of time and need some good luck, as well searching for the rarest of events; you are in competition with all the other people who are also seeking to break records. As with fishing, you will need to know your quarry and how it is likely to arrive and then watch, watch, watch as much as you can. If you are seeking a signal by some short-term mechanism such as sporadic E or meteor scatter you will increase your chances by leaving a receiver running with the aerial aimed at the target station and with some alarm device connected to either its AGC line or the synch lock detector.

4.12 The Way of the Future—Digital Television

The very first experiments with television were conducted with a picture resolution of 120 lines; the first regularly broadcast programs had a resolution of 405 lines and ever since experimenters have dreamed of providing finer resolution of small details by increasing both the number of lines and the bandwidth of the transmitted signal to make pixels ever smaller. In the late 1990s, experimental systems were running at line counts up to 1250 with bandwidth in the 20-MHz range. There is not room, however, for systems with 20-MHz bandwidth to be unleashed on the radio spectrum except in the microwave range so at the present time resolution is limited by available bandwidth and 7 or 8 MHz is about the widest the spectrum managers will allow.

Digital signal processing with file compression offers a way of overcoming this limitation, and there is a great deal of research and development going on at present which will continue for several years into the twenty-first century in a number of parts of the world to define the best possible standards for broadcasting of television by a digitally compressed bit stream. There is a very large measure of redundancy built into a normal television transmission, for most of the pixels for most programs, the specification of one pixel is exactly the same as the previous one and the one before that and so on. Every so often the specification changes and that is seen on the screen as an edge of a shape. There are also a very large number of cases where the definition of a whole line is precisely the same as the one above it and some times where a whole scene is the same for several frames at a time.

The power of digital compression is in these repetitions, and the benefits in potential bandwidth reduction are dramatic. A glimpse of what is possible comes from the technology of data transmission over telephone lines (bandwidth 3 kHz) as are used for connection to the Internet; still images can be transmitted using only a few kilobytes of data, and a 30-kb file is capable of transmitting pictures for a short sequence of movement. (Remember, there may not be a direct relationship between kilobytes of file length and any particular bandwidth, but a file being transmitted at the rate of 30 kb/s would only require bandwidth of hundreds of kiloHertz instead of the several megaHertz required for the same information in analog form.)

The success of digital television will depend very largely on finding the best compression algorithm for the purpose, and it will probably be several years into the twenty-first century before the ideal algorithm is defined and even more years after that before testing for the best one will be completed. At the end of the twentieth century there were several countries using official or semiofficial specifications for compression algorithms and transmission standards and planning is in progress to introduce digital television on-air in the 1st decade of the twenty-first century.

In Australia, a decision was made in late 1998 to use the European DVB-T encoding standard which is based on a modulation scheme described as "coded orthogonal frequency division multiplexing" in which between 800 and 1000 separate carrier waves are used to carry data in the parallel mode. Pixel encoding and digital compression will be in accordance with the MPEG2 protocol. This transmission standard is very adaptable, data is arranged in packets, and different packets can carry data with entirely different meanings.

The following list of possibilities will give some idea of the versatility of the overall system:

- one high-definition television channel (about 20 megasamples per second data rate) using the full bandwidth of the channel and all the available packets

- several lower-definition channels as, for instance, the outputs of all cameras showing the action at a sporting event from several different angles
- a lower-definition transmission which uses only part of the available packets with the spare packets being used for another purpose such as downloading of a computer program or a block of data—the computer information does not need to have any relationship at all to the television program

The standards allow for reconfiguration during transmission; for instance, a particular transmitter on a particular Saturday afternoon may be used to show a sporting event as several standard-definition channels, then later in the evening show a gala theatrical performance as a single HDTV program, and very late in the evening be used for a single standard-definition TV channel with the spare capacity being used for interactive computer games.

With digitally compressed television there will be an absolute time delay in the transmission process because the compression and expansion are both basically batch-processing functions, and both batch-processing operations must be completed before the screen can be displayed. The most comprehensive compression algorithms are likely to be the ones showing the longest delays so there may be a tradeoff between degree of compression and permissible delay. Any programs that are being played back from a recording delay of up to a couple of seconds or so will be of no account, but for live broadcasts, particularly sporting events, delays of more than a couple of tenths of a second could be significant.

Australians will be particularly interested in the fringe-area performance of digital television because of our relatively low population density and many of our citizens have great interest in sports so real-time transmission will be important to us. It is possible that the best algorithms for fringe performance may not be the same as those for maximum rejection of interference or that the algorithms which offer best compression may be too slow for our use. The adaptability of the DVB-T standard is a point in its favor when consideration is made of all these tradeoffs. The technical effect of propagation disturbances, noise, and interference on digital television will be related to but different from the effect on an analog signal. With the error-correction system used on the transmitted signal we can reasonably expect that if the signal can be resolved at all it will probably be seen on the screen as an almost perfect picture, and there will be a fairly sudden complete loss of picture as the signal strength is reduced, similarly to the "capture effect" of a frequency-modulated signal. At times when the signal is fading there will tend to be either a picture with little or no visible degradation or a blank screen.

In the 1990s in Australia the implications for servicing digital television are not fully known; it will not be possible to display the received signal on an instrument such as a spectrum analyzer and reconcile particular aspects of the display with particular features of the receiver, picture and sound as can be done with analog TV. One factor that is already known is that the requirements for flatness of frequency response across the passband will require closer tolerances than is presently needed for analog. When color TV was introduced to Australia there was a need to redesign some aerials to ensure that the color subcarrier was presented at the correct level relative to the luminance carrier. When digital is introduced we will probably see another round of the same process working to even closer tolerances.

The digital receiver will have the ability to process the signal to remove ghost images. A technical innovation that will be possible with digital television is that several translators can be operated at the edges of the service area of a major transmitter and all translators can be operated on the same channel. The receiver, if it happens to pick up the signal from more than one translator, will treat the second signal as a ghost image and reject it. With analog television a separate channel must be allocated to each retransmission, which results in very definite limitations on the number of translators that can be serviced by a major transmitter. This capability for reuse of the repeated channel probably means that in the future when the digital system is mature the coverage of each major transmitter and its translators will at least equal the coverage of the present fringe areas.

CHAPTER

5

Television Interference

5.1 Interference to Television Reception

The letters TVI stand for "television interference." The general subject of interference must be considered from several different aspects. The transmitter operator (of all voice and narrowband data services) has one aspect to attend to: he or she must not radiate excessive amounts of power on frequencies allocated to other services. In general, these spurious signals are harmonics or VHF parasitics, both of which are higher in frequency than the carrier of the genuine signal. The transmitters most likely to have trouble meeting a TVI specification are those which operate on lower frequencies than the TV channels and particularly those whose assigned frequency is one-half, one-third, or one-fourth of a channel which is used locally for fringe-area reception. Television transmitters may also interfere with each other by producing intermodulation products which generate adjacent channel interference; a problem due to that cause almost always indicates a fault in the power amplifier stages of the transmitter itself.

At the receiver, the aspect of harmonic and parasitic signals generated by transmitters can be off-loaded onto the transmitter operator once it is proven to be due to that cause, but that is the only type of interference that can be treated that way. Manufacturing industries are gradually realizing that equipment manufactured for other purposes can generate radio interference and in the late 1990s "electromagnetic compatability" legislation was drafted and enacted in many countries. The EMC laws, however, only set standards for new equipment and it will be many years into the twenty-first century before electrically noisy equipment already in service is worn out and scrapped.

Another consideration for receivers is that in fringe areas, when attempts are made to use the weakest of signals, the sensitive receiver is also more sensitive to interference and there may be instances where equipment which complies with the EMC standards is still too noisy for the television signal. This chapter describes the possible causes of each observed type of interfering signal and indicates some of the directions for action to cure problems. It focuses attention first on what the receiver owner or operator can do to his/her own equipment and later addresses the problem of interference from a neighbor's installations.

The social and political implications of TVI from a neighbor are acknowledged, but a full set of instructions for resolving all disputed claims is beyond the scope of this book. One valuable principle is, however, that if you have done all you can to make your own equipment as good as possible you are socially on much stronger ground when it comes to asking someone else to clean up their activities.

5.2 Interference from Equipment You Own

This section deals specifically with the situation of a rural or pastoral property or perhaps a mining camp where all human-made items within the immediate vicinity are under the control of the owners of the television set. Sections 4.3 to 4.5 deal with those cases where the limit of sensitivity of the receiver is set by thermal noise generated internally and is therefore unavoidable; the only things that can be done to improve those situations are to increase the signal strength presented to the receiver input. This chapter deals with cases where the limit of sensitivity is set by interference generated outside the receiver and in those there may be some opportunities to reduce or eliminate that interference. This section considers the case where the complete homestead or camp can be treated as one installation under one ownership and whatever needs to be done to isolate sources of interference can be done without the social and political issues that arise when someone else owns the equipment causing the interference.

The simple way to identify which equipment is causing interference is to have someone watching the screen of the television receiver while you switch particular items off. If the pattern disappears when a particular item is switched off then that is the source of interference. If after you have gone around and switched off all those items which can safely be switched and the pattern is still there then look for items which are permanently operating such as clocks, timers, and gas igniters. If the pattern has disappeared when you have switched everything off but it reduced in stages then several items are contributing to the problem and you must treat each as a separate source of interference and apply the appropriate cure to each.

For items which must be kept running, if you have access to a field-strength meter it can be used with a sniffer aerial to test individual items of equipment. A sniffer aerial can be made as a half-wave dipole for a specific frequency or it could be a loop with a suitable connection for the meter input. The general arrangement of a loop is shown in Figure 3.12; for this purpose, however, it is best used untuned (connect straight across where the tuning capacitor is shown) and the size should be about 100 to 200 mm on each side for VHF channels or about 30 to 50 mm each side for UHF. The total length of wire used must always be less than half a wavelength to avoid stray resonances.

You may need some common sense about interpretation of the results. There is detectable radiation from practically anything electrical if you get close enough to it. When you find what may be a source of interference back away from it a couple of meters and watch the decrease in the meter reading as you move. If the reading drops by about 10 dB or so in the first half-meter and then drops significantly again when you move out to 1 m separation the signal is an induction field and will probably not reach the aerial. If the signal is still significant at 1 m and only drops by about 6 dB when you move from 1 to 2 m it is a radiation field and should be taken notice of. A comparison between radiation and induction fields is shown in Figure 5.1.

The next two sections deal with interference in a lot more detail so I recommend you continue reading for suggestions on how to identify particular classes of interfering signal and how to reduce their effects. That information is particularly valuable when you must deal with

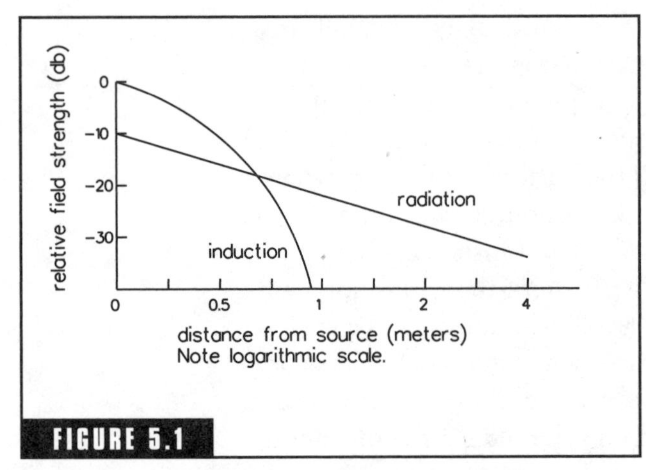

FIGURE 5.1

Graph of field strengths versus distance comparison.

interference from neighbor's equipment, which is a subject treated later in this chapter.

5.3 Recognizing Types of Interference

You can often gain a fair bit of information about the source of a particular interference by looking closely at the pattern it produces on the screen of the receiver. The information in this section is not a comprehensive list of all possible types of interference pattern but provides some examples of what you may be able to deduce from the picture on the screen.

First, concerning ghosts, they may look like a form of interference, but ghost images must be due to a signal which originates from the same transmitter as the direct signal but arrives at the receiver by another (time-delayed) path. Ghosting therefore is technically due to a propagation defect of the signal not interference from another signal. Section 4.7 deals with the technical aspects of ghosting.

Second, concerning random noises, the normal indication on the screen of a receiver which has no signal input at all is a random and ever-changing display of black and white spots and short horizontal lines. This thermal noise results from a multitude of very tiny unrelated events, each of which generates its own tiny burst of signal. When interference comes from many unrelated sources it may give a display which is hard to pick out from thermal noise. One important source of these signals is the ignition system of petrol-driven (or any electrically ignited) motors. A single engine or vehicle will give a display of fairly random short, dark horizontal lines which at some particular engine speeds may show some suggestion of a vertical or oblique alignment. When more than a couple of separate engines and vehicles contribute to the signal, the display will become more random and no patterns of any sort will be seen.

Third, consider displays showing definite horizontal bands. Signals that show horizontal lines or bands on the screen are periodic signals that are occurring close to the frame rate (usually either 50 or 60 Hz depending on which country you live in) or a harmonic of it. If the horizontal band is absolutely steady, the source is somehow related to the television signal itself and the prime suspect is probably an electronic fault internal to that receiver. If the band of interference drifts slowly, it is a signal from another source and in most cases it will be related to electrical mains operated equipment; for instance, a small AC/DC motor which uses a commutator will give two bands of short, dark horizontal lines because the interference is actually generated by sparks across the commutator, but their timing is controlled by the phase of the AC cycle. Single-phase devices will produce two bands of interference and three-phase equipment will give either three or six bands depending on whether the device is star- or delta-connected and whether the actual source of interference is directly connected to the AC mains or to a rectified component of it.

Another possible source of horizontal lines occurs when you are receiving signals on the same channel from two different transmitters. If both transmissions are received at about the same signal strength you may be able to identify the callsigns of both transmitters on test patterns but on normal programs the interference will be so severe that you may not be able to resolve a sensible picture from either. When one is about 20 dB stronger than the other, the stronger one will show a fairly clear picture over most of the screen; the picture information of the other will not be noticed but the synchronizing pulses will create shading of particular areas. The frame synch pulse will show a horizontal band several lines high and the line synch pulses will be seen as a vertical band of shading about one-twentieth of the total width of the screen. Drift rate will be very slow but the bands may be anywhere on the screen and drift may be observed in the form of the bands appearing at a different place each time they are seen.

Interference which shows as either vertical or oblique bands is usually due to a heterodyne. Heterodyne signals are capable of several different effects depending on how their frequency is related to the desired signal. If the frequency is within a couple of Hertz of that of the first line rate sideband (i.e., in Australia either vision carrier plus 15.625 kHz or vision carrier minus 15.625 kHz) it may upset the line synch detection and cause the picture to drift or jump sideways. For heterodynes which are close to those frequencies but in the range of about 70 to 200 Hz away from them, the picture may show the same horizontal movement but only in certain parts of the screen and the effect observed will be that the picture is torn into horizontal streamers. When the frequency of the interfering signal is separated from that of the vision carrier by between 30 kHz and about half a megaHertz the display will be seen as vertical or oblique bars; vertical bars indicate the actual frequency is very close to that of one of the sidebands due to the line synch pulses; other frequencies produce oblique patterns.

For heterodyne signals whose frequency is well away from the vision carrier if the frequency is more than 1.5 MHz below it, they will be out of band (in the Australian standard) and so will have no effect unless they are strong enough to overload the receiver, but if they are higher than the vision carrier they will produce a version of the vertical or oblique bars mentioned above that is so finely grained that individual bars cannot be distinguished. The signal then has the effect of an even gray screen over the whole picture.

Vertical or oblique bar patterns which show variation indicate that the interfering signal is modulated. Frequency and phase modulation on the heterodyne signal produce a wavy bar and amplitude modulation produces variations in shading density in the vertical bar. When the deviation of a frequency-modulated heterodyne is comparable with the line frequency of the television signal and the unmodulated frequency is not exactly related to one of the line frequency sidebands the oblique bars form a pattern of nested "V" shapes similar to the chevron pattern of heraldry or the shape of the ribcage of a fish skeleton. Because of its similarity to the shape of a fish skeleton its common name is "herringbone pattern." Figure 5.2 shows the sequence of events that form a herringbone pattern.

Heterodynes that fall in the frequency range of sidebands of the color subcarrier may cause false color in some or all of the observed picture. In the 625-line PAL system, this is the range of frequencies about 0.75 MHz either side of the subcarrier frequency, which is 4.43 MHz above the vision carrier. This is in the range of (vision carrier + 3.68) to (vision carrier + 5.18) MHz. Because the modulation on the color subcarrier has both phase- and amplitude-modulation components, the effect generated will depend on the exact frequency

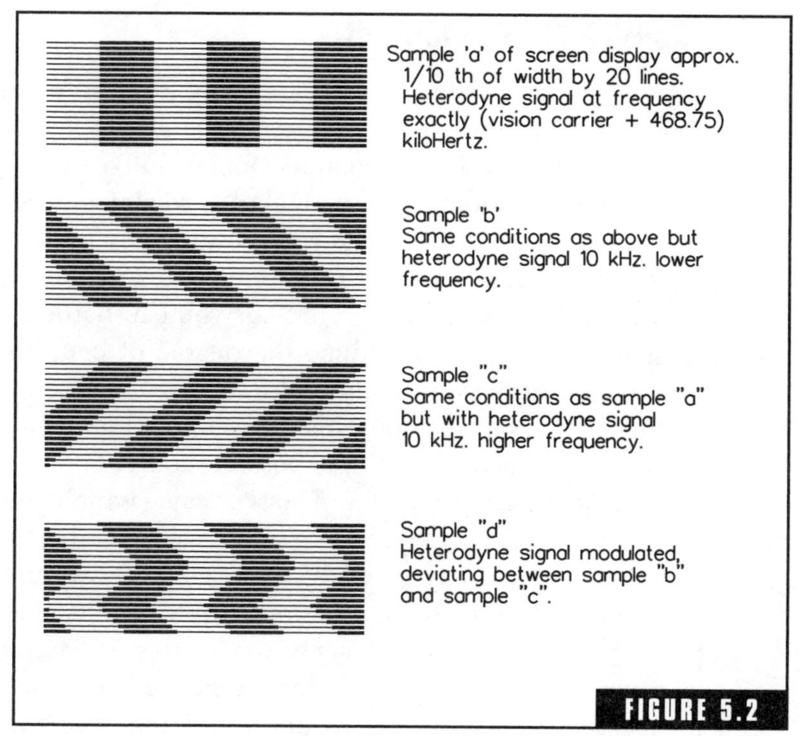

How a herringbone pattern is formed.

and strength of the interfering signal and so cannot be predicted in advance. The effect of a heterodyne on the color subcarrier will also depend on the system used for transmission; the PAL system, for instance, compares the signals on two successive lines in a way that is designed to discount the effect of external interferences so a signal must be strong enough to disrupt this comparison process before it is noticed; systems which do not include this facility (for instance NTSC) are susceptible to direct effects of heterodynes.

The sound transmission of a television signal is a normal frequency-modulated transmission which could, if required, be received by a normal high-fidelity FM receiver without reference to either of the vision signals. Interfering signals whose frequencies fall within the passband of this FM signal will show the same indications as for interference to other FM transmissions. There is more information on this in Chapter 7.

5.4 Hardening Your Receiver

If a television receiver is showing signs of suffering from interference there must have been two quite separate processes taking place. The interfering signal must have been generated and it must have been coupled into the receiver. I admit this statement is so elementary that it is hardly worth writing, but it is the key to defining two quite separate strategies for preventing the effect of interference. You can either prevent generation of the signal at its source or you can harden your receiver against its effect; that is, reduce the degree of coupling or change operating conditions so that the interference has less effect. If you are the owner of an isolated homestead or camp where you own both the receiver and all the possible sources of interference, it is purely a matter of convenience which you do, but in a country town where you have several or many neighbors and any or all of them may be owners and users of sources of interference, the sensible place to start is to make the receiver as insensitive as possible to everything other than the desired signals. This section deals with the paths by which interference gets into a receiver and the next five take each of those possibilities in turn and describes how to reduce their effect.

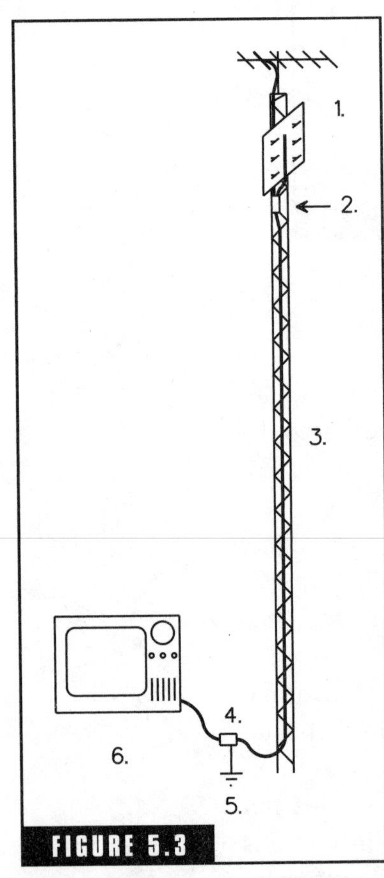

Typical features of a fringe-area television receiving station.

First, to ensure that all parties are thinking of the same items Figure 5.3 shows a typical arrangement. The numbered sections are as follows:

1. Aerial assembly which in some cases may be one broadband high-gain antenna unit or several single-channel units of which each is specific to a particular transmitter.
2. Combiner for feedlines (if needed) and/or masthead amplifier.
3. Transmission line (may be called a "feeder").
4. Power converter for masthead amp with surge diverter for lightning protection.
5. Earth for surge diverter.
6. Receiver.

There may be some installations that do not have all these units and there may be some that include other facilities. You may know some of these units by another name, in which case you are welcome to do a mental translation each time you see a reference to it in the following text.

Taking the case specifically of a multichannel aerial placed at the top of a high tower and connected to a masthead amplifier which is powered by low-voltage DC supplied from a ground-based power converter, the interfering signal could be any one of the following:

A radiated signal on a frequency within the passband of the receiver which is picked up by the aerial as part of its normal operation.

A signal on a frequency within the passband of the receiver which is being induced into the metalwork of the tower or the feeder cable and coupled into the input of either the masthead amplifier or the receiver. (These may be described as "common mode" signals.)

A radiated signal on some other frequency (for which the aerial has some sensitivity) of sufficient level to ride into the front end of the masthead amplifier and cause cross-modulation due to overloading; or a pair of signals which when combined are sufficiently strong to generate an intermodulation product; or a signal on some lower frequency which is being induced into either the tower metalwork or the feeder and coupled to the front end of the masthead amplifier with sufficient power level to cause harmonic generation by overloading.

A signal which is either on a frequency within the receiver passband or with a high-enough power level to be rectified and cause intermodulation products or cross-modulation which is entering the receiver directly via either the power mains or by being induced into leads to remote speakers or connections to other electronic equipment.

A signal on some lower frequency which is being induced into the tower metalwork, rectified by a rusty joint, and radiated to the aerial as a harmonic within the passband of the desired signal.

The effects of several of these mechanisms may be hard to differentiate but the measures required to prevent their effect are quite different from each other so it is important that both the point of entry and the mechanism of effect of the interfering signal are identified.

Step one in identification of the path of interference is to perform a signal-tracing operation. Start with the aerial connection at the back of the receiver (at room level), disconnect it, and note whether the interfering signal is still present. If it is, check all other connections to the receiver. (*Warning:* if you have a data connection of some sort, i.e., teletext or a remote tuning controller, for instance, there are some computer programs that do not take kindly to random disconnections of inputs or outputs.) Disable and disconnect them one at a time until the interference disappears. If it is, still present when nothing is connected except the mains power lead then that is the path of the interference, and you will be able to cure the problem by filtering that lead; refer to Section A5 in the Appendix for some principles of filter design.

If the signal disappears when the feeder cable is disconnected that indicates the signal is coming via the aerial system; the next section to test is the cable itself and the masthead amplifier. Climb the tower and, with the amp in its normal mounting position, disconnect its input while someone is watching the screen of the receiver. If the interference is still present then its entry point is either the cable or the masthead amplifier; this still includes a range of possibilities—refer to Section 5.6 for the next stages of the location process.

If the interference pattern disappears when the input to the masthead amp is removed, there is one more possibility to isolate. Touch the outer of the connector (if it is a coaxial cable) to a metal part of the case of the amp, or if the aerial connection is a pair of open wires, use the two wires combined. The point of this test is to check for signals being induced into the outer conductor of one of the coaxial feeds which is interrupted by the aerial connection being removed. If the interfering signal is still absent with the aerial removed but the outer conductor reconnected then the signal must surely be entering the system via the aerial.

If you have access to a field-strength meter and are able to distinguish the desired signal from the interference and measure them separately a refinement of the foregoing is possible which will give you information about the point at which harmonics or intermods are generated as a separate factor from the point where the signal enters the circuit. This test procedure also uses an attenuator which is specified as 10 dB but could be any convenient value between about 3 and 20 dB. The only requirement is that you can make a signal-level change and accurately measure its effect. Using the field-strength meter in its RF voltmeter mode, connect it into the line at a point as near the receiver as possible; a bridging connection at the receiver input would be ideal. Then try the attenuator at various points such as amp input, amp output, cable output, and so on as follows:

If you place the 10-dB attenuator in the line and both the signal and the interference drop by exactly 10 dB; this is a pretty good indication that the only source of both signal and the interference is nearer to the aerial than the point of the attenuator.

If the signal drops by 10 dB but the interference doesn't change that indicates the source of interference is somewhere later in the equipment chain.

If the interference drops by more than 10 dB this indicates it is being generated by overloading of the next amplifying device, and the cause is a substantial flow of power at its input. Remember that most harmonic generation or intermod production processes require several volts of signal across a nonlinear component so suspect first the places where power flow is highest, signal level is lowest, or frequency is highest. (Transistors for amplifying VHF and UHF have very narrow base regions so must operate on low voltages and are therefore sensitive to overloading.)

Once you have identified the entry path for the signal you can then reconsider more closely the list of possible coupling mechanisms at the start of this section. The next few sections give detailed information for that purpose.

5.5 In-Band Interference

Signals Received via the Aerial

Your opportunities to prevent these signals are limited to manipulation of the directions of nulls in the aerial pattern to reduce their effect. You may be able to slightly move the beam off the direct line so that the null is exactly pointed at the source of interference, or you may be able to change the aerial for one which has nulls at the required angle to prevent pickup of the interfering signal. Screen-backed or corner reflector-type aerials have better performance than Yagi types in regard to interfering signals from other than the desired direction. Using these and cunning choice of aerial pointing angle is about as far as you can go in minimizing the effect at the receiver from in-band signals received off-air. If these measures are not enough then you will have to attack the signal at its source, which may mean negotiating with the owner of the equipment responsible.

Common-Mode Signals

The sources of these signals are the input and output feeder cables and the power feed to the masthead amplifier. A coaxial cable has two conductors and ideally equal and opposite currents should flow in them, and the outer conductor should provide a shield for voltage fields which would otherwise induce signals onto the inner conductor. Anything which upsets the equality of the currents or allows a voltage to be induced into the inner conductor will reduce the effectiveness of shielding of the cable. Ideally these cables should be earthed at one point only; the circuit associated with the feeder cable is then treated as a loop with a constant current flowing at all points around the loop.

In the case where a folded dipole feeds signal to a balun, the dipole will normally be firmly bolted to the main beam of the aerial and the primary coil of the balun should be isolated from all other conductors. If the feed from the balun to the masthead amp is in coax, it would normally be earthed at the amp end. In that case there should be no conductive path between the primary and secondary of the balun and no earthing of the coax at the balun. Inductive coupling as shown in Figure 5.4 is used to transfer the signal.

A point worth looking closely at:

If one masthead amp is being used with two aerials, how are they combined? Is there an earth loop because of the two inputs? A balun in the line to each antenna section may be sufficient, otherwise a separate amp for each section may be worth considering if cost is not prohibitive. This could be arranged either with combining at the outputs or with two separate feeders to the set and switching of the feeders. If all other factors are equal, choose switched feeders because in the case where two aerials are combined you will always have the situation where only one is supplying the wanted signal and both are contributing to background noise.

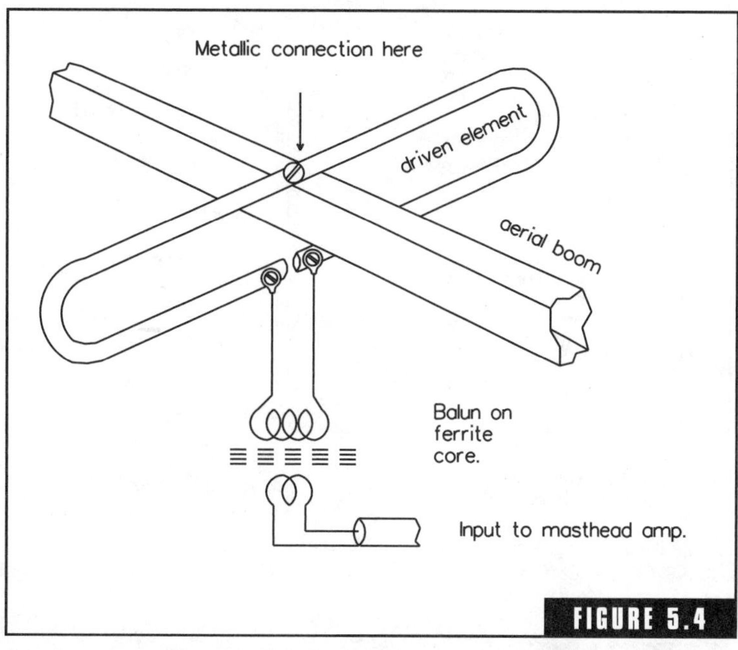

FIGURE 5.4

Antenna connection to minimize common-mode signals.

On the output side of the masthead amplifier the normal arrangement will be a coaxial cable with the outer conductor connected to the shield of the masthead amp and also earthed at the surge diverter and connected to the receiver chassis. There is presumably an earth connection through the mounting of the amp to the tower. There will be voltages induced in the tower from many different sources and also voltages induced in the outer of the coaxial cable. The aim is to make these voltages the same for both. To achieve that, run the coax down one leg of the tower to a very low height (ideally underground) then run it from the tower to the set with an earth bonding wire run following the same path as shown in Figure 5.5.

A tower that is 40 m tall is one-fourth of an approximately 170-m wavelength (frequency of 1.76 MHz) so it will have substantial pickup for frequencies down to the range of the broadcast band and have a resonance at every odd multiple of that frequency. Some of the implications of this will be referred to in later sections; the one that is relevant here is that it has the characteristics of a long-wire aerial (refer to Figure 3.23) for VHF and UHF signals so is capable of introducing lobes and nulls of sensitivity in a variety of unexpected directions.

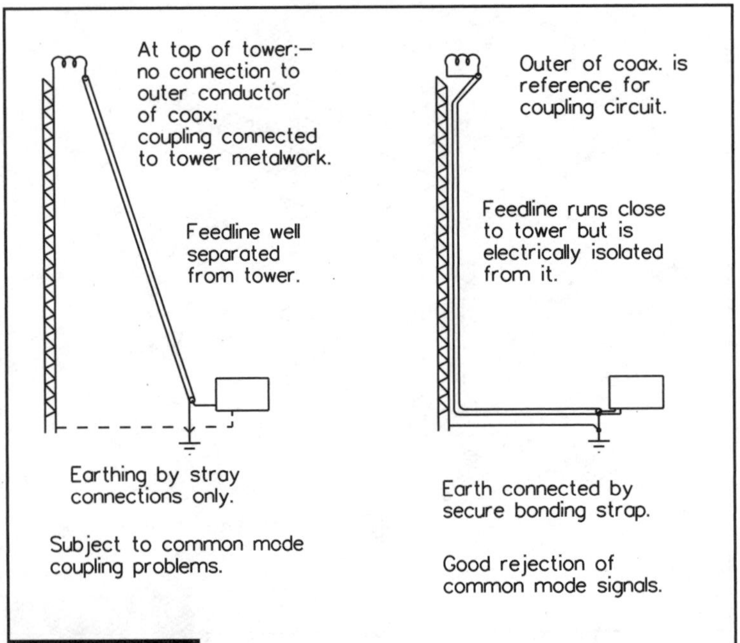

At top of tower:— no connection to outer conductor of coax; coupling connected to tower metalwork.

Feedline well separated from tower.

Earthing by stray connections only.

Subject to common mode coupling problems.

Outer of coax. is reference for coupling circuit.

Feedline runs close to tower but is electrically isolated from it.

Earth connected by secure bonding strap.

Good rejection of common mode signals.

FIGURE 5.5 Good and bad feedline arrangements.

Rejection of common-mode signals is primarily the responsibility of whatever is on the load end of the feeder cable (i.e., masthead amp or receiver) and depends on the construction of the input circuit of the equipment. If the common-mode rejection ratio is not high enough it may be possible to improve it by placing a 1:1 balun in the cable close to the input of the device arranged so that there is no DC connection across the balun and signal is transferred by inductive coupling.

5.6 Interference to the Masthead Amp and Cable

Front-End Overload Signals Arriving via the Aerial

A masthead amplifier operates in a particularly demanding environment; it is built to be as sensitive as possible but it must also operate broadband over the whole range of the spectrum used by television channels. That combination of specifications ensures that it will be predisposed to overloading if there is a sufficiently strong signal anywhere in its passband.

In a country town, signals from a nearby radio-communication base are often strong enough to cause front-end overloading of a masthead amplifier. Occasionally, it also may result from attempting to receive a more distant station in the presence of a strong local one or from nearby industrial activities that use electrical power in a way that produces arcs or frequent switching transients. The people most likely to be affected by this problem are the ones who live within 30 to 100 m of a VHF or UHF radio-communication transmitter (fire, police, and ambulance stations, some electricians and plumbers depots, or a mobile phone base), welding workshop, or light industrial plant.

If the interference is from only one transmitter on a single frequency you can place a rejection trap in the front end of the masthead amplifier. For signals on a range of the UHF spectrum (i.e., cellphone base), a band reject filter may be possible. Any trap or filter must be placed in the circuit before the first semiconductor; amplifiers and clipping diodes (lightning protection) are all nonlinear devices that will produce cross-modulation or intermodulation if overloaded.

A trap circuit can either be a series-tuned circuit placed across the input terminals or a parallel-tuned circuit placed in series with the input line. A stub of coaxial cable can also be used as a trap. The cable is connected as a T junction at the input of the amplifier and left as an open circuit at the end. At one frequency it is exactly a quarter-wavelength and the open circuit is reflected to the line connection point as a frequency-selective short circuit. The actual frequency of the trap can be adjusted by trimming small pieces from the open circuit end.

When more than one frequency must be reduced in level there may be a possibility of using more than one coaxial cable stub but there should only be one of them at the masthead amp input; the other one could be placed in the line nearer the aerial. The use of more than one trap can be unpredictable because the reactive components of the two traps will interact to produce stray resonances at frequencies neither of them are tuned to. Tuning and adjustment of a masthead amplifier with multiple traps would require a sweep generator and oscilloscope so that conditions over the whole spectrum can be watched all at once.

A degree of band reject function can be provided by a single trap using a parallel-tuned circuit with high inductance and very small capacitance. The edges of the stop band of a single-tuned circuit used in that way will have the shallow slope characteristic of a single-tuned circuit. If sharper selectivity is required, a filter using a combination of several L-C sections should be manufactured as a particular item. Filters like that are designed for a particular line impedance so matching of standing-wave ratio is important when they are used.

If trapping or filtering doesn't work and the owner of the set is sufficiently desperate it may be possible to move the aerial or the tower so that the source is in a null of the receiving aerial pattern. There is another mention of this aspect of the subject with more detail in Section 5.10 and again in Section 5.14.

Overloading Signals Induced into Feeder and Power Leads

This path of entry of interference is closely related to the subject dealt with in the previous section under "Common Mode Signals." If interference is evident due to this cause then whatever changes are needed to give the installation good rejection of common mode signals should be done before any more drastic changes are tried. When the station has good common mode rejection and the interference is not due to overloading of the masthead amplifier, the point of overload is usually the first semiconductor (protection diodes or amplifier) in the receiver itself. The same trap or band reject filtering circuits as were described for the masthead amplifier can be used at the input to the receiver.

The DC power supply to the masthead amp is a possible path of entry for interfering signals. If it is via the coax there is probably nothing special you need to do to it; filtering, if required, will have to be done in the power converter at a point where the DC supply is still separate from the signal. If it is via a separate pair of wires you may need a radio-frequency filter close to the amplifier. To check if that is needed, disconnect both legs of the DC feed and temporarily power the amp from a string of torch batteries with short leads to the amp. If the interference pattern is no different then filtering is no advantage. If filtering at the DC input to the amp is required Figure 5.6 shows how it would be arranged.

The RF filter on the DC feed, if one is needed, should have substantial rejection of all frequencies from 500 kHz to about 250 MHz. Figure 5.6 is a circuit that should have good performance with respect to common mode rejection. The aim of the RF filter is to divert currents from the DC lines to earth. It consists of some inductance in series with each conductor and capacitances from the lines to earth. The earth connection point should be the connection point of the case of the amp. The filter should be mounted as close to the amp as possible although up to half a meter or so of wire between them should not be too much of a problem.

5.7 Interference via Power Mains or Other Connections

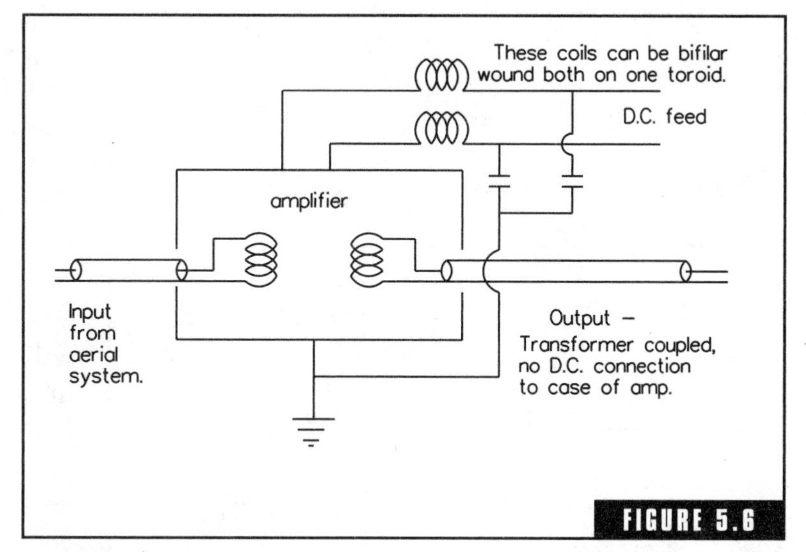

Power-supply filtering.

Interference arriving by this means has two most common sources. Overhead power lines can act as long-wire aerials and pick up substantial signals from a host of possible sources; or computers, other digital devices (such as machine controllers), or arc-generating machines can induce either data signals or harmonics of the clock or deflection signals or RF noise into any of the leads connected to the receiver. Within the receiver the method of coupling from the lead to the signal input is indirect; a coupling loss of 30 to 40 dB at least is to be expected so any signal strong enough to cause interference by this means must be a reasonably strong one on the lead itself. All the power and output connections to a television receiver carry signals of much lower frequency than the receiver input so comparatively simple filters should solve most of these problems.

There are a couple of catches to be aware of in the design of the filter. The capacitive legs are intended to shunt the interfering signal to earth but in many cases the physical size of the receiver chassis is comparable with a quarter-wavelength of the radio-frequency signal so an "earth" attached to the wrong part of the chassis may not be any protection at all and, in fact, in some cases can provide a path by which the interfering signal can bypass the filter and ride straight into the input of the receiver. If you have positively identified a particular lead as the source of interference but filtering does not have the expected effect, try some different spots for connection of the earth reference to the receiver. The layout and circuit of Figure 5.7 shows one example of a coupling path around the filter. The quarter-wavelength

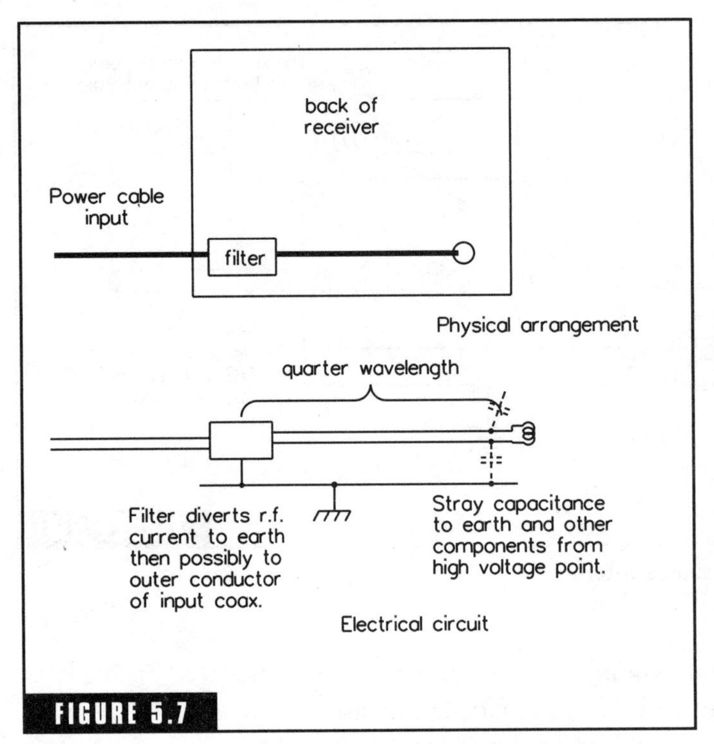

How position of earth connection may affect filtering.

of wire and chassis combine to form a lecher-line-tuned circuit and the stray capacitances carry part of the circulating current of that tuned circuit.

If the receiver is configured with a "double-insulated" power connection then a single-component filter such as this will breach the conditions of the double-insulation standard. In those cases, either change the power feed to an earthed connection (refer to an electrician if you are not one yourself) or add an isolation transformer which will provide the function of the second component and then place the filter on the equipment side of the transformer. Note in this case that the power mains safety earth will not be of any value as a reference for RF filtering; ideally the "earth" for that should be as near as possible to the physical point at which the signal enters the receiver.

Loudspeaker leads can be treated exactly the same as power leads in the case of earthed chassis receivers. In the case of double-insulated sets, the earth connection of the filter ***must not*** be allowed to touch the metal of the receiver chassis unless the power feed is permanently connected through an isolation transformer.

The treatment of data leads depends a bit on the type of data. The type of data usually used by computers complies with the NRZ (non-return to zero) standard, and for that the circuit must preserve the DC connection and have bandwidth from 0 Hz to about 10 times the clock rate of the data. For RZ data which has a reference level set at the end of each bit, transmitted DC continuity is not needed but the maximum frequency of the bandwidth required will be at least twice as high as for NRZ data at the same rate. The output of a modem can be treated as an audio signal usually with the bandwidth of a telephone line (300 Hz to 3.4 kHz).

5.8 When the Tower Speaks

Rusty Joints

This is usually only a problem where the receiver is physically close to a reasonably powerful transmitter. It is the plague of people who live within a kilometer or so of a high-power (10 kW and above) MF broadcasting transmitter. It is unlikely to be a problem in a town that does not have its own local broadcasting station, but if you suspect it, there are some ways to identify it as a source of interference.

First, its effect should change with the weather; normally it will disappear shortly after rain starts, reappear from a few hours to a couple of days after the rain stops, and may actually be worse in humid and slightly misty weather than when the tower has been hot and dry for several days. Also, if the transmitter is modulated you should be able to match some of the varying characteristics of the interference to the off-air program. If you suspect this problem, have someone watch the screen for changes to the interference pattern while you give the base of the tower a good hard thump with a sledgehammer—hard enough to rattle the whole tower. Changes of any sort indicate a fault due to rusty joints.

If you can identify a particular bolt, the cure is to either completely insulate the joint with fiber washers and sleeves around the bolt or to make a permanent metal-to-metal contact. If there are several bolts possibly contributing to an overall problem and the tower is not so old that it is unsafe to climb, the simplest solution may be to use an arc welder to make a small welding fillet across every bolted joint on the tower. (Read Section 3.12 before you do that.)

Harmonics of Low-Frequency Signals

Systems with poor common mode rejection are more prone to this problem than others so if you suspect it you should start by checking and correcting all the factors mentioned in the second part of Section 5.5.

The source of the interference will initially be a signal which is a subharmonic (i.e., one whose frequency is, for example, one-half, one-third, or one-fourth of the desired signal) that is somehow being coupled into a rectifying device (could be either a semiconductor or a rusty or corroded joint) with sufficient power level to swing the rectifier into a nonlinear region of its characteristic. The method of cure depends on where the nonlinearity is. Signals in which the harmonics are being generated in or near the transmitter can only be treated as in-band signals received off-air, as in the first part of Section 5.5. The note above which deals with rusty joints on the tower is relevant when one of those is the harmonic generator. For signals received off-air at the subharmonic frequency when an early stage of the masthead amp is the harmonic generator a trap similar to that described in Section 5.6 but tuned to the subharmonic frequency should help. When looking for the location of a harmonic generator don't forget the surge diverter, which is necessary for lightning protection; they usually include diodes across the signal path in a clipper circuit.

Broadband systems are more susceptible to overload problems than are single-channel receivers. If common mode rejection is good, there are no rusty bolts, and overload is still a problem, a change to a separate single-channel aerial for the channel that is affected with provision of its own masthead amp and feeder with switching of the feeders at the receiver will reduce the problem.

5.9 Sources of Interference in a Country Town

In modern times there are a multiplicity of sources of interference in even a quite small urban area or village—probably, on average, at least three or four for each inhabitant; if the village is big enough to have its own street lights and reticulated water and sewerage, then those facilities are other potential sources. Problems are exacerbated by weak signals; an interfering signal which causes severe disruption of a 100-μV desired signal would hardly be noticed in the background if the signal strength was several millivolts.

The following list of items are some of the most common sources of interference:

- radio-communication transmitters
- computers

- arc welders
- machine controllers
- small electric motors
- fluorescent lights
- lamp dimmers
- loose connections in electrical wiring
- dirty insulators
- anything and everything which switches or controls electric current

The next three sections provide some information on the items in that list. The list of possibilities is almost endless and these notes are by no means exhaustive but will give some information on typical common mechanisms which produce interfering signals. If you are involved in finding and eliminating interference remember that in practice most of these are point sources of radiation so the strength of their signal rapidly becomes weaker as the distance between the source and the entry point into the receiver is increased. Check first all those items which are closest to the receiver.

5.10 Radio-Communication Transmitters

These are low-power (~25 to 100 W) single-channel transceivers such as used for two-way communications by businesses and emergency services. There are two principle mechanisms by which they can interfere with television signals:

The transmitter may generate a harmonic which is radiated on a frequency within the channel of the television signal.

The degree of coupling between the radio-communication aerial and the television receiver aerial may be high enough to induce sufficient power flow into an early stage of either a masthead amplifier or the receiver to cause overloading.

The aspect of front-end overloading has been dealt with in Section 5.4 ("Hardening Your Receiver") and Section 5.6 ("Interference to the Masthead Amp and Cable") so if you are attempting to receive a fringe-area signal in a location which is within 100 m or so of the site of a radio-communication base station of any sort please read and apply the relevant parts of that section before you proceed to any more detailed investigation. Also if you are more than 100 m from the radio-communication site but your problem appears to be due to cross-modulation or intermodulation you should check for overload mechanisms (including rusty bolts on the tower, etc.) at the receiver before you look further afield.

The technical operator of the radio-communication transmitter will aim to reduce harmonic production and radiation as much as possible. Equipment used in this service in Australia is required to meet a specification for harmonic production which prescribes that the total power of all harmonics produced by the transmitter must be better than 40 dB (10,000 times the power) lower than the intended output of the transmitter, and this ratio is increased to 47 dB at the aerial output. When compared with the 50 W allowed output of a base transmitter for a single-channel fixed-and-mobiles network, the 47 dB represents total power of 1 mW. Some harmonic production, however, is unavoidable; that 1 mW will be shared between several frequencies but there may be 200 to 300 μW radiated on a particular frequency. The problem is that even 200 μW can be detectable over distances of several hundred meters in those cases where a major lobe of the polar diagrams of both aerials happen to line up (refer to Figure 5.8) and the interference is being compared with a desired signal close to the limit of sensitivity of the receiver.

Note that for out-of-band signals the angular position of nulls will not necessarily be the same as the published figure for the operating frequency; you will need to swing the receiving aerial to find an angle where the strength of interfering signal is minimum and then try to find a position on the block where that null angle for the interference is achieved at the same time as the main beam is pointing at the desired signal.

Figure 5.8 shows two possible polar diagrams of the same aerial with a comparison between the pattern for the fundamental frequency and a pattern for a harmonic. This is not intended to be the exact state of affairs of any particular brand of aerial, but one typical possibility of many. Note that lobe and null directions for harmonic frequencies are not something that aerial manufacturers test in any exhaustive fashion, so for a particular site you will need to start by assuming that at the harmonic frequency, major lobes in any direction are possible until proven otherwise.

If harmonic radiation from a radio-communications transmitter is the cause of the trouble, then the following will occur:

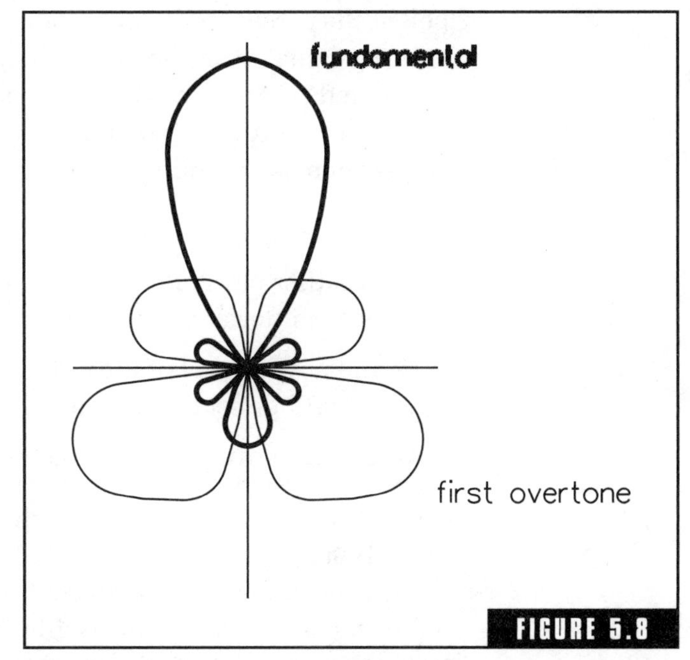

FIGURE 5.8

Comparison of fundamental- and harmonic-frequency polar diagrams.

- many receivers in the immediate neighborhood will be affected
- the interference will be specific to a particular television channel and will show a similar-looking herringbone pattern on all the affected sets

If you observe interference that has these characteristics, you are justified in suggesting to the operators of the radio-communications service that work at the transmitter is needed to cure the problem.

5.11 Signals from a Computer

The computer's clock oscillator is a generator of square waves at the frequency determined by the crystal. In some of the earliest computers that frequency was in the range of 1 to 5 MHz, but in recent years much research and development has been done by the manufacturers to make it higher; in recent years clock rates of 20 to 30 MHz have become common. The square-wave signal produces harmonics at all odd multiples of the clock frequency so there are potentially a lot of radio signals floating around. The only ones that are of concern, however, are those which happen to fall within the RF band of the desired channel. If the interference were from one particular computer you could look at the screen and make a fair estimate of the frequency of the interfering signal from the herringbone pattern (refer back to Figure 5.2) as detailed in Section 5.3, particularly in relation to heterodyne signals. The energy radiated by each harmonic depends on several factors such as the following:

- the harmonic number
- whether particular sections of the computer circuit are resonant at the interfering frequency
- the power level of the circuits being driven by the clock oscillator and deflection circuits
- any shielding that may be built into the computer

The last three have received much design attention in the latter years of the 1990s.

Most of the energy is concentrated in the lower-frequency range. For instance, for a circuit oscillating at 1 MHz to interfere with an Australian channel 10 signal, even though there is, in theory, a harmonic every couple of megaHertz, the energy in harmonic numbers 209 to 213 is so low that it would probably never be noticed. On the other hand, if the clock oscillator frequency was 30 MHz, its seventh harmonic would be at 210 MHz and would be readily noticeable. Figure 5.9 shows diagrammatically why that is so.

Finally, before leaving the subject of clock oscillators and their harmonics, during the latter part of the 1990s there have been computers marketed as having clock rates of several hundred megaHertz. This is not technically correct; the processors in these machines use parallel processing. A parallel processor has several (may be up to about 100 in some devices) relatively independent channels so that several sections of a batch of work can proceed at the same time. From the user's point of view the work gets done as fast as if it was a single-channel processor running at a clock rate of several hundred megaHertz but the actual clock rate that would be measured by placing a frequency counter on the oscillator output would only be in the below-30-MHz range. From the interference point of view the frequency of interest is the actual frequency of the oscillator output.

There will also be radiation from the deflection circuits of the monitor. With these, the power level is much higher than for the logic cir-

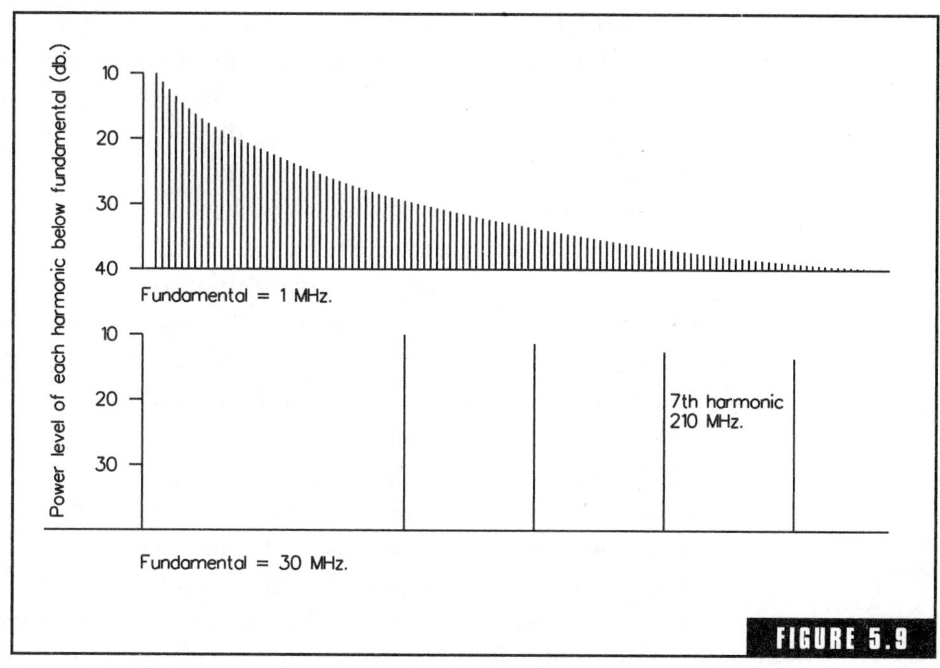

Energy distribution in a typical spectrum of harmonics.

cuitry; the most powerful radiator of the whole system is the horizontal deflection circuit and its spiky wave form produces a harmonic at every multiple of the line frequency. The rapid voltage transition of the flyback time segment as illustrated in Figure 5.10 is particularly related to the high-frequency components of the spectrum of harmonics. The fundamental frequency, however, is in the range below 50 kHz so the proportion of the total energy radiated in the TV band is very small. On the other hand, some of the lengths of wire in and around the monitor are comparable with the length of a half-wave of some television signals, so the final result may be something of a lucky dip.

If the interference you observe is due to harmonics from the deflection circuits of a computer (or group of computers) then there will be multiple frequencies present so there will not be any clear pattern such as a herringbone on the screen. The display may show an effect very similar to that of motor vehicle ignition interference but there will not be the variation that occurs when a motor vehicle is changing speed—it will be more related to the effect of a stationary engine which operates at constant speed for long periods of time. There may be a slow vertical drift or a slow regular pulsing of the pattern due to the frame rates of the television and the computer being very close to the same frequency but not phase-locked.

A signal from computer deflection circuits that is strong enough to be noticed as interference will probably also cause false colors, but because the color depends on the exact frequency and strength of the signal its effect cannot be predicted in advance. Systems that use PAL or similar encoding of the chrominance signal will be less affected by that than standards such as NTSC which don't use that encoding.

In systems where the television sound carrier is frequency-modulated, there may be no effect at all from the interference, courtesy of the capture effect of an FM system. However, if the interfering signal is received at the same (or slightly stronger) level as the desired program, it will take over and the capture effect will work in favor of the interference. If the source is the clock oscillator or any of the data-processing circuits of a single computer, the audio output in the television receiver will be a single clear tone and each different TV channel will have a different pitch of tone. Interference from the deflection circuits will be composed of several tones so will make a howling or wailing sound and will be approximately the same pitch

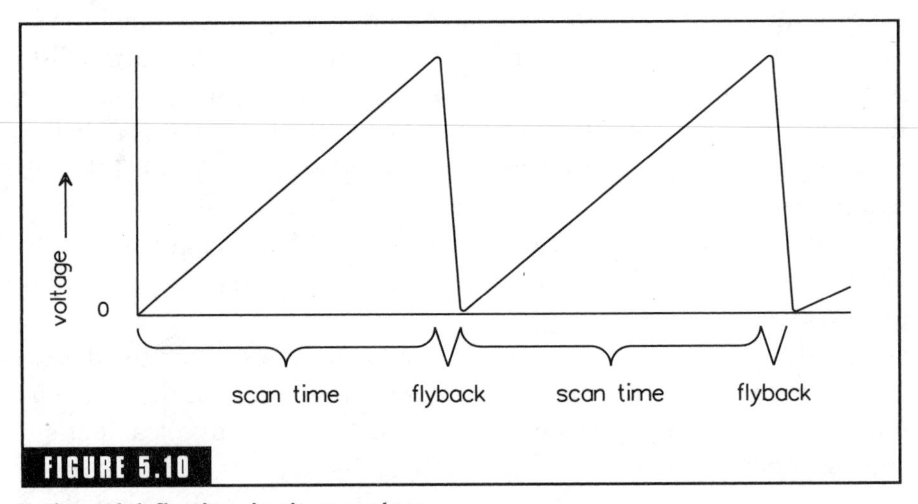

FIGURE 5.10

Horizontal deflection circuit-wave shape.

on all affected channels, and the power of the interfering signal will be greatest on the lowest-frequency channels.

Computer peripheral devices (automatic weather stations, stream gauges, etc.) can create interference if they are close enough to the receiver aerial. The effect created will in all cases have the same characteristics as signals from the clock oscillator or data-processing circuits described previously in this section. If the telemetry device is connected to the computer by a radio link then the interference may be due to the link transmitter and will have the characteristics described in the previous section under "Radio-Communication Transmitters."

If several or many computers and data devices are contributing to a combined interfering signal the observed effect would be very similar to random thermal noise but at a higher level than would be expected from a receiver running at maximum sensitivity with no input. There will be a general environmental pollution of many signals unconnected to each other.

Computers are not intentional radiators so the leakage from them will usually only be of low level and from a point source. For modern equipment that leakage should not be detectable over more than about 10 to 20 m. Older equipment is liable to be more noisy. If only a few computers are close enough to be suspect and they are all less than 7 or 8 years old the most effective cure is to check that shielding and common mode rejection of the receiver is up to par and then mount the receiver aerial high enough to be at least 20 m from the nearest one and if possible with nearby computers sited so that they fall close to the nulls of the aerial radiation pattern. For older computers a problem with television interference will probably be solved by upgrading to more modern equipment.

There will be some occasions when a radio-frequency signal is coupled from the computer onto either the mains lead or wiring to peripheral devices. This must be prevented at the source by filtering so that the RF signal is confined to the inside of the computer's case. A later section, "Signals from Mains-Operated Devices," has more detail on how these filters are used.

When many computers contribute to a general high level of environmental pollution and several neighbors are affected that is an indication that a CATV system or repeater may be needed. See Sections 4.8 to 4.10 for details.

5.12 Signals from Other Electronic Equipment

In the modern world electronic devices are everywhere and come in a multitude of different forms. Audio amplifiers and record players, control devices for all sorts of machinery, motor vehicle electrical systems, and airport and military radar units are a few examples that show the range of different functions that are performed by electronics. The electronics industry has a basic division into two major branches: digital or analog.

Digital equipment uses circuits that switch voltage states. From the point of view of the voltage waveform digital devices cause severe distortion, but that doesn't matter to them because the information they process is related to the timing of the switching transition. When radiated interference is being considered any device that uses digital electronics can be treated the same way as is described for computers in Section 5.11.

Analog devices are all those which handle a particular voltage waveform and preserve it intact with little or no distortion. Sine-wave oscillators and audio power amplifiers are examples of analog equipment. For an analog device the potential to radiate interference depends at least partly on the power level at which it operates and the physical size of the wiring. As a first approximation the majority of items which are battery-operated and self-contained into one small box are unlikely to be the souce of a major TVI problem. An audio amplifier which has remote loudspeakers connected to it by wires may need to be checked even if it is battery-operated.

A photographer's electronic flash unit has the required power level and its signal is in the form of a short rise-time spiky waveform but is usually not objected to because for most people its duty cycle is so low. The interference is only one pulse every few minutes and even at that rate the session may only last a couple of hours, and sessions are separated from each other by months or years. If an industrial plant used a continuing series of such pulses at a rate of once per second for the time of a working shift each day the potential to interfere with local television reception may be a severe-enough design constraint to cause the plant to be relocated to a less populated site.

Any analog equipment that includes amplifiers operating at a power level of more than a couple of watts and which has any connection to pieces of wire whose lengths are equivalent to at least a quarter-wavelength of the television channel has the potential to radiate interference. This, of course, includes almost all mains-operated audio and radio equipment and the television set itself. In most cases, interfering signals from these sources will have the form of unmodulated or amplitude-modulated carriers, which cause a heterodyne interference pattern. (Refer to Section 5.3.) The most likely possibilities for radiation of TVI of this type are as follows:

- harmonics of switch mode power convertor switching waveforms
- local oscillator radiation from receivers of all types
- faulty amplifiers which cause production of high-order harmonics
- loose connections in loudspeaker leads causing arcing
- VHF parasitic oscillations

High-order harmonics of power supply switching can be suppressed by filtering as is dealt with in the next section and fault conditions or loose connections in wiring will improve the operation of that equipment when they are removed so they can be referred to the equipment owner for correction.

The licensing authority will have attempted to minimize the chance of interference from other receiver's local oscillators in the initial planning of channel allocations, and in a well-planned band most of the problems can be avoided. Problems that do occur are most likely in the fringe areas where signals are weakest, receivers most sensitive, and unplanned combinations of channels may be used. When a problem does occur due to radiation from another receiver's local oscillator it must be treated as described in Section 5.5 ("Inband Interference Signal Received via the Aerial"). If several neighbors are all having trouble and the cause is all due to one particular set it may be in order to ask the owner of that set to install a trap tuned to the offending local oscillator frequency in his/her aerial line.

VHF parasitic oscillations can occur in either of two forms with very similar causes. There may be sine-wave oscillations or there may be "squegging" oscillations which have a sawtooth waveshape. For a circuit to generate a sine-wave oscillation the following conditions must be present:

- there must be a resonant circuit
- there must be a feedback path which gives positive feedback at the natural frequency of the resonant circuit
- the gain (in decibels) of the amplifier must be larger than the losses in the feedback path

Whenever these conditions apply, the output of the amplifier will contain a continuous tone at the frequency determined by the resonant circuit. In a high-fidelity audio amplifier the lengths of wire and printed circuit tracks around the power amp stage can form a resonant circuit with frequency somewhere in the VHF range. High-fidelity amplifiers include a negative feedback circuit to reduce distortion of the intended output. A circuit that is designed to give negative feedback for audio frequencies may not necessarily be negative for all frequencies; it may, in fact, rotate the phase of the feedback so much that it becomes positive in the VHF range. Careful design is required to ensure that amplifiers do not become oscillators and even with well-designed equipment there are a whole range of component faults that can cause the phase of a feedback circuit to be changed in the VHF range.

In valve-operated amplifiers VHF parasitics are usually suppressed by circuits which dissipate power in the VHF range in a way that loads and damps the resonance of stray circuits. Figure 5.11 shows the places where these damping components may be added. The "anode suppressor" consists of a small number of turns (about 15 to 30) of enameled copper wire wound around the body of a suitable $\frac{1}{2}$- or 1-W resistor; it is designed to appear as a short circuit for low frequencies and a lossy inductive impedance in the VHF spectrum. The "grid stopper" may be a resistor of several megaohms and relies for its operation on the fact that the grid appears as an open circuit for low frequencies but has input capacitance at higher frequencies, and in the VHF range transit time effects can make the input impedance appear quite low. Both these components are best placed physically very close to the valve socket.

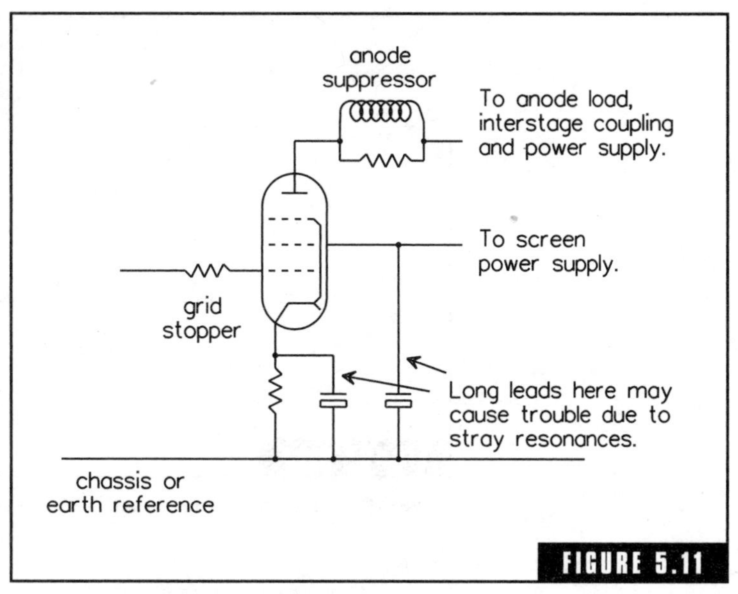

VHF parasitic suppression in valve amplifiers.

For solid-state amplifiers, the operating conditions are somewhat different than for valves; impedance conditions at all points of the amplifier are much lower and often the transistors or integrated circuits used as the amplifying elements will have significant losses of gain and efficiency in the VHF spectrum. In many cases, parasitic oscillations depend on common impedance in DC power and earth connections; the impedance may be a small fraction of 1 Ω and be too small to measure with a multimeter. These faults can be found by setting the device in operation and then, by using an oscilloscope or other instrument that can detect the parasitic signal, looking for places where there is a voltage across what should be a short circuit.

Squegging oscillations require all of the same conditions as are required to produce a sine wave but in addition must have the following:

- a rectifier, reservior condenser/time-constant function in the circuit
- the loop gain (amp gain minus feedback losses) must be much higher than the mininum required for a sine wave

Figure 5.12 shows the essential features of a squeg waveshape.

The spectrum of a squegging oscillation may show energy at many frequencies. The principal ones are listed below:

- the basic squegging frequency and its harmonics
- the sine-wave oscillation frequency

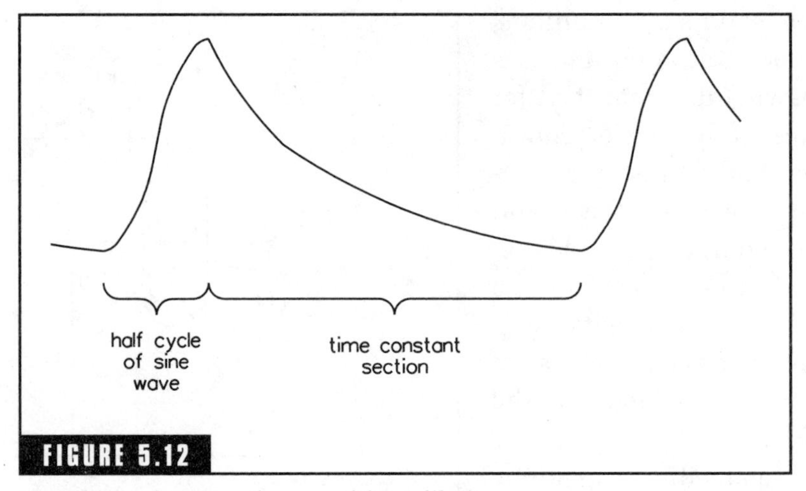

FIGURE 5.12

Waveshape of a squegging parasitic oscillation.

- a range of sidebands either side of the sine-wave frequency due to modulation by the squegging and its harmonics
- frequency modulation sidebands on all of the above. Both the squegging and the sine-wave frequencies are pulled by changes in voltages around the circuit so the genuine signal that the amplifier is intended to carry will cause FM sidebands to be added to the parasitic frequencies

From the amplifier operator's point of view parasitics are fault conditions. In a high-fidelity audio amplifier they create the effects of the following:

- high noise level in the intended output
- reduction of the available output power
- increased mains (or battery) power usage (the extra input appears as heat)

For devices such as machinery controllers using op amps the high noise level may not have any effect on the function of the device but all the other effects will be present. If you detect that a particular piece of equipment is radiating a parasitic signal you are in position to offer an advantage to the owner of the offending equipment in the process of getting it fixed.

5.13 Radiation from Mains-Operated Devices

Electromagnetic energy is radiated whenever a pair of contacts with current flowing through them is opened or an open pair of contacts with voltage across them is closed. The items with the greatest reputation are arc welding operations and small motors which use a commutator, but a multiplicity of other sources are possible. Power controllers with a thyristor or SCR as the control element are also common potential sources.

DC welders, motor vehicle starter motors, and the like, produce random noise over the whole spectrum which will show on the screen as random short black or white horizontal lines the same as when the set is tuned to a blank channel. All items driven directly by AC will produce the same random-lines pattern but only in horizontal bands across the screen. The radiation is only generated at particular times of the AC cycle (usually the voltage peaks) and because the frame rate is very close to the 50 Hz (in Australia) of the mains, the effect is seen as horizontal bands which drift slowly up or down the screen. That slow drift will only be seen on those services where the frame rate chosen for the television system is the same as the frequency of the local AC power mains. When those frequencies do not match, there will probably be a sequence of patterns which repeat at a rate related to the difference between the two; in those cases the observed effect may need some calculations to reconcile the observed pattern with a cause. A radiation source connected to a single-phase supply will give two bands. A three-phase supply could be expected to give either three or six bands depending on whether it was star- or delta-connected and whether the radiation is related to the direct mains or a rectified component of it. A pattern with six bands would be hard to pick from the pattern of a DC source mentioned above. Note that all items connected to the same phase will give a burst of radiation at the same time so will be seen as one source.

Solid-state switching devices connected to the mains (i.e., lamp dimmers and rectifiers) can be treated the same as devices with commutators, but the pulse of radiation occurs at a different part of the cycle. If you were seeing interference from two sources on the same phase of which one was a commutator and one a lamp dimmer, you would probably see four horizontal bands down the screen, and the spacing between the bands could be varied by adjustment of the dimming control.

Resistive loads such as lamp filaments or heater elements will not produce radiation unless there is a high-resistance connection causing arcing somewhere in their circuit. The same applies to purely inductive items such as transformers or three-phase motors. In cases where a capacitor is used on the mains it may be connected in series with the load; the capacitor itself generates no radiation but the associated device may do so if it includes switching or commutating contacts and in that case the radio-frequency signal will pass straight through the capacitor onto the mains wiring.

When a capacitor is connected across the line either for filtering or phase correction it should normally reduce the total radiation from the circuit, but in some cases, such as shown in Figure 5.13, capacitors c2a and c2b, the reduction may be less than expected because all they are doing is diverting the radio-frequency energy from the active and neutral lines to the earth lead, and if that goes halfway around the building to a meter box or subdistribution point before it is truly earthed then there is plenty of length to radiate the signal. In some cases (such as in Figure 5.13 capacitor c1) if the capacitor is physically placed so that the path length of the circulating current is close to half a wavelength of the desired signal it may cause concentration of energy into the interfering harmonics. Note that in cases of this type the size and/or construction of the capacitor may have almost no effect on harmonic production; it is the length of wire that determines which frequencies are selected.

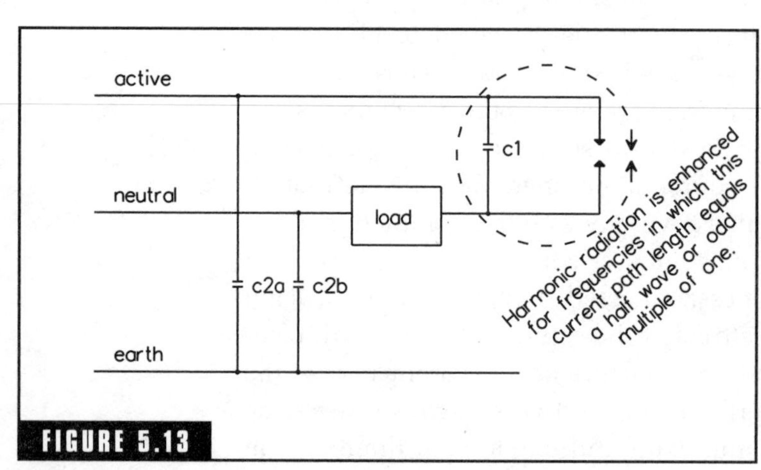

FIGURE 5.13

Capacitors on mains equipment.

Fluorescent lights and gas-discharge lamps are potentially powerful radiators of interference, but most have an inductive ballast and a phase-correction capacitor which combine to act as a filter circuit so the radio-frequency current is confined to the short pieces of wire connected to the lamp itself. However, in the case of a 40-W fluoro which is about 1 m long, this by itself will have the effect of a half-wave dipole for signals in the 2-m wavelength range (150 MHz). The actual frequency of maximum radiation due to this cause will be somewhat lower than that because the current is carried by ions which move significantly more slowly than the speed of light.

5.14 Solving the Problem after You Have Identified the Source

This is where your skill in delicate political negotiations will be used to its full extent. If you can identify a particular item as a source of interference and that item belongs to a neighbor you must approach that neighbor and advise them of the situation as soon as you are sure of the fact. In the first contact be friendly not accusative; your mission is to pass on information to your neighbor of which you assume he/she is not aware. At that time you must be able to demonstrate the effect of the interference and give the other person scope to prove the case for him/herself. Your neighbor may call into question the operation of your equipment and you must be prepared to demonstrate that it is working within its designed specifications, and it will be a help in that case to have worked through the "hardening" measures described in Section 5.4. Once you have demonstrated that there is a problem you must allow your neighbor reasonable time to fix it; ***do not*** under any circumstances offer to fix it yourself even if you have all the appropriate qualifications to work on equipment of that type and you work in a field related to it—let your neighbor arrange his/her own service work if nothing else to get a second opinion on your information.

If you are an electrical or electronics serviceperson called to solve an interference problem of that type, your actions will depend largely on what type of equipment is in question. In general, check first that the item is operating within its specifications; this may lead you to a defective component or nonstandard adjustment which only needs to be corrected to solve the problem. If a modification or extra function is required the circuit change you make will depend a lot on what the equipment or machine is actually doing.

For electrical machines where the source of radiation is a commutator or switching contact or electronic switch, an appropriate filter will solve the problem in most cases. Remember that for TVI, lead length is more important than a high number of decibels in the stopband of the filter and that the reference point to which you are trying to divert the RF current is not an earth, but the other side of the contact or switching point. An appropriately-sized capacitor placed directly across the contact will do more good than the most efficient filter on the market placed a half-wavelength along the line.

Figure 5.14 shows three different ways in which capacitors can be used to reduce television interference. In the top diagram the aim is that currents causing radiation of VHF and UHF signals should circulate within the small loop formed by the contact and the capacitor and if the lead length is kept short enough this circuit will be too small to radiate. The value of capacitance will also have an effect on the degree of arcing, as shown in Figure 5.15. In the second diagram the essential feature is that the attachment point must be to the inside of the metal box. If that is done then skin effect will ensure that the signal frequency currents will stay on the inside of the box so that interference is not radiated. Note that caution is needed in adding capacitance to signal lines; high data rate signals rely on voltage transitions within a few nanoseconds so the capacitor must be small enough to have no effect on the bit error rate of the signal. If an earth connection is needed for filtering, the bottom diagram shows how it must be done. For TVI this

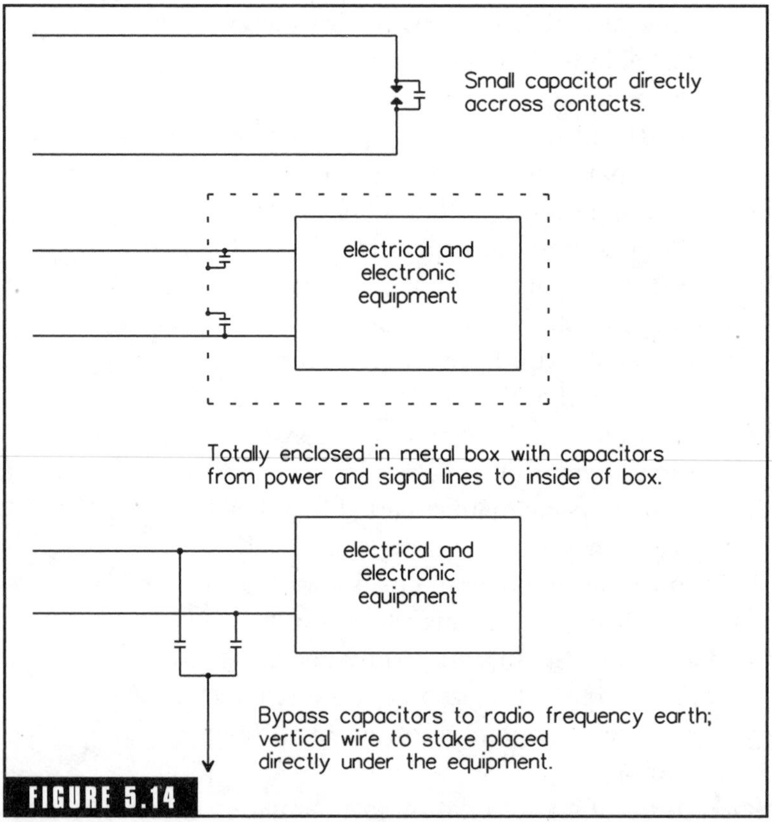

Small capacitor directly accross contacts.

electrical and electronic equipment

Totally enclosed in metal box with capacitors from power and signal lines to inside of box.

electrical and electronic equipment

Bypass capacitors to radio frequency earth; vertical wire to stake placed directly under the equipment.

FIGURE 5.14

Uses of filtering capacitors.

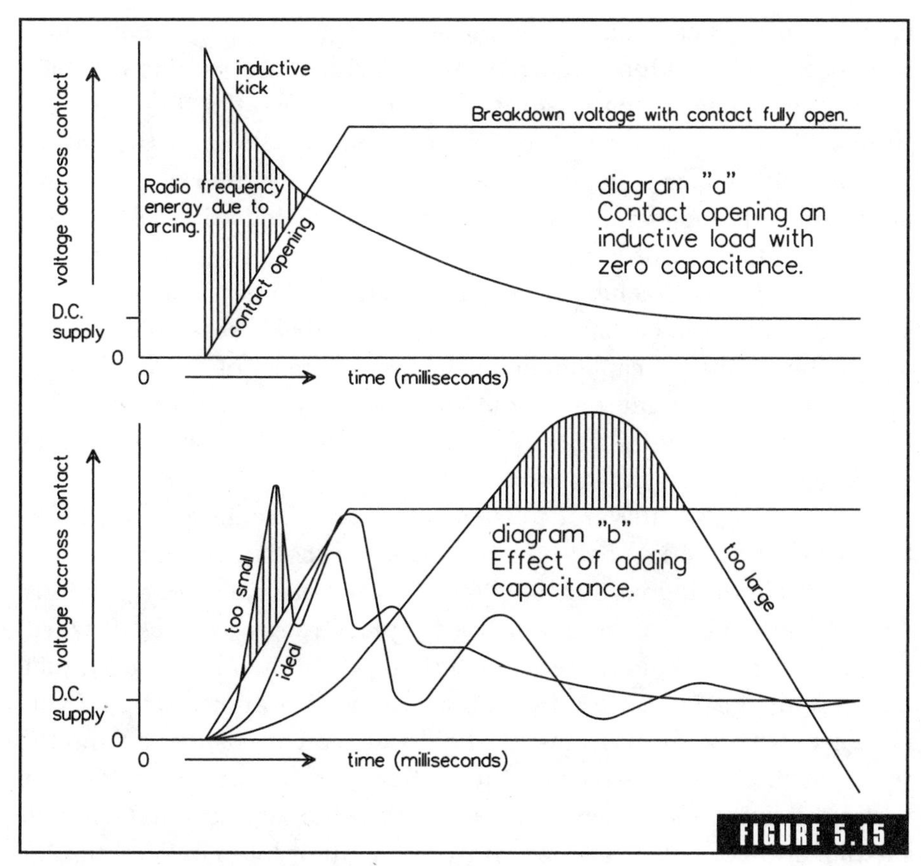

Waveforms of voltage across an opening contact.

will only be relevant for ground-floor locations, for upper-story uses there is some information on earthing in Chapter 3 (Section 3.3), and the overriding principle is that lead length should be kept as short as possible. Note that the electrical safety conditions must not be compromised; if an earth is required, then the cabling earth must be provided in the normal way and for double-insulated equipment, the TVI bypass system must have two mains-rated capacitors in series.

There is a correct value of capacitor for each contact, and that correct value will actually reduce visible arcing of the contact and extend contact life when compared to either bigger or smaller capacitors. The theory is that arcing is caused when the contact is slightly open but not wide enough to have sufficient insulation breakdown voltage for the applied power. The problem is most acute when the contact is breaking an inductive circuit, as shown in the upper diagram of Figure 5.15, because the reducing current in the inductance results in voltage across the contact being raised substantially above the figure for the supply. Capacitance across the opening contact will make the circuit resonant and larger capacitance will make the risetime of the initial pulse longer but also make the damped train of oscillation last longer. The lower diagram of Figure 5.15 shows the three possible states. In some cases a series combination of a relatively large capacitor and a small resistor, called a "snubber circuit," will give best results. Choosing the best value of capacitance can be the subject of a considerable portion of research and development time when new designs are being produced, for individual items in the field tuning for least-visible sparking of the contact is a good starting point.

If the strongest interference is coming from a small range of directions as seen from the position of the receiving aerial, you may be able to choose a cunning vertical angle that places that range of directions near the null angle. Although the best signal strength will result when the center of the beam is directed straight at the incident signal, that may not always be the condition for best signal-to-noise ratio. If you have access to polar diagrams for the aerials you are using you should see that there are nulls in both the vertical and horizontal patterns at particular numbers of degrees off the line of the main beam. (There is a explanation of polar diagrams, nulls, and minor lobes in Sections 7.5 to 7.10 of *How Radio Signals Work,* pp. 96–104 in the original printing.)

Imagine you are at the center of the aerial looking along the line of the main beam. In the horizontal plane there is a null a particular number of degrees off that line in either direction. In the vertical plane there is a null at a particular number of degrees of downward tilt. There is, in fact, a line of null sensitivity in a circular or oval pattern much the same as a rainbow is formed by points in all directions that are a particular visual angle away from the direct line from the observer to the Sun. When the aerial is mounted on a tower, that line of directions of null sensitivity laid out on a map (for example, Figure 5.16) would be drawn roughly as a parabola with the center of the curve at the point indicated by the vertical polar diagram and the more distant parts of the line tending to the directions indicated by the horizontal polar diagram. The pair of thick lines in Figure 5.16

are intended to indicate the approximate boundaries of the zone of maximum attenuation; outside those lines there will also be a wider zone where signals are weakened to a lesser extent.

Aerials of different types have the direction of that first null differing numbers of degrees off the line of the main beam, and you may find a comparison of two assemblies who both quote the same forward gain may give different performance in respect of signal-to-noise ratio because of one happening to have the null placed in a better position for that location.

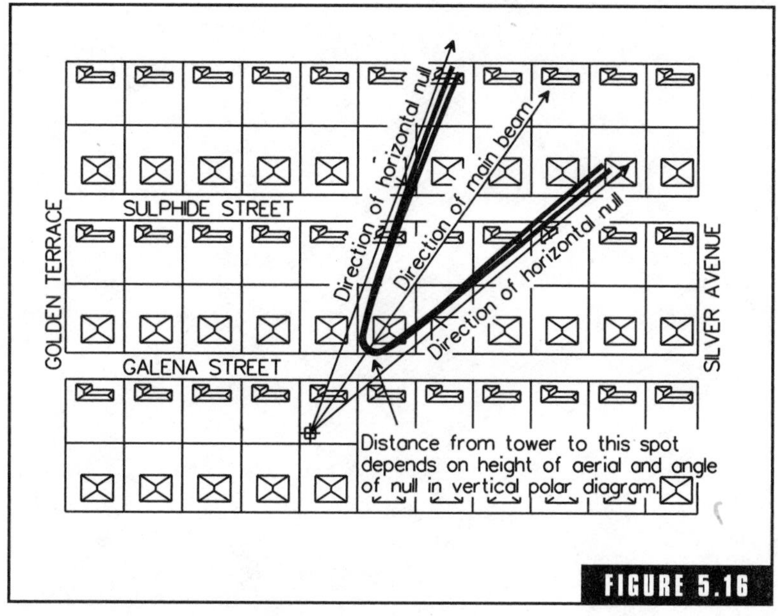

Polar diagram nulls laid out onto a map.

CHAPTER 6

The International Short-Wave Bands

6.1 The International Scene

Short-wave radio in the period from the end of the Second World War to about the end of the 1960s was the most important means of getting information from country to country. Since that time artificial communications satellites and broadband submarine cables have added facilities and partly eclipsed the importance of direct international broadcasting by HF radio. The former medium is still very important, however; it is still the only means by which a signal can be delivered to a private individual in another country without the use of facilities in that country which may be subject to censorship and/or propaganda. It is also the only means of long-distance communication which does not need an infrastructure which may be destroyed by a natural disaster.

Short-wave broadcasting services are almost entirely paid for and operated by national governments. This is not because of any embargo on private or commercial use (there is at least one famous station operated by a religious organization with worldwide interests) but because the stations usually are not well-received at locations close to their transmitting site; thus they have little or no local audience in the country of their origin. Commercial organizations usually seek a local market so are not interested in a service that has a substantial skip zone and whose only listeners are several thousand kilometers away. Governments are, however, interested in broadcasting messages to other countries as part of their international diplomatic efforts and foreign-affairs interests.

High-frequency radio (which is the part of the spectrum used for international short-wave broadcasting) is not a medium offering to carry a large volume of traffic; the total spectrum space available is not much more than that of a single undersea cable or a couple of transponders of a satellite. On the other hand, it is the most economical medium for occasional transmission of a single-voice channel or a few bits of information over long distances. Because of these properties HF radio tends to be mainly used in sparsely settled or remote areas where traffic density is low and distances are great or by people living expatriate who want news direct from a home which may be on the other side of the world.

There are not enough channels in the short-wave broadcasting bands for each country that wants one to have its own exclusive channel, and in fact a particular country claiming exclusive rights to a particular channel worldwide would be a very inefficient way to use the bands. Broadcasters aim their signals to a particular area. Directional aerial arrays are normal at transmitters, frequencies are chosen to place skywave reflections at the correct distance for that time of day, and the broadcast is commonly done in the vernacular language of the target country. Each of the nations that has a short-wave broadcasting service may have an interest in providing programs for several different countries at any one time, and if each is in the language of the target country each must have its own channel. Because the choice of frequency is so critical for effective skywave propagation it is common to "bracket" the optimum working frequency with at least two and possibly up to half a dozen or so transmitters all carrying the same program intended for the same target area. All these factors combined imply that there are a lot more transmissions being made on a worldwide basis than there are channels available; each channel in the broadcasting bands has maybe from a dozen up to a hundred transmissions competing to be heard.

There is a need to share the HF part of the radio spectrum between many different classes of service of which short-wave broadcasting of voice programs is only one, and each of those classes of service requires access to a range of frequencies to select the most appropriate for skywave reception at each particular distance. The HF spectrum is divided into many narrow allocations, and because of the long range expected of the signals the allocations must be coordinated internationally. Figure 6.1 shows allocations for broadcasting in Australia which is typical of the Pacific Ocean region and closely aligned to allocations in other regions of the world. (For information on the mechanics of skywave reflection and the reason it dictates choice of operating frequency refer to Sections 5.8 and 6.5 in *How Radio Signals Work*).

Transmitter operators each seek to make their own voice the loudest on the channel so transmitters are made as powerful as the state of the art will allow. There are technical limitations due to the massive production of heat in a small volume of components and the often-conflicting requirements that the physical size of the final RF stage must be a small fraction of a wavelength of the signal against the need for track length across insulators to be adequate for the breakdown voltage required. Transmitters capable of producing $\frac{1}{2}$ to 1 MW of radio-frequency signal continuously amplitude-modulated to 100% with a clipped sine wave were about the most powerful possible in the late 1990s. Aerial directivity adds to strength of signal in the target area with forward gain in the range of 18 to 22 dB being commonly claimed. A 1-MW transmitter feeding an aerial with 20 dB forward gain indicates effective radiated power of 100 MW in the center of the beam.

From the receiver operator's point of view there are not too many places on the Earth's surface where the short-wave broadcasting bands are other than crowded with very strong signals. The most sensitive possible receiver is sometimes little advantage, but the ability to separate signals from each other which are closely spaced in the spectrum is important.

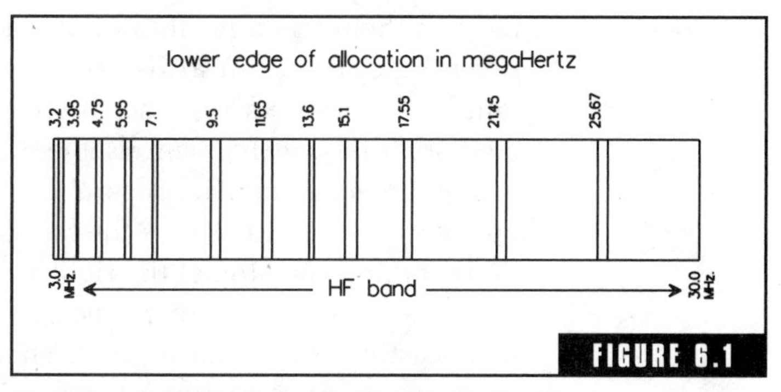

FIGURE 6.1

Short-wave broadcasting allocations in the Pacific Ocean region.

6.2 Features of a Good Short-Wave Receiver

The major factors determining the worth of a receiver for short-wave broadcast reception are as follows:

- adequate sensitivity
- good ability to handle strong signals
- adequate shielding
- very good adjacent channel selectivity
- variable or selectable IF bandwidth
- an accurate display of received frequency reading to the nearest 1 kHz
- continuously adjustable tuning with ability to reset to a particular displayed frequency
- effective noise limiter
- capability for independent sideband reception (refer to Section 2.6)
- provision for diversity connection (also dealt with in Section 2.6)

There are a multitude of receivers sold which are basically MF broadcast band receivers with tuning of one or two HF bands added. These can be used in situations where the wanted station is the strongest in that part of the band but have little value for resolving weaker signals. Their principle limitations are that they are not shielded at all, most of them use a ferrite rod aerial and the dial mechanism covers several megaHertz in a display which is usually less than 100 mm long with very poor reset accuracy. As an aid to making such a rudimentary dial mechanism usable the IF bandwidth is made wide enough to accommodate several channels with wide skirts so that as operators are tuning across the band they can in a sense hear the stations coming and stop at the right spot.

For a serious attempt to receive weak signals on the short-wave broadcasting bands the receiver should be enclosed in a metal box, and in many good receivers there will be other boxes within the outer case which internally shield sections of the circuit from each other.

The aerial connection of such a receiver will be a coaxial socket and if the receiver is operated at full sensitivity with nothing connected to that socket it should be possible to tune over the full operating range and hear no signals except the hiss of thermal noise. Tuning of a good receiver will either offer a mechanical adjustment mechanism with a very large reduction using special antibacklash gears or a frequency synthesizer and digital readout using a counter circuit reading from the local oscillator.

The next few sections will focus attention on each of the features listed above as individual items.

6.3 Maximum Usable Sensitivity

At the 3-MHz end of the HF band there will almost always be more noise entering via the aerial than is generated within the first amplifying stage of the receiver itself. At the 30-MHz end that condition will apply in most cases but it would be possible for a receiver operated with a restricted size aerial in a very low noise environment to suffer a marginal restriction of sensitivity due to internally generated noise. A situation where this might apply would be a small diesel-powered motor launch operating in a remote area well away from thunderstorms (polar regions for instance). If the vessel was a sailing yacht it would have a tall-enough mast for a sensitive aerial, or if it was petrol-driven (i.e., with electrical ignition) or operating within radio range of a city or thunderstorm the noise from the aerial would overcome internal noise.

The amount of noise actually produced by the loudspeaker or earphones will depend on IF bandwidth and the characteristics of the AGC section of the circuit. If the set is adjusted for reception of single-sideband communications its total bandwidth could be in the range of 2.1 to 2.4 kHz, and sufficient overall gain to give full output on a signal of 0.5 μV could be usable. On the other hand, reception of an amplitude-modulated signal with treble response to 10 kHz requires total IF bandwidth of 20 kHz, and sensitivity for signals smaller than about 10 μV is not usable. For any particular amplifier internal noise becomes less as operating frequency is lowered, and at the low end of the HF band it is technically feasible to build receivers with sensitivity down to about 0.16 μV for a 10-dB signal-to- (internal) noise ratio, but that extra sensitivity can only be used if the receiver must operate with a very restricted aerial system such as a mobile whip.

6.4 Strong Signal Performance

Sections 2.4 and 4.1 give some information relevant to the subject of dynamic range. Dynamic range for a good HF voice-channel receiver will be at least 60 dB and more likely in the range of 90 to 100 dB. In most cases, the signal that is strong enough to cause overloading will be in the range of 1 to 100 mV, and the exact figure will depend partly on how far off-tune the interfering signal is. The 100-mV figure is relevant to signals so far off-tune that they can be rejected by the aerial coupling and RF stage-tuned circuits. For signals closer in frequency to the desired signal, the RF amplifier and/or mixer will have some gain, and a higher level of interference will penetrate to the early IF stages and cause nonlinear operation in those stages so the overload threshold is reduced.

The graph shown in Figure 9.7 in *How Radio Signals Work* indicates how adjacent channel signals may overload a receiver. If it illustrates the effect of a single tuned circuit combined with a transistor amplifier stage as would be typical of the RF amplifier stage of many moderately high-performing short-wave receivers the bandwidth to the zero net gain point could be several hundred kiloHertz wide and a signal on a frequency 30 to 50 kHz off-tune may only be reduced by the tuned circuit between 6 and 10 dB. When that is followed by an amplifier with 30 dB gain, for instance, a 1-mV signal at the aerial terminal will correspond to 10 to 20 mV at the input to the next stage. For receivers which use multiple IFs and in particular those in which the first mixer is an up-converter delivering signal to a bandpass filter, that net amplification may apply to several stages before the signal is finally removed by an adjacent channel filter. For receivers of that type, there may be certain combinations of tuning settings and frequencies for which a signal of only 50 to 100 μV is sufficient to cause overloading.

The AGC characteristic of a receiver will have some bearing on the maximum signal it can handle. A good AGC circuit in an HF communications receiver should convert the range of input levels from 1 μV to 1 mV (60 dB) into a range of output levels of less than 6 dB. Good receivers may also have two separate AGC circuits; the one that controls gain in the RF and first mixer stages will be designed to cover the wider bandwidth of those stages and be very fast acting to protect later stages from overloading but may not have a great range of control. The IF AGC has the function of providing the constant level required at the input to the detector stage.

One factor that eases the strong signal situation is that for most receiving sites all the signals will have propagated via a skywave reflection and there will be a tendency for all the signals in a particular band to have originated from about the same distance away (± about a 50% tolerance of course) so signals stronger than a few millivolts will be rare.

6.5 Shielding

The requirements for a good HF receiver in this regard are the same as those for a high-performance receiver for MF broadcast reception (see Chapter 2). It should be possible to set up the receiver with everything connected except the aerial feeder and adjust tuning over the whole range available without hearing anything other than internally generated thermal noise. The arrangement of Figure 3.1 is relevant to this test if the short-wire aerial is left disconnected altogether.

The stages which handle low-level signals and where circuit impedance is high are the ones where shielding is most critical. In valve-operated communications receivers it was usual to build the whole circuit inside a metal case and use several internal barriers to divide the instrument into sections. Transistor circuits operate at lower impedance and fit into smaller boxes so coupling to a general radiation field is less but the lower impedance makes earth impedance (distance to a reference point) more critical. Shielding of transistor receivers may take the form of an individual small metal can or box for each stage that is at risk with all earth references for that stage made to the inside of that box.

6.6 Adjacent Channel Selectivity

This factor and resolution/resetability of the tuning dial and adjustment mechanism are the ones that separate the men from the boys (or women from girls if you happen to think of your receiver as having a particular gender) in regard to HF receivers. There have been a multitude of designs and different operating principles tried over the years; all so far have had some side effects or limitations as the following list of principle variations will show.

Adaption of an MF broadcasting receiver using 455 kHz or similar IF with addition of an RF amplifier and extra RF-tuned circuits to give the required image rejection with mechanical reduction gearing to give tuning bandspreading.

Defects: RF-tuned circuits must all be switchable and tunable so are very difficult to keep aligned. Strong signals close in frequency to the local oscillator frequency may ride into the mixer stage and pull the frequency of the local oscillator. Mechanical gearing prone to inaccuracy due to wear.

As above but with higher frequency for the IF channel, frequencies in the ranges of 1600 to 1700 kHz and 10.7 MHz ±50 kHz have been commonly used.

Defects: Hard to tailor the response of the IF filter. Mechanical problems same as before.

Use more than one IF.

Defects: Each extra IF requires an extra local oscillator which generates interference signals at intermod frequencies (each additional oscillator doubles the number of spots). There is a problem of where to place all the IF channels in the spectrum so that they don't mask desired signals. Same mechanical problems as at first.

Use a fixed-tuned or switchable first local oscillator and tunable first IF.

Defects: Only works well over limited ranges. (A broadcast band's only set could be made this way but it is no good for general coverage.) Harmonics of tunable local oscillator sweep over the RF ranges at several times the tuning rate of the genuine signal.

Use a main tuning dial combined with a crystal calibrator to set the general coverage front end to a particular frequency then use a secondary "bandspread" tuning dial mechanism to offset from the calibrator frequency to particular desired channels.

Defects: Complex procedure to adjust tuning. Requires the operator to have a considerable degree of understanding of the technical processes involved to get good results from the equipment.

Have a frequency in the low-VHF range as the first IF to avoid problems from image frequency responses and oscillator harmonics.

Defects: Frequency drift of local oscillators; requires either a crystal-controlled first LO or a drift cancellation system such as the Wadley Loop, which uses a lot of components to produce a drift-free tunable oscillator source. Limited dynamic range because there are several relatively broadband amplifying stages before the filter, which determines adjacent channel selectivity.

Despite its limitations this last option is the one that is commonly used for really high performance receivers and transceivers at the present state of the art (late 1990s).

6.7 Variable IF Bandwidth

In the vast majority of cases, reception of signals in the HF band is accompanied with reception of some background noise which is a broadband signal. There is an advantage in limiting the receiver's IF bandwidth to only just sufficient to receive the required signal. The actual figure may vary from 20 kHz total bandwidth as needed for high-fidelity reception of a strong clear broadcast signal down to a minimum of 100 Hz bandwidth as required for reception of a morse code signal from an emergency rescue system. The circuitry required to make bandwidth adjustable over that range is incredibly complex so bandwidth adjustment is usually provided in switchable form. One point of practicality relevant to the selection of a receiver is that when a narrow bandwidth is selected ideally the shape factor of the filter should stay the same; i.e., the skirts of the filter characteristic should be at least as steep for narrow bandwidth as for when it is wider, as shown in Figure 6.2, but to achieve that, a separate filter for each selection of bandwidth is required and that means a very expensive IF amplifier.

Note that it is normal terminology for bandpass filters such as are used in an IF amplifier to express the bandwidth of the filter as the range of frequencies between the 6-dB rejection limits. The frequencies at which the filter achieves 60 dB rejection indicate the stop-

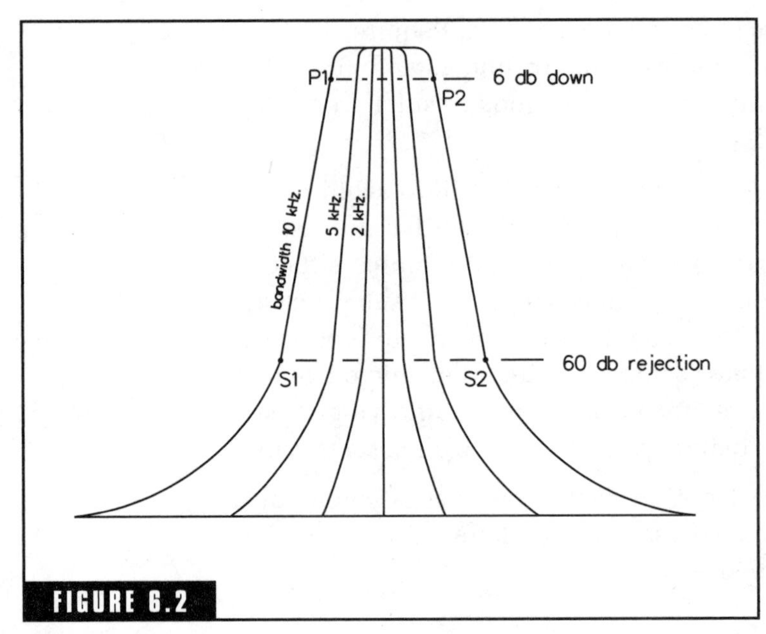

FIGURE 6.2

Ideal IF bandpass response curves.

bands. Filter designers aim to make the ratio between the frequency range for 60-dB rejection compared with the bandwidth for 6-dB rejection (this ratio is named the "shape factor") as small as possible.

The points P1 and P2 in Figure 6.2 are respectively 5 kHz lower than the design center frequency and 5 kHz higher so the bandwidth of the filter is those figures added, which is 10 kHz. A ratio of 2:1 is a good shape factor to aim for, and to achieve that the points S1 and S2 must be no more than 10 kHz above and below the center frequency, for a total of 20 kHz. For the 5-kHz bandwidth, the P points are 2.5 kHz above and below the center and the S points should be no more than 5 kHz away. For a 2-kHz bandwidth the P points are 1 kHz either side of center and the S points should be no more than 2 kHz away.

The diagrams of Figure 6.3 show the effect of economizing on the ideal to the extent of using a tuned circuit instead of a filter for each selection of bandwidth. The situation is improved when more than one tuned circuit is used for each bandwidth, but there really is a need for at least three and possibly four circuits for each selection; the practicality is that a piece of equipment with four tuned circuits will cost about as much to produce as one giving the same performance with a filter.

6.8 Tuning Rate, Resetability, and Resolution of Display

For high-quality MF broadcasting receivers it is technically feasible to manufacture tuning capacitors and mechanical dial mechanisms which can be adjusted over almost a 3:1 frequency range and display

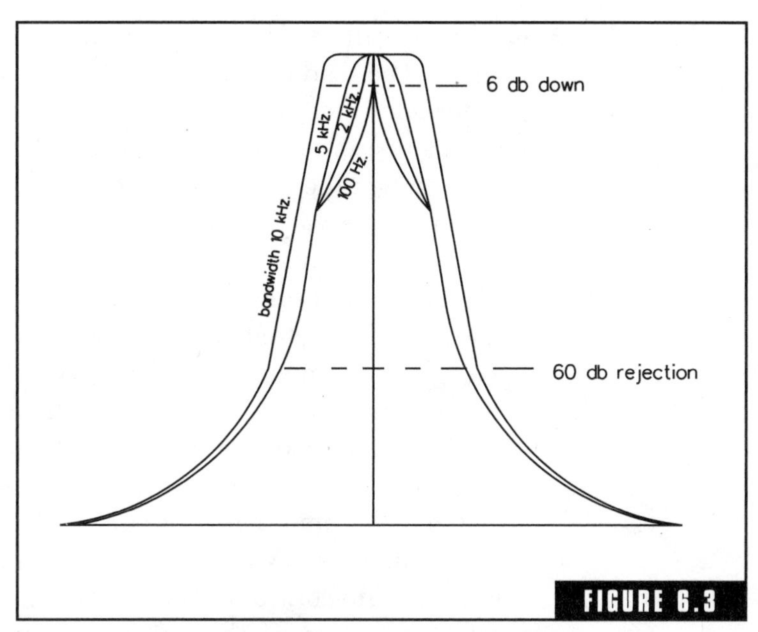

Response curves of "economy" IF bandpass selection.

the 1 MHz of the band over a dial readout of 100 to 200 mm length with sufficient accuracy to be able to reset to a particular channel just by visual alignment of the pointer. The tuning rate of such a mechanism is between 5 and 10 kHz/mm of dial, and it only requires a mechanical system with less than 1 mm of total backlash to achieve the above. When a manufacturer attempts to scale up that system to cover the HF band, the 3:1 frequency ratio may be from 10 to 30 MHz and the 1 mm of total backlash can account for up to 300 kHz of tuning. In addition to that, reception of limited bandwidth for improved noise immunity makes the accuracy requirement much greater.

A mechanical dial mechanism for an HF receiver should ideally spread the displayed dial readout at a rate of 1 to 2 mm/kHz of tuning adjustment with the operating knob geared so that each revolution of the knob corresponds to about a 7- to 10-kHz change of tuned frequency with backlash of that mechanism equivalent to less than 100 Hz of tuning. For general coverage, this implies a dial scale between 30 and 60 m long with 200 to 300 revolutions required to tune from 10 to 30 MHz. Mechanical systems of this type are normally only practical for receivers designed for a particular class of service in which each allocated band is only a few hundred kiloHertz wide, and in that case the range of adjustment of the tuning is electrically limited by padding and trimming capacitors to the limits of those allocated bands. This electrical padding and trimming makes the mechanical specifications much easier to meet reliably.

The mechanical difficulties can be side-stepped in a general-coverage receiver by replacing the dial mechanism and tunable local oscillator with a frequency synthesizer and counter with a digital readout. The readout directly shows the tuned frequency in kiloHertz and for good performance should read to the nearest 0.1 kHz. There must also be provision for trimming the exact frequency for continuous coverage; the basic operation of the synthesizer is to switch the frequency in a very large number of discrete steps, and the receiver must include provision for adjustment to fill in the gaps as is required, for instance, to resolve a single-sideband transmission.

If you are selecting an HF receiver for a high-performance general-coverage requirement there are a couple of points worth checking on synthesizer/counter systems. The digital display uses square-wave signals operating in the multiple hundreds of kiloHertz range and these can cause an increase in the internally generated noise level which may become significant under some conditions (diesel-powered motor launch receiving near 30 MHz in arctic waters for instance). The electronic control of tuned frequency is usually accompanied by electronic tuning of the RF-tuned circuits (using varactor diodes) which may imply a limited capability to handle strong signals. Neither of these potential problems are insoluble and digital readouts of frequency are becoming increasingly common in high-performance HF receivers, but for equipment which appears to be low priced for the performance it offers those specifications of dynamic range are the ones most likely to have been economized on.

6.9 Noise Limiting

The signal arriving at the detector of an HF receiver is a modulated carrier wave combined with a large number of random impulses which will be heard as noise. The impulses come from thunderstorms and similar high-energy electrical discharges. At the receiver input the impulses are many times stronger than the desired signal and very short in time (a microsecond or less), but in the process of passing through the IF amplifier and its filter the energy of the pulse is smeared in time so that the pulse height is less and the duration longer.

The simplest and least-effective noise-limiter circuit is an audio clipper at the output of the detector biased so that it just allows signals of the level of all components of the wanted signal but clips the energy

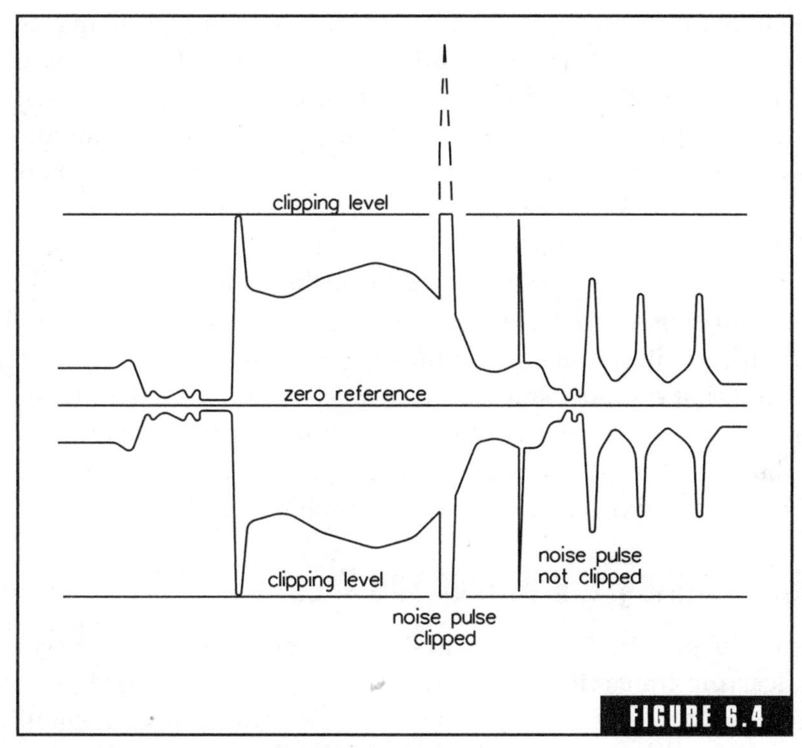

Audio clipper-type noise-limiter function.

of all signals of higher level. A diagram of the output of such a device is shown in Figure 6.4. Simple clippers of this type can give improvement in signal-to-noise ratio but often to have any worthwhile effect they must be set to clip so tightly that they cause distortion of the audio waveform.

The display of Figure 6.4 is a drawing of the envelope outline of a couple of syllables as would be displayed by a storage oscilloscope set to a slow trace time taking perhaps $1\frac{1}{2}$ to 2 s to cover the screen; it would approximately correspond to the sound "-ta-ttt" in that time.

Noise limiters can be made more effective by moving the function toward the front end of the receiver where the noise impulse is still a short-duration high-amplitude pulse, but that introduces practical difficulties of other kinds. Often the signal has not been amplified enough to operate a clipper circuit and there may not be a fixed reference level for the clipper because that part of the circuit is before the stage at which the AGC circuit has applied its stabilizing action. Noise-limiter circuits can be applied to stages as early as the output of the first mixer by providing a separate broadband amplifier with its own AGC function and using the output of that to operate an electronic switch to remove a small time slot of the program that contains the noise pulse. A block schematic diagram of this function is shown in Figure 6.5. In relation to the diagram of Figure 6.4 noise pulses at the output of the mixer have the same total energy as the pulses shown but they occupy between 10 and 100 times less time and are the corresponding number of times larger in relative amplitude with the result that even the smaller ones are easily identified and removed by the switching circuit. The critical feature of a system of that type is that the switch must operate extremely quickly or else the noise pulse will have passed through before the switch responds.

6.10 Finding the Signal You Want

When you set out to tune in a signal from a particular short-wave broadcasting transmitter you will in almost all cases be able to hear something at the spot on the dial where the signal you want should be. There will not normally be large sections of the dial giving nothing except background noise as is common for sections of the MF broadcasting bands in the daytime. The task at hand is not primarily to boost sensitivity enough to make the signal audible but to remove the effect of other signals so the one you want can be heard clearly.

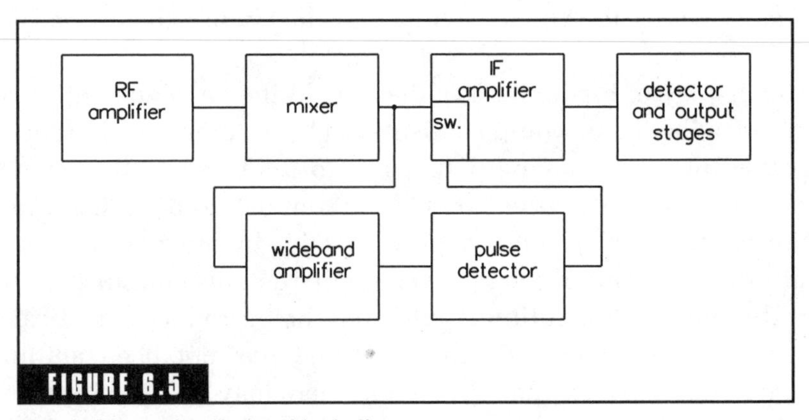

FIGURE 6.5

IF switching noise-limiter block diagram.

There are a range of measures you can take to improve your chance of picking a particular signal out of a field of interference. That range includes the following:

- check that you are listening at the correct time of day
- positively identify the signal
- select the best frequency of those offered
- use an aerial with directional properties
- select for best available sideband and the most suitable IF bandwidth
- check that your receiver's front end is not being overloaded

The next few sections of this chapter will address each of these points in detail.

6.11 Time of Day

The calculation of local time for two different time zones on the Earth's surface is something that looks easy, but there are enough hidden traps in the calculation so that most people get it wrong at least a few times in their life. It was a calculation that was never made by anyone until about the seventeenth century and then for a couple of hundred years was the special mystery known only to a few senior mariners and astronomers. The concept became public knowledge with the advent of intercontinental air travel and messages by radio in the first half of the twentieth century so collectively we are still learning the finer points of the technique.

There are several reference scales of time which at first do not appear to have a great deal of relevance to each other so it is important to fix very clearly in your mind the time scales you are going to work in. The ones that matter for intercontinental radio signals are "local time" and "mean solar time." Astronomers have a couple of other time scales that may be relevant to reception of satellites and for radio astronomy and these will be mentioned later in Chapter 8 ("Receiving the Extraterrestrials"); physicists use "atomic time" and a variety of short-term time scales to measure very fast actions, but none of these other scales are relevant to short-wave broadcast reception.

Both local time and mean solar time are derived from the rotation of the Earth on its axis. That happens at a regular rate once every 24 h and as a result each place on the Earth's surface in the tropical and temperate zones normally has one sunrise and one sunset in each 24-h period. Local time for each place on the Earth's surface can be defined in terms of local sunrise and sunset but daylengths vary between summer and winter so these are not constant references.

A more accurate reference is the exact moment the center of the Sun crosses the "meridian," which is the imaginary line from the point of true north on the horizon, vertically up to the point directly over your head then vertically down to the point of true south on the horizon. Even that solar reference is not absolutely accurate because the Earth's orbit around the Sun is not truly circular; the eccentricity of the orbit causes the Sun to appear to slightly speed up and slow down in its passage across the sky when compared to a series of days each of an even 24 h long. Mean solar time is the time scale that would show the real position of the Sun in the sky if the Earth's orbit was an exact circle. At certain seasons of the year the actual position of the Sun in the sky may be up to about a quarter of an hour fast or slow when compared to the position indicated by a year of exactly even 24-h periods.

Mean solar time is referenced to the local time at Greenwich, which is a suburb of London in England. That is purely the result of the historical fact that British seafarers were the first ones to successfully use astronomical time scales to fix their position in the east–west direction and the Royal Astronomical Observatory happened to be located at Greenwich at the time. There is no reason to change it though. There needs to be a reference location and every point on the Earth's surface is as good as any other. Greenwich mean time is mean solar time for longitude 0.00, which is the meridian of the Greenwich observatory. It is the same as local clock time for England in the English winter and would be for the summer as well if they did not have "daylight saving" in England. In the modern scientific world a more stable time scale has been developed by adapting Greenwich mean time to a standard derived from an atomic clock which is called the "Universal Time Coordinate" (UTC). For radio broadcasting purposes GMT and UTC can be treated as the same thing, they are always within 1 s of each other.

Local time is different for each different north–south zone of the Earth's surface (refer to Figure 6.6). Lunchtime in central Europe is an hour or so before lunchtime in England. Lunchtime in the eastern edge of North America corresponds to late afternoon in England and for the western seaboard of America lunchtime corresponds to mid- to late evening in Europe. Conversely, lunchtime in England (approximately 12:00 noon GMT) corresponds to early afternoon in Europe, mid-morning in eastern America, and a bit before sunrise in California. On the other side of the world 12:00 noon GMT corresponds to about sunset in India, early evening in Singapore and Western Australia, late evening in eastern Australia and northern Japan, and close to midnight in New Zealand.

One of the potential points of confusion in working out the relationship between local times in two different time zones is to be very sure of the standpoint you are viewing the problem from. You can either place yourself at a particular spot and work out differences related to that spot or you can imagine yourself following the Sun around the Earth so that the part you are flying over is always a constant local time. In both cases when you attempt to work out differences between two places the numbers of hours will work out the same but in one case differences have to be added and in the other they must be subtracted.

For international operations such as air traffic control and long-range radio signals Greenwich Mean Time (or UTC) is commonly used as a worldwide reference with each operator making his/her own calculation of the difference

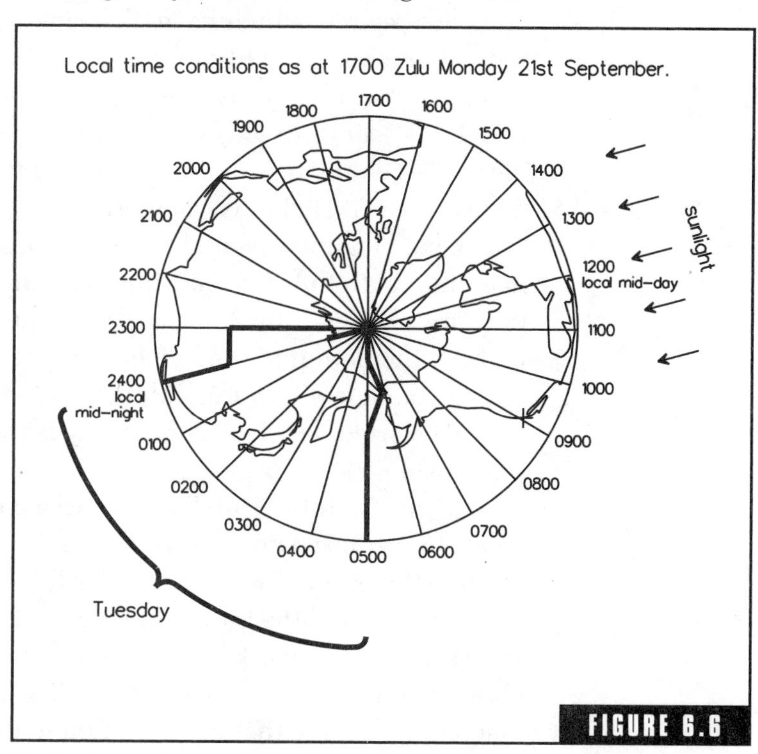

FIGURE 6.6

The Earth showing local dates and times.

between GMT and local time for that particular location. Greenwich Mean Time is commonly called "Zulu" or "Zed" time because it is often written using the four-figure 24-h scale with the numbers followed by a capital Z to distinguish it from local times expressed in the form of four figures to denote 24 h.

The following examples should show how the system works:

In western Australia (which does not have daylight saving)
00:00Z = 08:00 local time.

In eastern Australia in the summer season
01:00Z = 12:00 local time (clock noon).

In California without daylight saving
08:00Z = 00:00 local time (midnight).

In the English summer
12:00Z = 13:00 local clock time.

The date can also cause confusion. At most times the world contains two regions which have different dates. There is an International Date Line which roughly follows longitude 180.00 but has some deviations so that it is all over sea rather than partitioning some nations. Longitude 180 is directly opposite the longitude of Greenwich so midnight GMT is noon at the date line and vice versa. For an instant each day at noon GMT the whole world has one date but if we compare the time zone of Gilbert Islands (near the equator a little to the west of the date line) with the adjacent time zone of Winslow Reef we would find they are separated by 23 h not 1. A minute before the instant of midnight at the date line the Gilbert Islands are preparing to farewell the old day but a few hundred kilometers to the east, where Winslow Reef has just started its new day, that day is the one that the Gilbert Islands are finishing.

For each point on the Earth's surface the date changes at midnight local clock time so one of the boundaries between the two date regions is a line that sweeps around the Earth in step with local midnight. The other boundary is the International Date Line. The diagram of Figure 6.6 shows how this works; it is drawn from the point of view of an observer high over the North Pole. The conditions under which this diagram works are that the direction to the Sun stays constant and the Earth rotates under the observer counterclockwise once every 24 h. It is actually shown at the instant of clock noon at New York on the day of the Equinox and at that time the Sun is rising in the Pacific Ocean and setting in the Mediterranean Sea. If the radio signal you wish to receive crosses either of the date change lines there is a chance that you could look for the program on the wrong day. If, however, the signal crosses both the date change lines then it is back to the same date as it started from.

6.12 Identifying the Signal

In contrast to the shorter range services such as MF broadcasting, television, and so on, short-wave broadcasting transmitters do not stay on a fixed frequency continuously. A particular transmitter may have its operating frequency changed up to a dozen or so times in each 24 h of operation and be switched to directional aerials in different directions each time the frequency is changed. The transmission is not replaced by a beacon. When the transmission is finished on a particular frequency it is switched off and there is no further radiation of power on that frequency until near the time of the next scheduled program.

From the receiver operator's point of view if you wish to hear the program from the start you must be able to set your receiver accurately to the frequency where the transmitter will appear before there is a signal to tune to. That is the reason why it was recommended in Section 6.8 that your receiver be capable of tuning to a particular kiloHertz figure just by dial reading alone.

The transmitter operators want you to hear their signal so they make your job a bit easier by bringing up the transmitter on the new frequency a few minutes before the official program starts and radiating a "tuning signal" which is a few minutes of easily recognizable modulation. A tuning signal may be a few bars of the country's national anthem repeated over and over again, or it may be the call of a bird which is a famous representative of that country, or it could be a famous piece of music which is widely known to be related; the aim is to make a sound which a receiver operator tuning over the channel can recognize instantly. Tuning signals are required to be interspersed with voice announcements of callsign at least once every 5 min.

If your interest is focused on receiving programs from a particular country it is worthwhile to memorize the tuning signal of that country well enough so that you can identify a particular couple of notes of it as you tune across the dial or you can recognize it even if it is suffering interference from a much stronger station. Program schedules are often available from the consulates, High Commissions, or embassies in the national capital of the country you live in and in a few cases some regional consular offices may be able to supply the information. If these people do not have the program schedules available to distribute they can usually give you an address to write to for one to be supplied. Propagation conditions on the HF band change regularly so schedules are changed with the seasons; if you have a continuing interest in following the program try to get your name added to a mailing list for automatic posting of each changed schedule.

If you have missed the time of the tuning signal and must identify the station from its program content the task gets a bit more tricky. The program guide is a help to match what you hear with the expected type of program but also if you have ever received that program successfully before at that time of day and you have recorded such things as actual dial reading, IF bandwidth, and aerial pointing angle you should be able to simply adjust your equipment to the previous settings and hear the program, it may be improved by some final trimming but it should be at least detectable if you duplicate the previous settings.

6.13 Selecting the Best Frequency

For each path over the Earth's surface and each time of day there is a range of frequencies from the "absorption limiting frequency" to the "maximum usable frequency" which is usable for communications. For explanation of these terms see Section 5.8 in *How Radio Signals Work*. Normally each short-wave broadcast program is transmitted on at least two or three frequencies within that range and there may be a considerable difference in clarity of reception between them. One variable is the path loss due to absorption; signals at a frequency very close to the MUF are stronger than those of lower frequency. Another variable is the risk of drift of MUF; signals on a frequency even slightly above it are not heard at all. As another consideration each frequency channel has its own "flora" of other users and the significance of interference for your receiving location may be different from all other locations.

A phenomena particularly associated with signals on the HF band is "selective fading." Section 5.3 in *How Radio Signals Work* describes the mechanism of fading in general, and selective fading occurs when the signal has two valid propagation paths which give signals of about equal strength at the receiver and one or other of the signals has a component of its received phase depending on the exact frequency of the signal. Selective fading causes frequencies in a small part of the received channel to be exactly opposite in phase and therefore severely attenuated. For an AM signal, this can cause degradation of the frequency–response curve if the nulled frequency happens to be in one of the sidebands, but there can be severe distortion if the nulled frequency band is close to the carrier. Multipath propagation is associated with frequencies low in the usable range, in general, channels near the MUF have only one propagation mode so the presence of selective fading is an indication that the channel in use is lower than ideal for that time of day and that distance of path.

On the program guide there should be information on the frequencies and aerial directions of all transmitters carrying the program you want and while the tuning signal is on you should make a quick check of all the advertised frequencies. Do not waste a lot of time searching carefully for weak signals; it is quite possible for any of them to be either buried under interference or be above the MUF on one particular hop and in those cases you may not hear a signal at all. This procedure should only take a couple of minutes and once you have done it you can select the strongest and clearest frequency and concentrate tuning refinements on that channel.

If you have a continuing interest in a regular program such as a daily news broadcast you should repeat the comparison of all channels at least once per week and be able to do a quick check every time reception becomes difficult on the channel you first chose.

6.14 Directional Aerials

Chapter 3 gives a lot more information on aerials for the MF and HF bands in general. This section focuses on the particular aspects that apply to shortwave broadcast reception.

If the signal you want is suffering interference from another station on the same frequency there is nothing you can do to improve it by tuning the receiver, but if the interfering station is coming from a different direction you may be able to make a considerable improvement by building directional properties into the receiving aerial. Note at the start that directional aerials do not have to be rotatable (and therefore expensive); a wire dipole has directional properties, and if you happen to have two trees for support at the right compass bearings you can have a directional property in the aerial for little more than the cost of a length of coaxial cable and a balun.

The directional properties you require of an aerial are something you need to think about at the planning stage of the installation. If your interest is in short-wave Dx-ing and monitoring of signals from several different countries you will have very little alternative but to install a rotatable beam with elements cut to favor the broadcasting bands. On the other hand, if your interest is news from home of a particular overseas country, then it may be feasible to install fixed wire aerials with quite definite tailored directional responses for your particular need.

If you are setting up a fixed wire aerial for receiving from a particular direction remember that the maximum response of a dipole is a broad lobe but the maximum rejection at the null is a very sharp point. If there is only one other principle transmission causing interference you will get best results by adjusting for maximum rejection of the interfering station.

Remember the vertical angle; maximum rejection will occur when the dipole is aligned and tilted to be exactly end on to the interfering signal. You will be able to calculate to within a couple of degrees the required azimuth and elevation figures, but for the wavelengths of short-wave broadcasting the point of exact null is only a few centimeters across so some method of trimming adjustment will be required. The diagram of Figure 6.7 shows one possible method. When adjustment by this method is exact the interfering signal can be reduced in level by about 25 to 30 dB, but in practice with long-range signals changes in the ionosphere will cause the angles of arrival to change from day to day. If the adjustment is set once and then left fixed, the long-term average will be a lower but still very worthwhile figure.

For initial setting up of the aerial shown in Figure 6.7 use the receiver tuned to the interfering signal as the sensor. Adjust IF bandwidth to the narrowest available and tune exactly to the carrier fre-

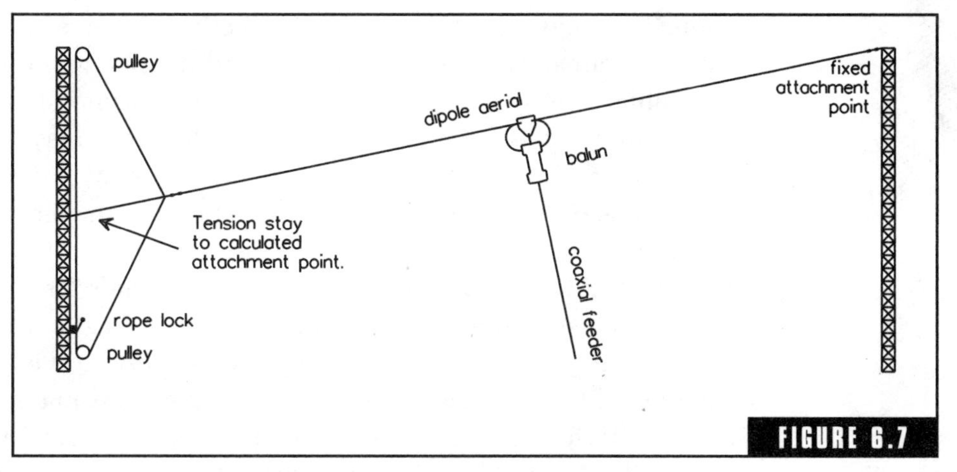

Dipole aerial with adjustment for exact nulling of interference.

quency of the interfering station. If the desired station and the interference are on exactly the same frequency, look for a time of day when the wanted station is off-air and use that time for setting up the aerial. If the receiver has an S meter, use that for signal-strength indication; if not, switch to manual gain control or disable the AGC (and use a suitable input attenuator if needed) then use a multimeter to measure audio voltage across the loudspeaker as an indication of power output. Fading on the signal will cause the meter reading to wander so you will need to make an adjustment then watch the reading for at least a minute or so to estimate an average then repeat the "adjust and watch" procedure to see whether the average is lower or higher.

It is probably not practical to attempt to null out two sources of interference by this method; the variations due to fading of both signals acting independently will make it impossible to establish when the setting for minimum interference has been reached. When there is interference from several sources the use of directional aerials should be aimed at maximizing forward gain in the major lobe and directing that exactly at the arrival angle of the desired signal.

Chapter 3 gives more information on directional HF band aerials. Almost all the techniques described for MF broadcasting band directional aerials will work equally well for the HF band with suitable scaling of size. One difference is related to the earth reference of a vertical element. A quarter-wave vertical for 25 MHz is only about 3 m long so it may not be particularly effective if used at ground level in a built-up area. In cases like that a ground-plane reference at the top of a suitable pole will make an effective aerial. A tuned loop (as in Figure 3.12) for frequencies in the high end of the range may be a bit small to be a good signal collector; a loop tuned for 25 MHz, for instance, would be only a small fraction of a meter in diameter.

From the other direction there are some opportunities for scaling up the techniques used for VHF television aerials for use on the higher frequency short-wave broadcasting bands. A Yagi cut for a frequency in the range 20 to 25 MHz is not excessively unwieldy and can be made to fit on a rotator. There are rotatable Yagis made and sold commercially for use on the amateur bands which have multifrequency capability and some with inductive loading of the elements can be used on frequencies as low as 7 MHz. Screen-backed dipoles and corner reflectors would be too awkward to fit on a rotator, but for fixed direction use they could very simply be given multifrequency capability with very good front-to-back ratio.

While considering the subject of directional aerials there are a couple of points worth noting in relation to the calculation of directions and arrival angles:

The direction required is from the location of your receiving aerial to the location of the transmitting aerial. Location of the originating studio is not relevant. This point becomes particularly important when a program is relayed; you will need to identify whether you are receiving the original transmission or the relay.

The signal always uses a great circle path. For example, London and southern Canada are on about the same latitude but the path of a signal from London to Vancouver starts and finishes about 45° off the east–west line and goes as far north as Greenland and Baffin Island.

There are several factors contributing to the calculation of vertical angle so, if possible, obtain a predicted figure for the required frequency and transmitting and receiving locations. If you do not have access to predictions remember that the maximum practical length of one hop is about 3200 km and the height of the reflecting point is between 300 and 500 km in altitude for F-layer reflections, with frequencies close to the MUF being reflected at the highest altitudes. Unless you have a computer program that takes in all the factors and makes exact calculations, the surest way of using these figures is to draw a scale model of the curve of the Earth on a very large sheet of paper, draw in the ray lines, then measure the resulting angles with a protractor.

Refraction (the "bent-stick" effect) will make the signal appear to come from higher in the sky than that diagram indicates so add a couple of degrees to the vertical angle for hop lengths between 1000 and 2000 km and perhaps up to about 6° to 8° for hops over 3000 km. If you have chosen the correct number of hops the accuracy of the result should be within 2° to 5°, which should be close enough to use as the initial attachment point of the tension stay shown in Figure 6.7. The ionosphere varies with time so the actual angles for both direction and elevation will vary a few degrees from the calculated figures.

The design of aerials for particular directional properties is an enormous subject so if you want to delve deeper into it there are books devoted entirely to aerial design which give many examples. One good one on that subject is *The American Radio Relay League Antenna Book.* That book gives designs specifically for the amateur bands, but in many cases, they can be adapted for nearby short-wave broadcasting bands just by scaling the measurements. There are probably also sites on the Internet that have relevant designs and these can be found by delving through listings on the search engines.

6.15 Selecting for the Best Sideband and IF Bandwidth

Receivers which offer a selection of IF bandwidths are intended to make best use of the tradeoff between treble response for a high-quality clearly received program and reduced noise and interference when conditions are less ideal. The use of single-sideband reception has been dealt with in Section 2.6. If your receiver does not have single-sideband capability you may be able to gain a portion of the advantage of ISB if your receiver has a fairly flat passband characteristic for the IF filter and you juggle the tuning and bandwidth selection as in the following example.

Assume the signal you wish to receive has a carrier frequency of 12,000 kHz and is amplitude-modulated with frequencies up to 9 kHz. There is severe interference from a signal on 12,005 kHz amplitude-modulated with treble response to 5 kHz and significant interference from a station on 11,990 kHz. These signals are shown in the spectrum diagram of Figure 6.8. If your receiver has a flat-topped portion of its bandpass characteristic you may be able to gain considerable improvement by setting the tuned frequency 2 to 3 kHz lower than exactly 12,000 kHz and adjusting the IF bandwidth to about 5 to 7 kHz.

The most offensive part of the interference is the strong heterodyne tone from the carrier signal on 12,005 kHz and the effect of that can be made perhaps 15 to 20 dB less by carefully placing it on the sloped part of the IF filter characteristic. The limit of how much you can gain by this means is set by the flatness of the passband of the IF filter. If the carrier level of the desired signal is reduced then when the program has high volume in the mid-range of frequencies there will be modulation over 100%, which will sound severely distorted (compare with Figure 2.13.)

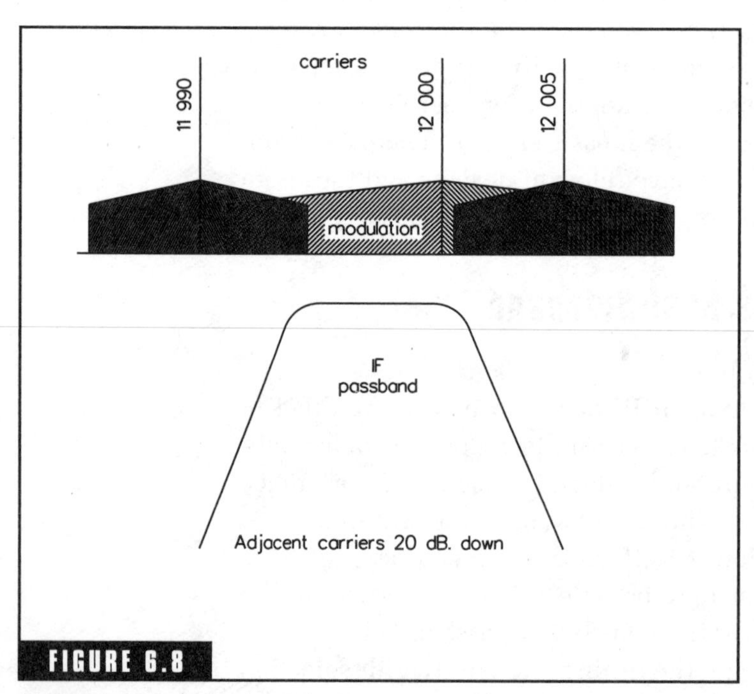

Setting IF tuning and bandwidth to minimize interference.

If the passband of the IF filter is flat-topped, as is typical of a crystal filter, you may be able to set the tuning up to about 3 kHz off the theoretical point with good effect. If the passband is more rounded, as would be expected of a series of tuned circuits, then you may only be able to move the tuning a few hundred Hertz before distortion sets in. If your receiver has a facility for "enhanced carrier reception" using a "beat frequency oscillator" then distortion will be no problem and you will be able to set a receiver with almost any-shaped passband up to several kiloHertz off-tune with good clear reception. The BFO (beat frequency oscillator) is an oscillator circuit built into the IF amplifier whose frequency can be trimmed to any particular setting within the passband of the IF filter. In a receiver designed for single-sideband reception the BFO is usually called a "carrier insertion oscillator." For enhanced carrier reception the BFO is accurately set to zero beat with the incoming carrier of the desired signal with the result that the signal presented to the detector always has a very low modulation percentage so there is no distortion due to overmodulation.

6.16 Checking for Receiver Overloading

HF receivers can be overloaded by strong incoming signals and the overloading signal may not necessarily be detectable within the received passband. Overloading causes cross-modulation and intermodulation, and in the crowded short-wave broadcasting bands there may be quite a few extra signals present before the operator notices any effect other than a general increase in the crowding of the band. The effect of limitations on dynamic range of the receiver is described in Section 2.4 and in relation to television in Section 4.1. The same principles apply on the HF bands but in a less restrictive form; the available dynamic range of a good HF general coverage receiver should be 60 to 90 dB instead of the 30 dB common for television. If you suspect that a particular interfering signal is an intermod or cross-modulation product, temporarily include an attenuator in the aerial feeder. Attenuation of the incoming signal by 20 dB or so will completely eliminate most overloading signals and significantly affect the wanted signal to interference ratio in all cases. Note that an attenuator in the feedline will not affect intermods produced by corroded connections in the aerial wiring or rusty joints on the tower or surrounding metalwork. (Even be suspicious of gutters on neighboring buildings.)

To check for intermods produced in the aerial structure use a modification of the test shown in Figure 3.1. The modification is that the temporary aerial should be cut to a length close to a quarter-wavelength of the frequency under test then rigged in a clear space as near vertical as possible. With that test all genuine signals should be detectable although maybe weak and noisy, but intermods generated in the joints of the aerial system should be reduced in level by at least 30 to 60 dB.

6.17 Monitoring and Reporting Procedures

When a local or regional broadcasting service (MF band or television for instance) is being established part of the commissioning procedure is a field-strength survey. A technical officer of the installation crew takes a field-strength meter to representative locations and makes a number of measurements; those measurements and their locations are plotted on a map and, by interpolation and extrapolation and checking by theoretical calculations, lines of equal field strength (isopleths) can be drawn in a similar fashion to the way isobars are drawn on a weather map.

For a number of reasons, some technical and some economic or political, that procedure is not feasible for short-wave broadcasting. Short-wave broadcasters rely on monitoring stations and listener's reports for information on the coverage of their transmissions. Each of the national governments that are themselves involved in broadcasting also operate monitoring stations which will regularly survey particular short-wave broadcasting bands and refer all the information collected to the International Telecommunications Union, who then sort out which reports apply to which stations and pass on the information.

A typical working schedule for an official monitor may be to survey a particular band, for example, the 15-MHz short-wave broadcasting band which covers from 15,100 to 15,600 kHz. An observation is made for every 5 kHz of the band with notice taken of whether that frequency is occupied by a carrier or by sidebands of an adjacent channel. If a carrier is present, a report is made in the SINPO format of the quality of the received signal and if identification of the transmission is possible, that is included.

That process may be repeated every half-hour, 24 h per day for a week. The next week the same overall process will be repeated but a different one of the broadcasting bands will be surveyed.

Information supplied to the ITU includes the exact location and technical details of the aerial with the reports of all the observations and particularly records of any positive identifications made. The ITU can combine similar reporting from many monitoring stations in many different countries; extract and collate all the reports relevant to a particular member broadcaster and advise them of the results. These official monitoring reports are also used to identify possible clear channels when a member requires to change the frequency of an existing service, commence a new service, or give warning of possible interference problems if a truly clear channel is not available. Official monitoring and reporting provides a reliable backbone of information but it is only a skeleton; there may only be one or two monitoring stations in each nation and that does not give enough points on a map to plot a chart of coverage area. The broadcasters encourage and actively solicit listener's reports to fill in the gaps between official stations.

For a report to be useful to the broadcaster there is certain technical information about the receiving station that must be included to make it possible for the user of the report to reconcile it with all the others. The list of details they will need to know includes the following:

- the exact location and type of aerial used for reception and the height above local ground level
- any directional properties the aerial may have and the direction of the major lobe used for the reception reported
- the type of receiver and if relevant the detection mode and IF bandwidth used.

The actual report of reception should use the SINPO format, which is an acronym for strength, interference, noise, propagation, overall. Each of these factors is given a rating between 1 and 5 which the listener estimates according to the following scales:

Strength

S1 = signal barely detectable
S2 = just strong enough to have a slight effect on the receiver AGC
S3 = moderately strong signal
S4 = strong enough for comfortable listening
S5 = so strong that any further increase in strength would make no difference to the signal

Many receivers have an S meter which displays information on the voltage present on the AGC line which is usually calibrated in a scale of nine "S units" with some room for overrange indications. The nine-unit scale is used by amateur radio operators and if properly calibrated for their use the scale would correspond to the following:

S9 = 100 μV at the aerial terminal

Each S unit below that = 6-dB reduction in signal level.

These S meters rarely stay calibrated for very long, but if it is relevant the meter scales could be roughly related to SINPO numbers by taking meter S9 and overrange as SINPO S5 and then counting two metered S units as one SINPO unit. Be aware, however, that your particular receiver may have a substantial zero error in its calibration and that zero error may be different at different parts of the spectrum.

Interference

This scale only measures the effect of other intentional transmissions carrying intelligent (?) signals. All spurious emissions and signals from nonhuman sources are described as "noise."

I1 = wanted signal unreadable

I2 = interference strong enough to reduce the readability of the wanted signal

I3 = interference strong enough to cause difficulty but signal still readable

I4 = interference detectable but having no practical effect

I5 = no significant interference

Noise

The difference between noise and interference is due only to the intelligence or otherwise of the source. Noise is all nonintelligent signal inputs to the receiver. The scale for noise is the same as the scale for "interference."

Propagation

This factor includes fading, selective fading, intermodulation effects caused by rapid fluttering, and the effect of all factors of that type.

P1 = signal generally unreadable

P2 = variations sufficient to reduce readability of signal

P3 = significant variations but signal still readable
P4 = slight variations due to fading, etc.
P5 = absolutely steady signal

Overall

The overall rating is a "stand back and survey the finished product" type of assessment after all the individual ratings have been decided. It should have some of the functions of a checksum in that if all the individual factors were recorded as a particular number then the overall rating would be expected to be close to that number. There are, however, other factors such as the overall readability and possible technical problems in the transmission itself which will not be assessed by any of the individual factors.

O1 = signal generally unreadable
O2 = defects noticeable enough to affect readability
O3 = noticeable defects but signal still readable
O4 = defects detectable but no practical effect
O5 = quality as if you were in the studio

If you have collected all the technical information required for a useful report on a particular transmission there are basically three options open to you depending on what your interest is.

You could be someone who seeks to gain confirmed reports of reception of as many different stations as possible in a competitive, record-breaking manner. Most people who do that have some postcards printed called "QSL cards," which may have all the technical details of the receiver and aerial and the operator's return address printed on the card so that all that has to be done is to add the SINPO figures, a few words about the program content, the broadcaster's address, and a stamp and post it. If that describes your interest then what you are really interested in most of all is the broadcaster's confirmation of your reception which also will normally be in the form of a QSL card. "QSL" was originally a telegraphic shorthand for "confirmation of message received," but it has moved out into the wider field of radio communications of a variety of types.

On the other hand you could be a migrant from a particular country and have a continuing interest in that particular country. You may be able to offer to send a regular series of reports. Broadcasters are very grateful for these but do not normally offer payment; they would add you to their mailing list for regular program guides and some may send you some stationery (i.e., blank QSL cards) so that your reports to them are in a standard format but the actual collecting of the information and writing of the report would be done on a voluntary basis.

More casually than either of these you may be someone whose real interest is in general knowledge of faraway places of all types and you just happen to have heard their broadcast. If you send a short letter to the contact address given in the program you have heard to advise them that your real interest is in the country, in general, rather than specifically just their programs, they will welcome that contact as well and may be able to send you some "tourist brochure" level of information in addition to their standard acknowledgment of your report. You will need to include the receiver and aerial technicalities and the SINPO report in your letter. Broadcasters encourage these casual reports and welcome advice of what your real interest is; they pass that information on to program directors and reporters as a guide for the direction of future programs.

Short-wave broadcasting is an established technology; it had its heyday in the 1930s, and from then until the early 1970s was the most significant means available for transmission of broadcast information over national boundaries. For many smaller and developing countries the possession of a short-wave broadcasting service was a major and possibly the most important thrust of their foreign policy at that time. In the 1990s there were alternatives, and the broadcasting organizations widened their focus so that now HF broadcasting may be supplemented (and programs integrated) with programs via geostationary satellites or with images on a Website delivered via the Internet.

CHAPTER 7

FM Stereo and Television Sound

7.1 Stereo Multiplexing

The starting point of design specification for a broadcasting system which transmits stereophonic information is that the receiver's IF output can also be demodulated without stereo decoding into a signal that gives all the program that would be normally received as a monophonic signal. This requirement for a dual role adds complication to the multiplexing scheme.

In a stereo audio amplifier two separate channels are provided; one channel is used to drive loudspeakers or earphones on the left-hand side of the listener and the other drives an output on the right-hand side of the listener. A monophonic signal equivalent to the combination of these, complete except for the removal of the stereo information, is obtained by adding the signals from both channels [mono signal = $(L + R)$]. The radiated mono signal can only be one channel so the baseband signal is required to be an $(L + R)$ combination. The signal carried by the subcarrier portion of the multiplexed signal is that portion which transmits all the information about separation of that mono signal into two stereo channels. It consists of an $(L - R)$ combination; the $(-R)$ component is produced by reversing the phase of that channel.

To reconstitute the original left channel exactly, equal levels of the two multiplexed channels are added together; the (L) signals combine but the $(+R)$ and $(-R)$ cancel each other. In similar fashion, the original right channel is obtained by adding the same two signals but with a phase reversal on one of them; in this case the components are $(L + R)$ and $-(L - R)$. The (L) and $(-L)$ components cancel and the right-hand channel components are (R) and $-(-R)$, which add.

When these components are modulated onto a radio carrier the requirements are as follows:

- phase relationship between main and subcarriers must be preserved exactly so single-sideband techniques cannot be used
- devotion of audio signal to radiation of continuous tones will waste modulator power

In order to make maximum use of the capability of an FM broadcasting transmitter, the stereophonic signal is compiled into a multiplex baseband signal composed of the following:

1. A signal in the frequency range of 50 Hz to 15 kHz which transmits the $(L + R)$ combination and which can be received by a monophonic receiver. This component is designated as the M channel.
2. A double sideband suppressed carrier signal referenced to a suppressed subcarrier on a frequency of 38,000 kHz. This transmits the $(L - R)$ signal in the form of amplitude-modulated sidebands in the frequency range of 23 to 53 kHz. If letter designations are being used, this is called the S channel.
3. A low-level pilot tone (typically 20 dB down on maximum mod. level) at 19,000 kHz.

The combined signal, as illustrated in Figure 7.1, is then used as a 50-Hz to 53-kHz input for the modulator of an FM broadcasting transmitter with a carrier frequency somewhere in the range 88 to 108 MHz and deviation of 75 kHz (if it is in Australia). That signal requires a channel allocation (bandwidth) of 200-kHz.

At the receiver the stereo decoder senses the presence of the 19-kHz pilot tone, doubles its frequency, and amplifies it to provide a carrier frequency for demodulation of the double sideband $(L - R)$ signal. The $(L + R)$ signal is recovered by simply low-pass filtering the

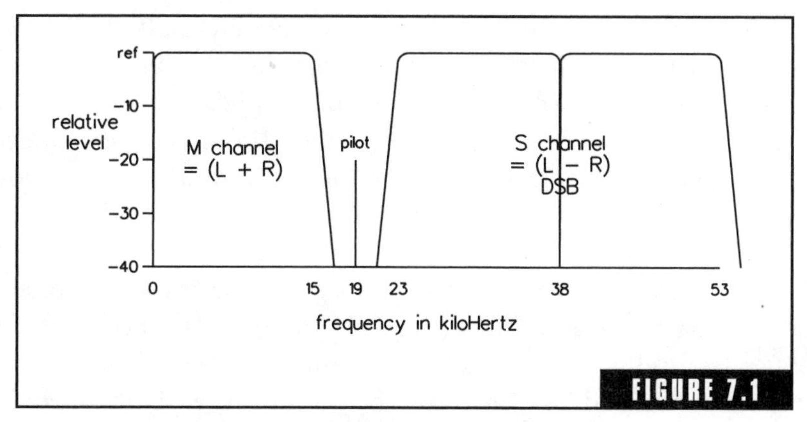

FIGURE 7.1

Modulating signal for an FM broadcasting transmitter.

output of the modulator. With those two signals available (there must be both positive and negative phases of one of them) the right and left channels can be derived by a simple resistive matrix. These functions are illustrated in the block diagram of Figure 7.2. Stereo separation is maximized by adjusting levels to be exactly equal and by providing phase correction; separation of about 25 to 30 dB is possible with a well-adjusted decoder. In practice, in setting up an FM stereo decoder this relationship is used the other way around. The decoder output is connected to equipment for measuring crosstalk between the chan-

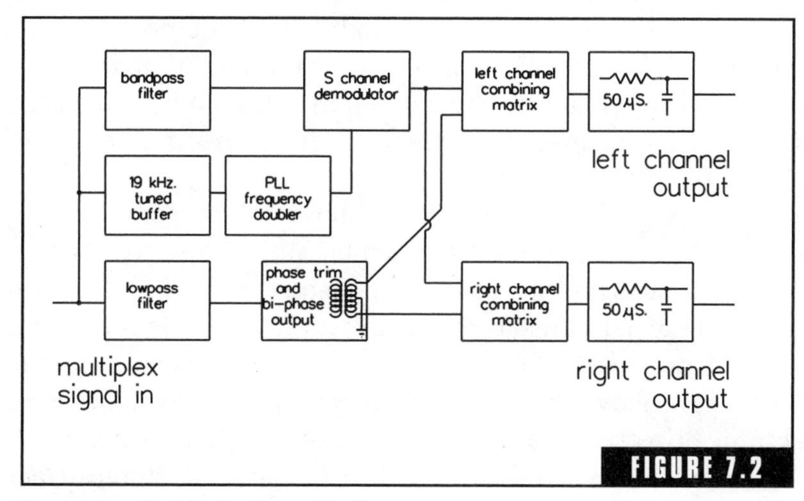

FIGURE 7.2

Stereo decoder block schematic diagram.

nels, adjustments are made so as to minimize crosstalk, and it is assumed when those adjustments are correct that the two components have been adjusted to exactly equal level and phase.

Preemphasis and deemphasis are used on the signals to reduce the effect of noise in the treble end of the audio-frequency response. Preemphasis must be applied to the left and right channel signals before encoding. After encoding, the frequency response from 0 to 53 kHz must be kept as flat as possible to preserve the relationship between main and subcarrier components. Deemphasis in the receiver is applied to the audio channels after decoding. These factors are specified as the effect of passing the signal through a resistor/capacitor network with a specified time constant. In Australia, the specified figure is 50 μs, which means that the transmission level of a 15 kHz note is 13.6-dB high relative to the level of a note in the bass register. At the receiver an equal and opposite change in frequency response brings the overall signal back to its original level and makes the effect of high-frequency noise up to 13.6 dB less.

In a modern solid-state FM broadcasting receiver all the functions shown in Figure 7.2 can be built into a single LSI integrated circuit, and in fact the one piece of silicon may accommodate all that plus some other functions such as a ratio detector and/or audio preamplifiers for use in those receivers where component count must be minimized.

In the 625-line PAL television system when the sound transmission is monophonic it is frequency-modulated onto the sound carrier, which is 5.5 MHz higher in frequency than the vision carrier at 10% of the power of the synch pulse tips (−10 dB). Deviation of that signal is 50 kHz, and audio frequency response is provided from 50 Hz to 15 kHz.

When TV sound is transmitted in stereo the power of that signal is reduced to 13 dB below the synch pulse tips, i.e., power is halved, with the same deviation and frequency response figures as before. That carrier is used to transmit an $(L + R)$ signal. Stereo information is transmitted by a second sound carrier whose frequency is 5.7421 MHz above the vision carrier at a power level 20 dB below synch pulse tips (i.e., 1% of the power of the vision carrier). When this second sound carrier is used for transmitting stereo it carries the right channel signal. In the receiver the left channel is decoded by subtracting (adding in reverse phase) the second carrier signal from the main ($L + R$) signal. The second carrier offers the option of transmitting the program with sound in two languages as an alternative to visual subtitles. To signal which form of transmission is being used a low-level pilot tone is added to the modulation of the second carrier. That pilot tone is modulated with signals similar to CTCSS tones (explained in Section 9.2) to identify whether the sound is monophonic, stereo, or dual-language. The difference in transmission method between FM broadcasting and television sound is worth underlining. For FM broadcasting there is only one carrier which is frequency-modulated by a signal which contains three components (main channel, pilot tone, and double-sideband multiplexed channel). For television sound two separate carriers are used, which if necessary could be received using an FM broadcasting receiver with extended band coverage. The observed signals that would be received in that way would be two quite separate monophonic signals with a frequency response of 50 Hz to 15 kHz.

Digital television does not require any encoding for transmission of a stereo signal. The MPEG2 digitizing protocol makes provision for several sound channels so left and right channels can be transmitted and received as separate signals.

7.2 Propagation of Stereo FM Broadcasting Signals

Both television and FM broadcasting services operate in the VHF and UHF bands so coverage is limited to "line-of-sight plus a little bit." Section 4.11 details some variations to that but the line-of-sight rule is a good rule of thumb for normal users. Sections 5.5 to 5.7 in *How Radio Signals Work* give information on the factors that control the size of the "plus a little bit."

The designers of an FM broadcasting service will have attempted to place the transmitter so that it covers the maximum population of a major city or provincial center and they would like to place the edges of the service area in places of low population density. That implies that fringe areas will be in rural areas which usually have lower levels of electrical noise than are prevalent in large cities. Under those conditions the limitation of sensitivity of an FM broadcasting receiver will be close to that set by the level of internal noise in the first amplifying stage of the receiver; for a monophonic signal with 15-kHz frequency response transmitted at 75-kHz deviation the weakest detectable signal is less than a microvolt.

The FM capture effect is enhanced in those cases where there is a large ratio between deviation and modulating frequency. In reference to the figures mentioned above (75 and 15 kHz) where there is a 6:1 ratio between them, the signal strength required to give a 40-dB signal-to-noise ratio at the audio output may only be 6 to 10 dB above that required for a just detectable signal. Fading due to variations in propagation conditions, will have some effect on signals in marginal conditions but the final result in many electrically quiet locations is that the minimum signal required for a good quality monophonic signal with no objectionable noise content is in the range of 3 to 10 μV at the input to the receiver.

A significantly stronger signal is required for stereo reception. Modulation bandwidth is wider, the ratio between maximum modulating frequency and deviation is reduced to about $1\frac{1}{2}$ to 1, and the various steps of the decoding process are all susceptible to the effect of noise. The result is that the signal usually needs to be about 20 dB stronger for stereo than is required for the same final audio quality of a mono transmission. Multipath fading is a significant factor affecting the stereo decoding process; its particular effect is to change the relative levels of the sidebands, which degrades the equality of level between the M and S channels of the signal into the decoder, which causes distortion and loss of stereo separation in the audio outputs. Both weak signals and multipath fading become more likely as distance from the transmitter is increased so for most FM broadcasting services reception of the signal in stereo is limited to the primary service area. Outside of that stereo-reception area there will usually be a substantial region where the signal can give good audio-quality reception in the monophonic mode.

Due to the capture effect the fringe area where signals may be heard but are weak and noisy will only be a narrow border on the outer edge of the monophonic region. For over-the-horizon reception where there is some semiregular fading the observed effect of marginal signal levels is almost a switching function between one state of a relatively clear signal and the other state in which the receiver output is thermal noise with no intelligent signal at all.

In a high proportion of cases the receiver of an FM stereo broadcast is in a motor vehicle. If that is relevant to your use of the service, check with Chapter 10 for information specific to the mobile environment. Continuous reception of a stereo signal while mobile requires that signal level be strong enough to give noise-free decoding in places where the standing-wave microclimate works to weaken the signal. Section 3.8 explains how the microclimate works. When the vehicle moves from a place of very strong signal to a more marginal region the first sign of the approach of a weaker signal area is intermittent operation of the stereo decoding; standing-wave effects also apply in that case so the effect may be a semiregular pulsing of the stereo function.

The disruptive effect of multipathing can be present no matter how strong the signal is; it can occur whenever the level of the reflected signal is within a few decibels of that of the direct one. The disruption of equality between the M and S channels of the multiplexed signal is greatest when the variation of relative signal level has a high rate of change in relation to the frequency spectrum. Reflections in the first few Fresnel zones (explained in Section A1 in the Appendix) result in broadband effects which only vary at a slow rate due to frequency changes; reflections where the Fresnel zone number is in the 100-to-500 range are much more disruptive to the stereo decoding process. The places where a reflected signal in that range of Fresnel zones arrives at about the same level as the direct signal are in locations within a kilometer or so of a cliff face or multistory building and the path for the direct signal is slightly obstructed (Figure 7.3). The conditions that cause troublesome multipath effects for FM broadcasting signals are identical with those that cause troublesome ghosting with television; refer to Section 4.7 for more information.

7.3 Fringe-Area Stereo Television Sound

The television stereo sound signal is two independent carriers, each frequency modulated, with a monophonic transmission of 50 Hz to 15 kHz (ignoring pilot tones for the present). The chance of disruption due to multipath fading is still possible but compared to the effect on an FM broadcasting multiplexed signal it is much reduced. Due to the intercarrier process, there is a chance that demodulation could be affected by selective fading of the vision carrier but the characteristic of

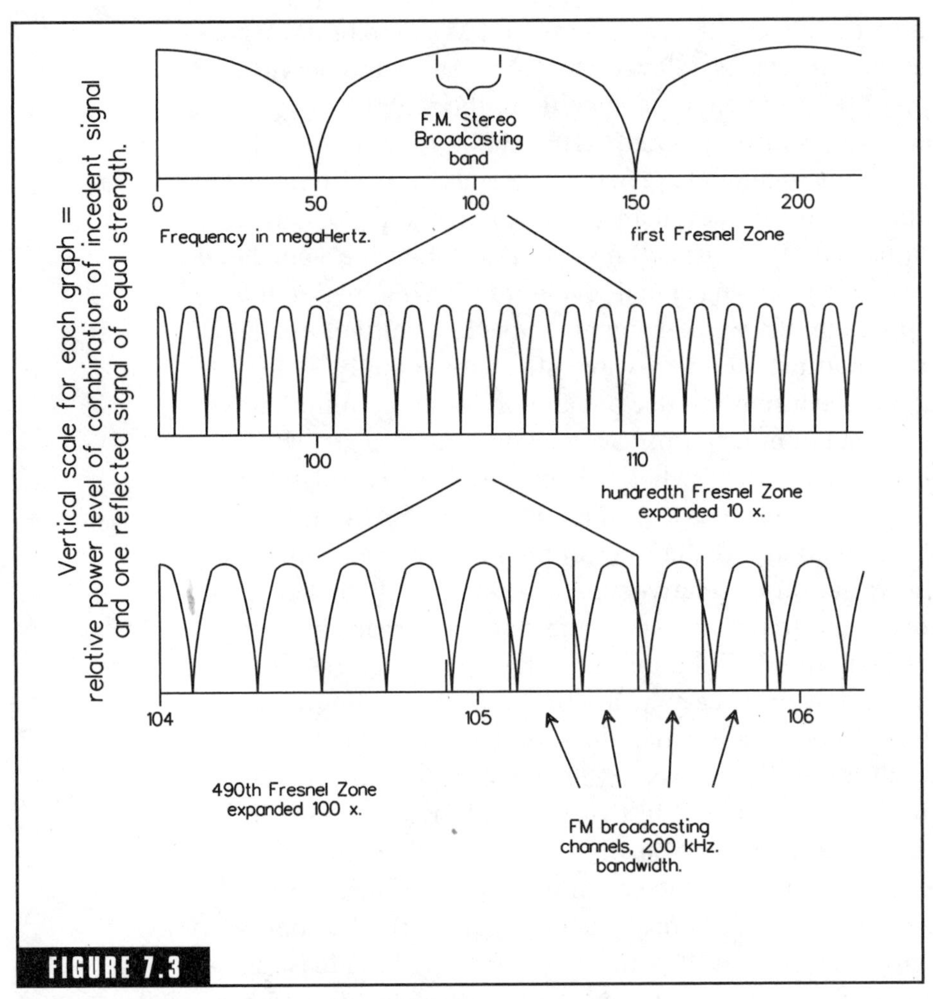

FIGURE 7.3

The effect of Fresnel zone number on frequency response.

the system is that so long as there is sufficient vision carrier there can be a quite wide variation due to selective fading with no observed effect. Both the sound carriers have less power than the vision carrier and the frequency response of the receiver's IF amplifier increases the difference in relative level between vision and sound carriers.

The IF amplifier response curve of Figure 7.4 is not that of any particular receiver but is typical of the general shape of many, there is very little true passband filtering; the bulk of adjacent channel selectivity is provided by traps on the possible carrier frequencies. The

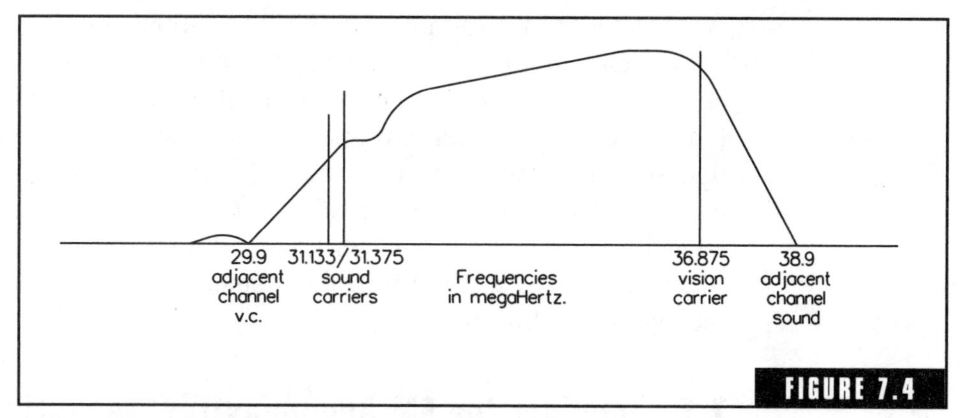

Television IF amplifier frequency-response curve.

sound carriers that are transmitted at 10, 13, or 20 dB lower power than the vision carrier are further reduced by the selectivity of the IF amplifier. Selective fading would have to reduce the level of the vision carrier by at least 15 to 20 dB before degradation of audio output would be noticed. If it does happen the effect is to cause "frame buzz," which is due to each frame-synchronizing pulse causing a momentary change to sound demodulation. Frame buzz more commonly occurs when the signal is detuned in the direction of reducing the relative level of vision carrier so if it is noticed the first check should be of the accuracy of tuning.

For a noise-free picture, a vision carrier signal of 1 mV at the input terminals of the receiver is ideal and a signal level of 100 μV is about the minimum usable signal for most receivers. The FM sound signal can give quite reasonable audio quality with a signal level of 1 μV if there is low electrical noise. When a monophonic television signal is of marginal signal strength and the signal is slowly decreasing, the picture will become unwatchable before the sound quality is noticeably effected. The same factor will apply to television with stereo sound in a reduced form because the second sound carrier is transmitted at a lower level.

In marginal conditions, the decision whether the sound is presented as a monophonic signal or decoded into a stereo pair will be made by the pilot tone decoder. The pilot tone is a frequency of 55 kHz added to the modulation of the second sound carrier at a level sufficient for 2.5-kHz deviation of that carrier. Coding of the pilot tone is done by amplitude modulating it at either one of two subaudible frequencies in a manner very similar to the CTCSS encoding of VHF and UHF radio-communication transmissions. An unmodulated pilot tone signals monophonic transmission so when the signal of the second sound carrier is lost in the noise the receiver ceases to detect modulated pilot tone and presents a mono sound signal. The carrier level at which that happens will depend on the selectivity, sensitivity, trip level setting of the pilot tone detector and the adjustment of the receiver's fine tuning.

7.4 Aerials for FM Broadcasting Reception

There are so many technical similarities between television and FM stereo broadcasting that the transmitters for those services are often found sharing the same mountaintop and there are many cases where the organization providing the transmitting function will provide a facility for both services. In those cases, both transmitters will be housed in the same building and aerials will be mounted on the same tower. The implication of that for fixed receivers is that the aerial for television reception and that for FM broadcasting reception will be the same type of structure pointing in the same direction. In the primary service area, the one aerial can cover both functions with a two-output passive splitter, as shown in Figure 4.12, to separate the feeds to the two receivers.

In weaker signal areas, the 6-dB loss of signal across the splitter would affect the quality of the television signal so that must have a direct feed from its own aerial with a separate aerial provided for the FM receiver. The whole band from 88 to 108 MHz (Australian FM broadcasting allocation) is wider than the 5% bandwidth offered by an unloaded half-wave dipole so some broadbanding will be needed. That aerial will have many of the same characteristics as a television aerial although the specification for flatness of response over a particular group of channels will have a much wider tolerance than is required for color television.

For receivers located in the secondary service area of a large city (roughly equivalent to the commuting range for city workers) a screen-backed dipole or corner reflector aerial would work well. Broadbanding could be achieved by building the driven element with large-diameter tubing or constructing it as a folded dipole with a couple of hundred millimeters between the conductors of the dipole. Polarization will be the same as is used for the local television channels although in some places where television channels are horizontally polarized, the FM signal may be transmitted with circular polarization because a substantial part of the station's audience will be in motor vehicles with vertical receiving aerials.

In country areas, the need for truly broadband receiving aerials may not be so urgent. There may only be two or three towns within radio range which are big enough to have their own FM broadcasting service. If you are planning the installation of a receiver in a rural area you should look first at the frequencies of the stations available locally and ask the question, "Could they all be fitted within the passband of a single Yagi aerial (nominally about 5%)?" If the answer is "yes" then the most appropriate aerial is a narrowband array with whatever directional properties and polarization is required tuned to a frequency halfway between the lowest and highest frequencies of the stations you wish to receive. If the transmitters are in different locations you will probably need to mount that aerial on a rotator.

For fringe areas, all of the measures already detailed for television in Chapter 4 are relevant to FM broadcast reception; the factors of aerial gain, aerial height, and length of feedline are all equally relevant here. Reflections that cause multipath fading will be more annoying for FM broadcast reception than for television; if your television receiver shows a visible ghost at all it is very likely FM reception will be degraded. Fading due to weather conditions will either not be noticed at all or have a greater effect than is seen on a television picture; due to the capture effect if the signal is strong enough to be detected at all it will be within a few decibels of the strength required to give a noise-free monophonic signal. For stronger signals, the stereo-decoding function will tend to follow the same pattern—it will either be switched on and work properly or be switched off, and the change in signal level required to perform the switching function is only a few decibels. The standing-wave microclimate may have an effect about equal in magnitude to the change required to make a just detectable noisy signal strong enough to be noise-free. At 100 MHz the wavelength is 3 m so in marginal conditions a change in position of the aerial of $1\frac{1}{2}$ m may be all that is required to change an installation from workable to unworkable or vice versa.

In many fringe-area installations the tower for the television aerial will also be used for FM broadcast reception. In a direct comparison such as that it will require less technical resources to produce a good FM signal than will be required for the same quality of reception for television. The performance of such a dual system will usually be best balanced if the tower height is planned for the requirements of the television receiver with the television aerials at its top; then the FM aerial is added close below at a distance just sufficient to avoid problems with interaction between the aerials.

When the FM receiver must stand alone the required height of tower and aerial gain can be calculated using the "link budget" form of calculation based on the published figure for transmitter station ERP and a signal level at the receiver of between 30 and 100 μV. The 30-μV figure is appropriate in noise-free locations if you estimate there will be little or no fading due to atmospheric effects; larger figures can be used to build in greater fading margins where the path is very obviously not in radio line of sight or in those cases where some local electrical noise interference is expected.

If a single-channel FM aerial is needed it will probably be more expensive than a broadband unit of the same complexity. Mass-production techniques can be used for aerials covering the full broadcasting band but tuning of aerials is achieved by changing the length of the elements. Making a different size of the same design is a different manufacturing process so the benefits of mass production are reduced for individually tuned products. If single-channel aerials are available commercially they will most likely be found in radio/electronics/electrical shops in the primary service area of the transmitter. If an aerial must be custom-made or home-built the designs for VHF amateur use can be adapted to the FM broadcasting channel by calculating the ratio between the wavelength of the FM signal and the wavelength for the center of the amateur band and using that ratio to scale all the dimensions.

7.5 Fault Types Peculiar to FM Systems

Providing that the wanted signal is a few decibels stronger than any other signals in the receiver's passband there are very few forms of interference that can degrade the quality of the final audio output. If the desired signal is not the strongest in the channel the interference will overcome it so that the capture effect causes the receiver to resolve the interfering signal and reject the desired one. If in a particular location one channel has two signals available with some fading on both of them the effect will be that the receiver output will switch from one to the other depending on which shows least effects of fading at that moment.

Selective fading due to multipath reception will cause distortion. That has been dealt with in Section 7.2.

Within the receiver, restriction of bandwidth and inaccuracy of tuning will cause distortion. The effect on the audio output of restriction of bandwidth of an FM signal shows exactly the same characteristics as overloading (clipping) of an analog amplifier which gives distortion due to restricted capability for voltage or power output. Mistuning causes distortion due to exactly the same mechanism but the effect is only on one side of the audio waveform so predominantly even harmonics are produced. Figure 7.5 is a graphical representation of the FM demodulation process showing how distortion occurs due to bandwidth restriction.

When the FM receiver is the sound channel of a television system, accuracy of tuning is attended to by the intercarrier method of reception; fine tuning of the television signal as a whole can be adjusted by several megaHertz without any distortion of the sound or disruption of the stereo decoding function apart from the possible generation of a certain amount of frame buzz.

Because of the risk of distortion and the presence of a capture effect the frequency response of an FM receiver is often quite broad with much less adjacent channel selectivity than would be required for an amplitude modulation broadcast receiver. The spectrum managers will have attempted to ensure that for each location strong signals are separated from each other by several channels and the receiver designer only needs to build enough selectivity into the IF amplifier to ensure that the wanted signal is a few decibels stronger than the others. Flatness of the passband over all the range of the sidebands is the critical design parameter for a high-fidelity FM receiver.

The FM receiver does not have an AGC circuit in the same sense as is used for AM or single-sideband receivers. The amplifying stages of the IF amplifier are limiters (voltage clippers) and variations in signal

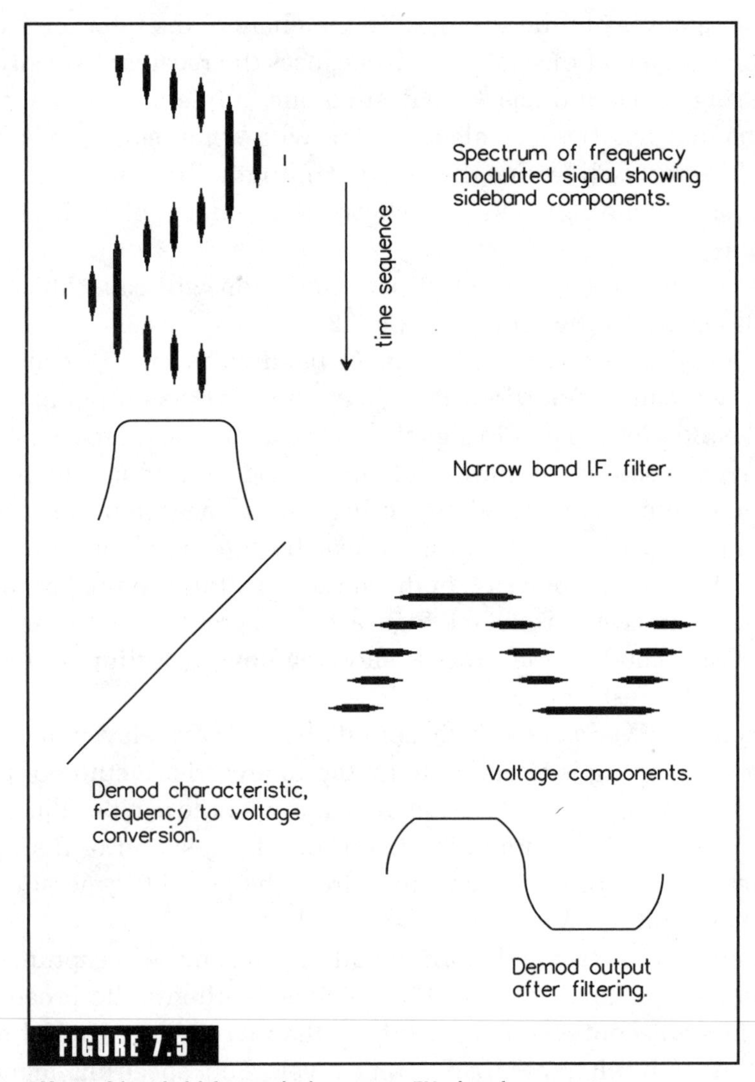

FIGURE 7.5

Effect of bandwidth restriction on an FM signal.

strength are absorbed by the limiting function so that a constant-level signal is presented to the demodulator stage. The overall gain of the IF amplifier in the no-signal condition is very high (may be over 100 dB) and so the receiver is possibly prone to parasitic oscillation. In some cases, the parasitic is damped out by the presence of an input signal and in some marginal cases it can disappear as soon as the offending circuit is loaded by the connection of a test probe. These parasitic oscillations can be either sine or squegging waveforms (refer to Section 5.12). A sine-wave oscillation may have the effect of an unmodulated carrier which causes quieting of the receiver; in other cases the parasitic oscillations will increase the noise level which will show in proof-of-performance measurements as a reduction in sensitivity.

A parasitic oscillation which only shows in measurements as a reduction in sensitivity and is damped out by the addition of a test signal or measurement probe can be ***very*** hard to identify. One way of possibly proving its existence is to use an AM receiver tuned to the FM receiver's intermediate frequency with a small sniffer coil to search for radiation or induction from the FM receiver's circuits. The catch is that this test is not definitive; if you find a signal you can definitely say that a parasitic exists, but absence of a radiated signal does not guarantee the FM receiver a clean bill of health.

When all is correct, a direct comparison between FM and AM broadcasting signals of comparable signal strength shows that the FM signal gives much higher fidelity. For an AM signal of 100 μV at the receiver input the signal-to-noise ratio is unlikely to be better than about 40 dB and the treble response will be limited to 9 to 10 kHz. An FM signal of the same strength can be received with a signal-to-noise ratio approximately equal to that of the signal leaving the transmitter (about 68 to 70 dB) and treble response to 15 kHz. The AM signal is also a lot more degraded by the effects of electrical noise. The workable range of the FM signal is shorter than for AM broadcasts in the medium-frequency band, but within the service area the FM signal gives a significant improvement in fidelity of the audio output.

CHAPTER

8

Receiving the Extraterrestrials

8.1 Of Stars and Satellites

This chapter deals with all signals that originate from outside the Earth's atmosphere. Artificial satellites intended for broadcasting and communications are included and the scope of the chapter extends to the basics of moonbounce and interplanetary radar experiments and radio astronomy. For each of these subjects there is an enormous wealth of information and whole books are devoted to each one. The present text provides only an outline of each with some suggestions for detail on each subject.

The common factor about almost all signals from space is that they have traveled over a very long path so all are very weak. (The only exceptions are radio noise from the Sun and occasional signals from the planet Jupiter.) For example, the signal strength available at ground level from a 30-W transponder on a geostationary satellite is liable to be significantly less than 1 pW/m^2; expressed in terms of decibels below 1 W/m^2, figures in the range of -125 to -130 are common. (A picowatt is 120 dB less than a watt.) Signals used by radio astronomers and the returns from interplanetary radar experiments are liable to be yet another 60 to 80 dB weaker. The most sensitive possible receiver is essential, and highly directional aerials are commonly used, but there are no really strong signals coming from those directions so great dynamic range is not required.

The next section gives some very basic information on finding your way around the sky. Signals from artificial satellites which can be placed in either geostationary or low earth orbits will be the subject of the following two sections; Section 8.3 deals with signals from geostationary satellites and Section 8.4 is devoted to low Earth orbits (LEO). Later sections deal with "Radio Astronomy" in relation to the large professional installations and "Moonbounce and Interplanetary Radar," which is of particular interest to amateur radio operators and research institutions.

The final section of this chapter, entitled "Domestic-Scale Radio Astronomy," is intended to show some entry points into the subject for practical experiments and observations by an interested amateur or private experimenter.

8.2 The Geometry of the Sky

If someone wishes to inspect a particular block of real estate they can travel to the site, find the surveyed corner pegs, walk all over it, and take a tape measure and define the exact positions of planned facilities such as fences, gates, and buildings in terms of the exact length from each corner peg. In the surveying of the Earth there is another system of measurements based on angles which is expressed in the form of degrees, minutes, and seconds of latitude and longitude. Distance measurements from surveyed pegs can be reconciled with the coordinates of latitude and longitude, but the calculation is often involved and requires an enormous degree of accuracy of measurement of the angles to get a resolution as good as even that of a relatively careless person with a tape measure. The ability to go to the land and walk around on it allows us access to the greater accuracy of measuring in reference to a nearby survey peg.

If, however, we were suspended above the land, the tape measure would be useless but the angle coordinates could still be used and distances on the ground could be measured in terms of the visual angle as it appeared to the suspended observer. Note that the observer does not need to know how high above ground he/she is; providing his/her position is kept constant, each observable point on the Earth's surface can be uniquely specified by angles in relation to the survey points. The visual angle measurements used in this case can provide a system for specifying the position of points in the surveyed area in exactly the same way as the tape measure can be used by a person walking over the land.

When we look at the sky we cannot go there with a tape measure, walk over it, and plot positions of objects by measuring distances; the only observing point available to us is the Earth, which in astronomical terms is so small that all our measurements are limited to observations from one fixed point. However, the visual angle method of establishing positions works equally as well as it did in the previous example, so for surveying of the sky all "distances" over the "surface" are specified in terms of visual angles as seen by an observer on the Earth. In surveying the positions of stars the concept used is that of a hollow sphere seen from the inside with the stars all given positions on the inner surface of that "celestial sphere." We do not know the radius of that sphere (and, in fact, modern measurements show the universe to be a three-dimensional object with stars being at a range of distances away) so we cannot use a tape measure to specify distances on its surface but we can use visual angles as seen by an Earth-based observer to specify positions of all the heavenly bodies.

There is a system of angular coordinates for specification of "positions" of heavenly bodies which is almost exactly the same as the latitude and longitude system for specifying positions on the Earth's surface. The references to the celestial coordinate system are based on the rotation of the Earth as illustrated in Figure 8.1 and described as follows:

There is a North Celestial Pole and a South Celestial Pole, which are located on the celestial sphere directly above the Earth's pole that each is related to.

There is a celestial equator, which is a line on the celestial sphere directly above the Earth's equator.

Every point on the celestial sphere can be specified in degrees, minutes, and seconds of angle north or south of the celestial equator. That specification is named "declination" and has characteristics very similar to the specification of latitude for a point on the Earth's surface.

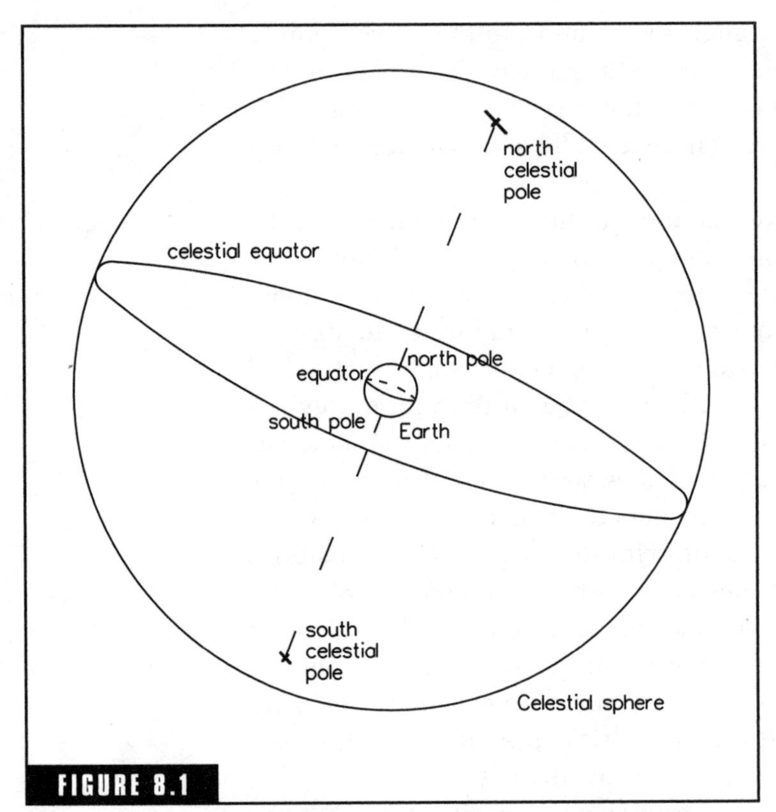

FIGURE 8.1

Basics of the celestial sphere.

In the east–west direction points on the celestial sphere can be specified by the sidereal time of day at which they are exactly overhead. The name for that parameter is "right ascension" and it is expressed in units of time (hours, minutes, and seconds). The Earth's rotation carries the local meridian across the celestial sphere at the rate of 15 degrees of angle for each hour of time. Note carefully that the minutes and seconds of right ascension ***are not*** the same size in the sky as the minutes and seconds of an angle measurement.

Specifying positions in the sky and tracking movement of objects is more complicated than the equivalent operation on the Earth's surface using latitude and longitude because the system based on the celestial poles and celestial equator is only one frame of reference, and several frames are possible. That one is generated by the daily rotation of the Earth itself. Another frame of reference is based on the plane of rotation of the Solar System and yet another is based on the rotation of the Milky Way Galaxy.

If the straight line from the Earth to the Sun were extended to the celestial sphere it would trace a line as the Earth moves on its orbit which would be similar to the celestial equator but would take a year to complete instead of a day as it does for the celestial equator. That line is called the "ecliptic." The planes of the orbits of the Moon and all the planets are close to the same plane as the Earth's orbit so the path traced out in the sky by the movements of all those bodies is always within 5° of the ecliptic. The ecliptic is inclined with respect to the celestial equator by approximately $23\frac{1}{2}°$ and it is that relative inclination between the two planes which produces the seasons on Earth.

Within the time scale of a human lifetime the fixed stars that form the background to the celestial sphere may be regarded as stationary, but, in fact, the whole Milky Way Galaxy is rotating about its center which is in the direction of the constellation Sagittarius. Our Solar System, as a whole, is part of that rotation and is moving at about 20 km/s in a direction toward the constellation Hercules. That plane of rotation about the center of the galaxy forms another frame of reference which is of great interest to professional radio astronomers.

Artificial satellites which move in orbit around the Earth can be tracked in relation to figures for azimuth and elevation as seen from particular points on the Earth's surface or they can be related to positions in declination and right ascension against the background of the fixed stars. Both systems of measurement can be used, and there are most appropriate uses for each one. To translate from one to the other can be complicated and must give due allowance for the rotation of the Earth and its associated time considerations. A satellite in geostationary orbit, for instance, maintains the same azimuth and elevation figures in relation to the Earth for very long periods but is moving in relation to the fixed stars so that it is carried around the celestial sphere once each day.

8.3 Characteristics of Geostationary Satellite Signals

In a geostationary orbit a satellite is at a particular altitude, approximately 35,840 km above the Earth's surface, such that its orbital period is exactly 24 h traveling from west to east at a particular spot over the equator. Because the orbital period exactly matches the rotation of the Earth its apparent azimuth and elevation as seen from a point on the Earth's surface stays constant so a high-gain aerial pointed at that spot in the sky can have a continuous signal from the satellite without the need for any tracking movement.

In practice, the synchronization between the satellite orbit and the Earth's rotation is not exact and is continuously being slightly degraded by tidal effects from the Moon, passing comets, asteroids, other planets, and the Sun. When the orbit becomes elliptical the observed effect from the Earth is a daily movement back and forth in the east–west direction. Errors in matching latitude with the equator are observed as a daily drift in the north–south direction. These two movements are combined in random phase so the combined effect as seen by an Earth-based observer can be approximately a circle, an ellipse, a line at any direction, or, more commonly, a figure-8 pattern traced out in the sky. Tiny gas jets are built into the body of the satellite and these can be controlled by a ground-based operator whose job is to make periodic small bursts of gas propulsion with the dual aim of making the error pattern as small as possible and husbanding the gas supply. Rate of use of the gas determines the life of the satellite.

Figure 8.2 shows how the resultant cyclic variation appears to a ground-based observer or a fixed high-gain aerial. The satellite driver is attempting to keep the pattern as small as possible but must not use excessive gas to do it. The accuracy required depends at least partly on the beam width of the earth terminal aerials that will be used for communication with the satellite, and the conventional wisdom has been that an aerial with high-enough gain to be used as a receive-only terminal giving an "end user" standard of signal should be able to produce that signal without the need for cyclic tracking of the satellite's movement. On the other hand, stations producing the stronger, clearer signals needed for rebroadcasting use larger, higher-gain aerials, and some tracking of cyclic movements may be required. Section 8.6 later in this chapter deals with tracking systems.

Path-loss figures from the geostationary orbit to Earth are generally a few decibels over 200. Transmitter power for a particular transponder is usually less than 100 W (more detail in Section 8.5) and a target the size of a continent looks quite small from 36,000 km away so highly focused beams are used and, in some cases, may be shaped in quite a detailed way to follow national boundaries.

There are many uses of geostationary communications satellites; broadcasting of television to remote areas and wide-area mobile communications are two of the major ones. The frequency bands mainly used for television are C band (4 GHz) and Ku band (12.5 GHz). Mobile communications generally use lower frequencies than those, many multichannel systems use the L band at 1.5 GHz.

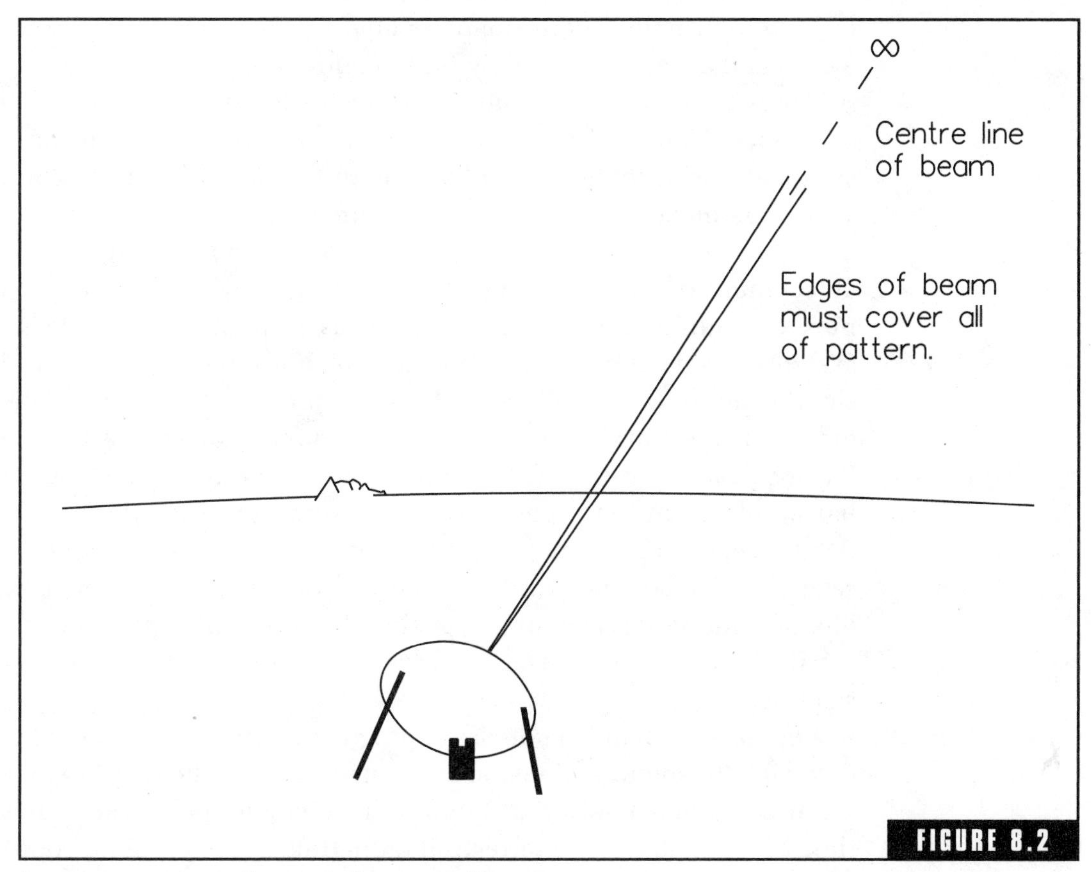

Apparent movement of a geostationary satellite.

A variety of transmission standards are used, as determined by the government of the country that owns the satellite. To achieve the aerial gain required for a television bandwidth signal, the frequency must be in the microwave range or else the aerial must be impossibly big. In some parts of the world, reception of C-band transmissions using a parabolic reflector requires a dish diameter up to 5 m even to receive an "end-user" standard television signal; frequencies lower than 3000 MHz are not generally practical for broadband signals from geostationary satellites. The L band can be used for narrower bandwidth services such as mobile telephone and radio communications based on 1 VF analog channels or digital bands at up to about 64 kb/s. Some aerial gain is needed so the aerial must either be equipped with the capability to track or be constructed in a cunning fashion to provide a cone-shaped pattern illuminating the correct elevation all around the vehicle (see Section 8.9).

Transit time may be a significant factor for signals in some types of service. The round-trip time for a signal from Earth to the geostationary orbit then back to Earth is in the range of 220 to 250 ms. That time makes an appreciable difference to the flow of conversation, but a two-way exchange of information is still possible. Telecommunications system managers have agreed not to allow duplex communication channels (such as ordinary voice telephone systems) to be routed via two-satellite hops (500-ms delay). The significance of that delay depends on the program type; some types of data transmission can operate but need to recognize and allow for it. In most cases two-way data communication is unusable. The delay has an effect on the type of error correction used for the data link. ARQ schemes which involve a return of information at a rapid rate are severely affected. "Forward error correction," in which enough extra bits of information are sent to enable the receiving equipment to reconstruct a corrupted word, works much better with geostationary satellites.

On the other hand, a television program is almost unaffected by the delay if both sound and vision are transmitted via the satellite path. If a watcher/listener set up a station to receive the image via a satellite link, but the sound via a terrestrial radio link, then the sound could arrive noticeably before the visible action or it could serve to counteract the time required for sound to travel to a microphone placed at the edge of an oval; the magnitude of the delay is roughly equivalent to sound traveling through air for about 100 m.

Geostationary satellites are enormously expensive to install so all are owned by either national governments or by the biggest of multinational corporations or cartels of large commercial companies. There is only one geostationary orbit on the line vertically above the Earth's equator and satellites are placed along it with reference to terrestrial longitude. In each operating frequency band there is room for one satellite at every 2 or 3° around the orbit and in the closing stages of the 20th century some parts of the orbit within range of the large centers of population are close to being full.

Sections 8.8 and 8.9 describe a typical TVRO station as might be used on an Australian cattle station or a Pacific Ocean island. Section

8.10 describes a satellite mobile telephone as required for use with a geostationary service.

8.4 Characteristics of Signals from Low Earth Orbits

In a low Earth orbit, the satellite can be at any altitude below the figure for a geostationary orbit providing it stays above the altitude at which the atmosphere causes drag. From the radio signal's point of view, the implication is that the satellite does not stay in a fixed position in the sky—it moves across the view of an Earth-based observer so high-gain fixed aerials cannot be used. The aerial system of an Earth-based receiver must either be omnidirectional or have the ability to track the direction of the incoming signal. Signals are strongest when the path length is short so satellites that fly just above the Earth's atmosphere can be used in combination with receiving equipment using dipole, loop, or quarter-wave vertical-type aerials to give a workable signal strength. LEO satellites commonly have lower aerial gain than would be used by a comparable geostationary service. The target is bigger in terms of visual angle so to cover it the beam must be wider.

The lowest possible path would be a circular orbit about 160 km above the Earth's surface, and a satellite would complete one rotation at that altitude in 96 min. There is still enough atmosphere there to limit the lifespan of such a device to a few months or a year or so. For longer life the altitude must be higher but not a great deal extra is needed; at 50% higher altitude the expected life of the satellite would be many years providing that there is only small eccentricity in the orbit. The life expectancy due to atmospheric drag is mainly determined by the altitude at the lowest point of its orbit. LEO satellites commonly have altitudes between 200 and 500 km above the Earth's surface.

Carrier frequency does not need to be in the microwave spectrum for use by LEO satellites. The practical low-frequency limit is about 100 MHz; below that, signals can be affected by sporadic E-layer ionization in the ionosphere often enough to be a nuisance. The small capture area of a nondirectional aerial limits usefulness of the highest frequencies. Carrier frequencies in the range 100 to 500 MHz are commonly used and some services may use up to 1500 MHz. Services that use directional aerials with tracking ability commonly use frequencies in the range from 1000 to 3000 MHz. In the modern world, a wide-enough clear spot in the spectrum may be a problem; the LEO satellite may eventually be above every point on the Earth's surface so whatever frequency is chosen must be permitted worldwide. If individual 1VF channels are to be allocated, then Doppler shift becomes significant (see Section 8.7).

Because the LEO satellite drifts across the sky then journeys on around the Earth a single satellite cannot offer a continuous communication service. A group (or cluster) of satellites can give continuous contact if they are well-spaced along a particular orbit. The number required to give continuous service depends on the altitude. At 200 km the horizon is a little over 1600 km away and 13 satellites spaced at 1600-km intervals around the orbit would be required to give unbroken coverage. At the other end of the scale, 3 satellites can cover the complete circumference of the Earth if they are at least a little over 6300 km above its surface and spaced around the orbit at 120° intervals. Figure 8.3 illustrates these conditions.

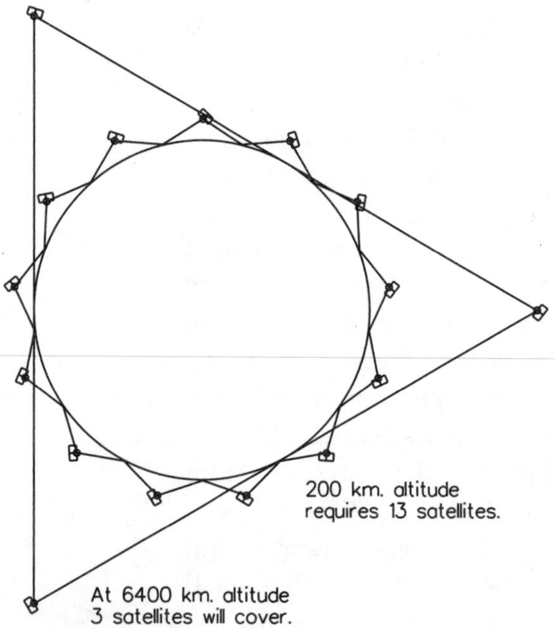

FIGURE 8.3

Continuous coverage related to satellite altitudes.

8.5 Satellite Power Supply

Transmitter power in the satellite is limited to what can be collected from the Sun. Communications satellites use banks of solar panels which are kept aligned for maximum pickup. The total power collected is usually in the range of 1 to 5 kW for about 22 h per day (geostationary orbit); in the other 2 h the orbit passes through the Earth's shadow. That power supply must provide for all satellite functions and also must provide a reserve to charge batteries for the shadowed time. Each satellite carries several transmitters so the power available must be shared between them. The actual RF output figure for each transmitter is usually in the range between 10 and 100 W.

When the signal from a geostationary satellite reaches the Earth it has spread to a beam that may cover a fair proportion of the area of a continent or an ocean so the signal strength in terms of power per square meter of the Earth's surface is very small. Highly directional aerials with a capture area of several square meters are needed to make the signal strong enough to overcome the internal noise level of the receiver front end. Actual figures, for example, are that for a TV receive-only station operating at 12 GHz a 1.2- to 1.8-meter-diameter parabolic dish aerial (gain in range 42 to 46 dB, beam angle less than 1°) is required and with that aerial a low-noise amplifier with a noise figure less than 1 dB is needed to produce a signal that is in the range 10 to 20 dB above the internal noise level in the receiver.

For LEO satellites, the field strength of the radio signal depends to a large extent on the altitude of the orbit. At one end of the scale, satellites close to the minimum altitude can be treated almost the same as distant terrestrial repeaters on very high mountains; for higher altitudes, the Earth terminal must use a gain aerial and because the azimuth and elevation do not stay constant, tracking aerials are needed.

The satellite transmitter output stage must produce the required level of RF power as efficiently as possible to minimize power supply loading but it must also have a very high power-to-weight ratio. In the future, microwave semiconductors may be developed that will do the job, but at the end of the 20th century the best equipment available used traveling wave tube amplifiers. Traveling wave tubes have a feature that has had a marked effect on satellite system design; they are nonlinear in their input/output characteristic to the extent that if a system really needs a linear analog signal out of the satellite then the power must be backed off approximately 12 dB from the maximum. If the modulation principle will tolerate nonlinearity then the full output up to the saturation level of the TWTA can be used.

If each transponder channel is only required to handle one signal then frequency modulation with deviation covering the whole bandwidth of the channel will do all that is required. When the transponder must carry many channels, multiplexing schemes have been developed which combine the best features of both FDM and TDM. The basic philosophy is to use a multiplexed channel with many carriers in a fixed relationship to each other with digital information shared between the carriers in a manner similar to that used for digital television transmission. The digitizing and encoding processes can be configured in such a way that a "capture effect" similar to that of a frequency-modulated signal is produced which enhances the signal-to-noise ratio of the output signals. Providing that the signal received off-air has at least a 10- to 15-dB signal-to-noise ratio the final figure for the output can be at least 40 dB.

8.6 Tracking Aerials

Figure 8.4 illustrates the principle of a tracking directional aerial. The system operates by comparing the signal strength from each of the four feedpoints; when all four are exactly equal the aerial is aligned so that the center of its beam falls exactly on the junction point at the center of the four feedpoints. If all outputs are not equal, the one that has the strongest signal indicates the direction of the movement required to bring the aerial back into exact alignment and a simple servo loop can be used to keep the aerial seeking for the strongest signal.

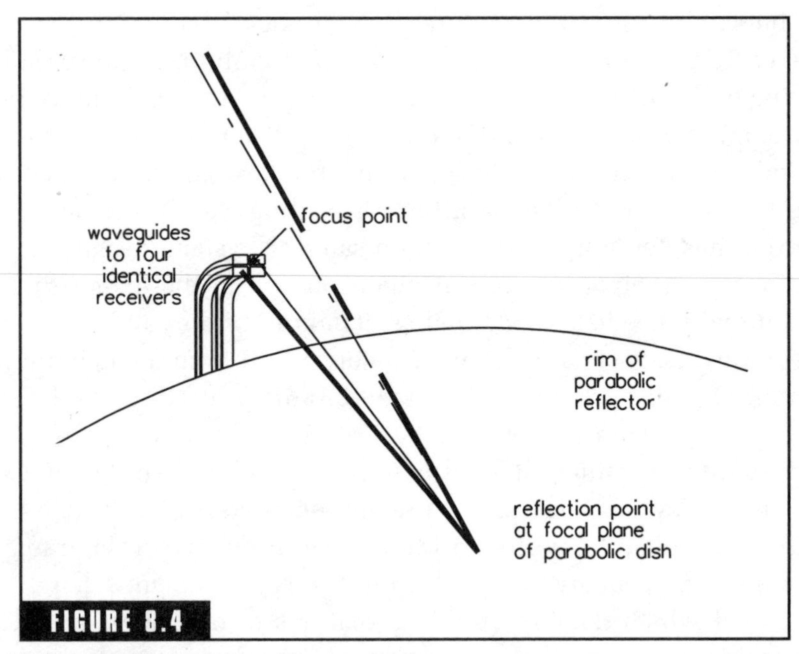

FIGURE 8.4

Principle of an aerial tracking mechanism.

In practice, tracking systems do not use four separate receivers; the practical mechanism only has one feedpoint to one receiver. At the feedpoint, a small motor rotates the signal collector in an eccentric fashion. If the feedpoint has linear polarization (such as a single dipole or a normal horn on the end of a waveguide) the polarization must stay aligned with a particular direction while rotating, with motion similar to that of a connecting rod big end bearing on a crankshaft. If the aerial has circular polarization the feedpoint can be solidly fixed to the rotary platform and move in the same fashion as the crankshaft journal. The radius of rotation must be equivalent to a movement in angular position of about half the beamwidth of the aerial.

At the receiver, if the aerial is not exactly aligned with the incoming signal, the received carrier level will show a cyclic variation which can be detected as an AC voltage or current component in an AGC-type detector or FM limiter stage. The frequency of that AC voltage or current signal will be determined by the speed of rotation of the motor, the amplitude of the AC signal will indicate how far off alignment the aerial is, and the phase of the AC signal relative to the rotation of the feedpoint will indicate the direction to be moved to bring the aerial in line with maximum signal. That information can be used as the input to a servo loop to control movement of the aerial. The servo loop will adjust the aerial position so that the RF carrier level of the incoming signal is steady and at that setting the center of the main beam of sensitivity is exactly aligned with the center of rotation of the feedpoint mechanism.

Figures 8.5 and 8.6 show the basic elements of a practical tracking aerial system. These diagrams are based on a feedpoint with circular polarization; the mechanism for tracking with linear polarization is much more complicated.

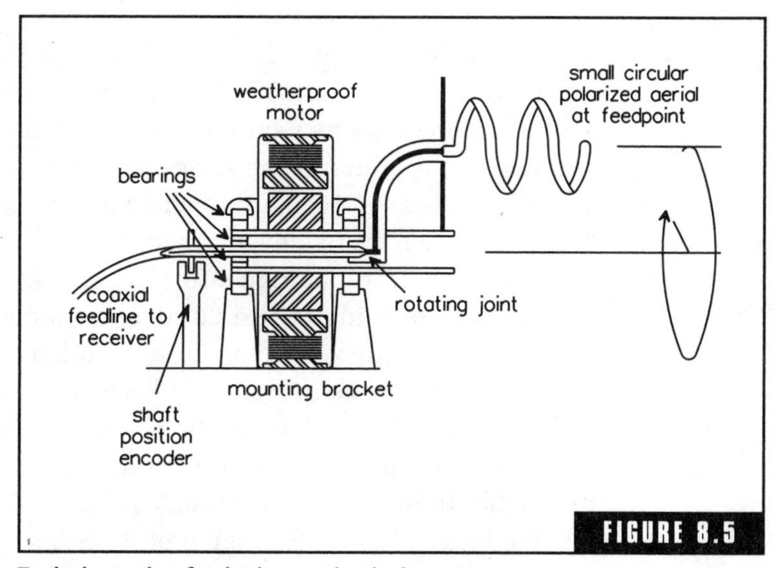

Typical rotating feedpoint mechanical arrangement.

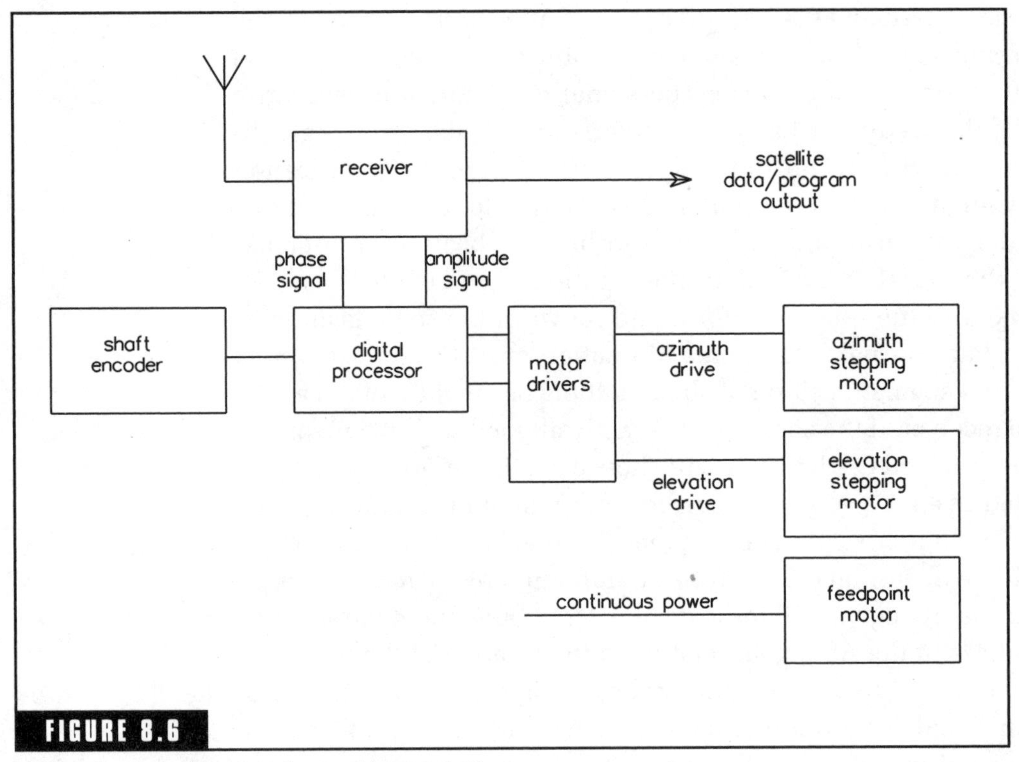

FIGURE 8.6

Block schematic diagram of a tracking controller.

The directions of the axes of the gimbal system the aerial is mounted on and the method of connection of the RF feedlines across the gimbal mechanism may have a significant bearing on the operation of the station. The simplest system would use a vertical shaft to give movement in azimuth and a horizontal shaft carried on that to allow movement in elevation with a flexible coaxial cable or waveguide with a strain relief loop sufficiently long to allow movement over the full 360° of azimuth and 90° of elevation as illustrated in Figure 8.7a.

That simple arrangement will work but the feedline connection ensures that all the movements have end stops on them so there will be times when a full revolution of movement is needed to allow the aerial to track past the end stop, with signal lost during that time. Azimuth/elevation mounting can be used with commutating joints in the feedline to get the signal past the bearing movements, and the majority of commercially built satellite-tracking aerial systems in which

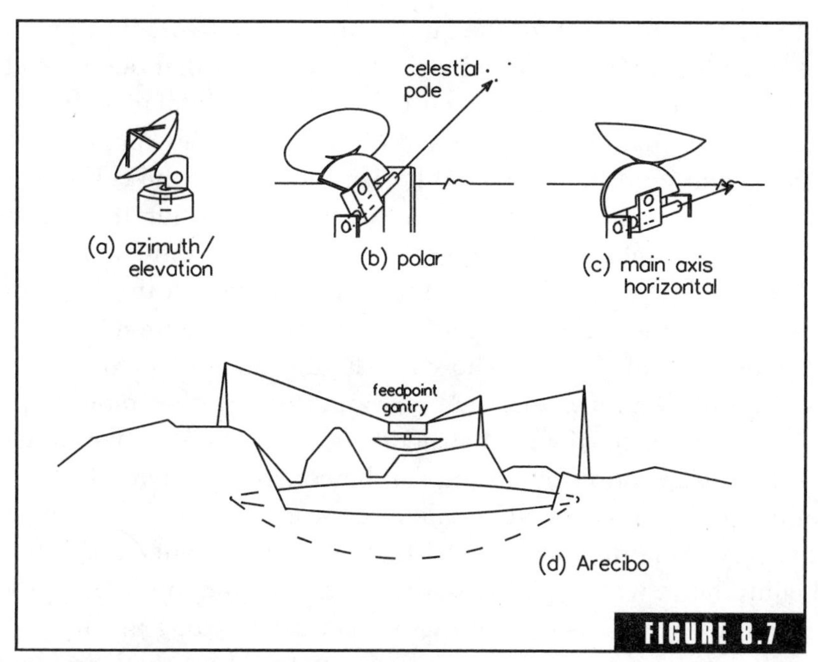

Tracking aerial mounting arrangements.

waveguide is used as the feeder use that mechanism. Another cunning way of getting past the end-stops problem that can be used with coaxial cable feedlines is to wrap the feeder for several turns around the azimuth-bearing shaft and allow sufficient total slack for the aerial to cover at least 720° of horizontal movement. At the start of a tracking session the aerial is set to the position that gives at least a full 180° of travel in either direction before an end stop is contacted.

Small radio telescope aerials, on the other hand, often use a polar mounting as shown in Figure 8.7b. The major bearing for that is on an axis aligned with the axis of the Earth's rotation; it points to the celestial pole. The second axis is at a right angle to the polar axis and gives movement to cover the sky in a north–south direction. With the machinery arranged in that way a particular spot in the star field can be tracked by a clock drive controlling only the motion of the polar axis.

For a mounting which is to be home-constructed or a one-off construction, a variation of the polar mounting may be easier to produce than an azimuth/elevation mounting. Commutating joints in coaxial cable or waveguide require very fine engineering to get them right and would be quite difficult for the home constructor. (Note that a rotating joint at the feedpoint is normally unavoidable, but that one is so close to the source that standing-wave problems are minimal even if the mechanical construction is not perfect.) If, however, the "polar" mounting is aligned with the terrestrial horizon instead of the celestial pole the whole sky can be covered with no limitations due to end stops. The aerial can track anywhere across the sky with the feedline being carried across the bearings just by strain relief loops in flexible sections of the feeder. This arrangement is shown in Figure 8.7c.

A mounting of the type shown in Figure 8.7c offers simpler programming of its movement than an azimuth/elevation mounting but requires a bigger mechanical structure for a particular size of aerial; the azimuth/elevation mounting can be constructed with all the mechanical parts approximately balanced over the central bearing, and that is very difficult to achieve with other arrangements. For commercial equipment, with repetitive production, programming of a processor is cheaper and easier to provide than extra-strong metal work so azimuth/elevation mounting with commutating joints and circular polarization is common for commercially made tracking aerials.

The largest radio telescope in the world, located near Arecibo on the Caribbean island of Puerto Rico, has a pointing/tracking system which is different from all of those previously described. The 305-m-diameter spherical dish is permanently fixed in a natural bowl between hills and the feedpoint assembly is mounted on a gantry structure supported on cables connected to towers at the edge of the bowl as shown in Figure 8.7d. Pointing is achieved by moving the feedpoint; positioning is controlled by a quite complex program and signal strength is only one of several inputs to the processor. There are limitations on the range of movement available with a mechanism such as that; the Arecibo telescope cannot cover the whole visible sky.

8.7 Doppler Shifts

Whenever a wave motion travels from a source to a receiver and there is relative movement between the source and receiver so that the distance between them is changing the observed frequency of the wave at the receiver will be different from the frequency generated by the source. That applies to all wave motions. For sound, it is the factor which causes the sound of a motor vehicle or railway train to change to a lower pitch as it passes you. For light, it causes changes of color, although because the speed of light is so great the relative speed must be several kilometers per second for the change to be measurable. That factor is very useful to astronomers for measuring the relative speeds between us and other star and galaxy systems when velocities of many kilometers per second are common. For radio signals, the Doppler effect causes carrier frequencies to be shifted. Figure 8.8 shows how a Doppler shift is produced.

The magnitude of change of frequency depends on the following:

- propagation velocity of the wave motion
- observed radial velocity between transmitter and receiver (i.e., that component of the total movement which determines the rate of change of distance between them)
- the carrier frequency of the signal

The relationship between these factors is quite simple, the radial speed divided by the propagation velocity of the wave motion gives a

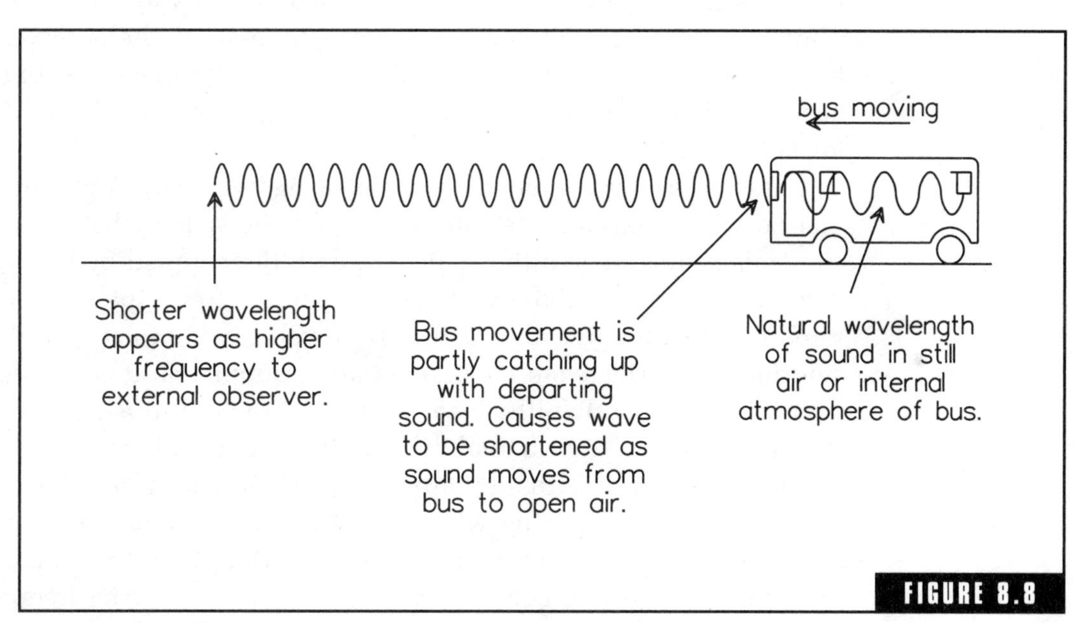

FIGURE 8.8

How a Doppler shift occurs.

fractional figure and the carrier frequency of the signal is changed by an amount equal to that fraction.

The direction of change is that when the two bodies are moving toward each other, the observed carrier frequency is raised and when the bodies are moving apart the observed frequency is lowered. When LEO satellites first appear over the horizon, they are moving toward the ground station with the greatest radial velocity; as they move across the sky, the radial velocity decreases to zero when the satellite reaches the zenith then starts to increase in the opposite direction as the satellite moves toward disappearing over the other horizon.

For an example with actual numbers, consider a satellite passing just above the Earth's atmosphere in a circular orbit whose actual speed along its orbit is 27,800 km/h and which is transmitting a signal on exactly 300 MHz. When first observed the satellite is not moving directly toward the ground station but is actually moving at an angle approximately 14.4° to the line of the observer. The radial velocity is 26,927 km/h or 7.480 km/s. That figure is 0.00002496 (0.002496%) of the speed of light and that is the proportion by which the 300-MHz signal's frequency is changed. The actual frequency of the satellite signal when first observed is 7.487 kHz above 300 MHz. As the satellite moves across the sky, the observed frequency drifts lower; at the instant at which the orbit is exactly at a right angle to the observer the frequency is exactly 300 MHz, and as the satellite moves on the frequency drifts down until, at the moment the signal is lost, it is 7.487 kHz lower than 300 MHz.

For terrestrial voice mobile services using frequency or phase modulation with maximum deviation of 3 kHz the spacing between 1VF channels is 15 kHz and the actual bandwidth occupied by the signal and its sidebands is about 9 kHz. The drift due to Doppler effect of a satellite signal totaling almost 15 kHz is well beyond the capability of a normal 1VF private mobile radio system. Doppler shift is less significant to wider bandwidth signals; when compared to a 2-MB/s data circuit the nearly 15 kHz total drift at 300 MHz is only about 1.5% of the occupied bandwidth and only about 1% of what would be an appropriate channel spacing when guard bands are included. On the other hand, higher carrier frequencies make Doppler effects more significant, the figure for equivalent conditions with a carrier frequency of 3000 MHz is ±74.87 kHz and for 12 GHz it is ±299.5 kHz, which amounts to nearly 60% of the occupied bandwidth of a 2-MB/s data stream. Signals from geostationary satellites are rarely affected by Doppler shifts; the radial velocity is only a few meters per second so the total shift even at 12 GHz is only a few Hertz.

For astronomers Doppler effects are the tools that have enabled them to measure the movement of stars and gas clouds. There are many chemical reactions which produce radiation of highly specific, accurately known frequencies. For observations at optical infrared and ultraviolet wavelengths these reactions produce dark or bright lines in the spectrum. For radio wavelengths the same spectral line phenomena occur but are detected in a different way—by tuning a very narrowband receiver across the range of frequencies close to that of the spectral line. A sharp change in level as the receiver is tuned indicates the presence of the line which signifies that particular chemical reaction. The exact frequency measured for the line can be compared with standard figures for that reaction and the difference indicates the radial velocity of the object in which that reaction is occurring.

8.8 A Typical TVRO Station

The letters TVRO stands for "television receive only." Direct broadcasting of television programs to viewers in remote and isolated areas has become one of the high-volume users of geostationary satellites. In different parts of the world there are a number of different services using different transmission standards and different frequency ranges. Historically, one of the first on the scene was a U.S. overseas armed-forces broadcast which used the NTSC standard on frequencies in the C band (3 to 4 GHz). When a communications satellite was introduced into Australia it was a general-use bird intended to support communication of all types equipped with frequencies in the Ku band with the space-to-Earth direction operating at 12.25 to 12.75 GHz. A transmission standard called BMAC specifically designed for satellite use was initially chosen for television broadcasting; the name means "type B multiplexed analog components," and it has been used by the system developers as a stepping stone to the newer digital TV standards. In 1999 Australian satellite direct broadcasting systems use the digital

MPEG2 standard for transmission. Refer to Section 4.12 for information on digital television.

In Australia a 12.5-GHz domestic satellite TV terminal has a parabolic dish aerial of between 1.2 and 1.8 m diameter which can usually be mounted on the roof of a building. The signal-collection equipment at the focus of that dish has a small aerial feeding to a waveguide which is only a few centimeters long. Equipment for that is arranged as shown in Figure 8.9.

The electronics at the feedpoint is a sealed unit called a "low-noise block down-converter" which contains a low-noise input amplifier and a mixer/local oscillator combination. The first amplifying stage in that LNB module is the major limiting factor for sensitivity of the whole receiving system, and the limit is due to noise introduced by that amplifier. LNBs are sealed, weatherproof modules that cannot be repaired economically so must be treated as components and replaced

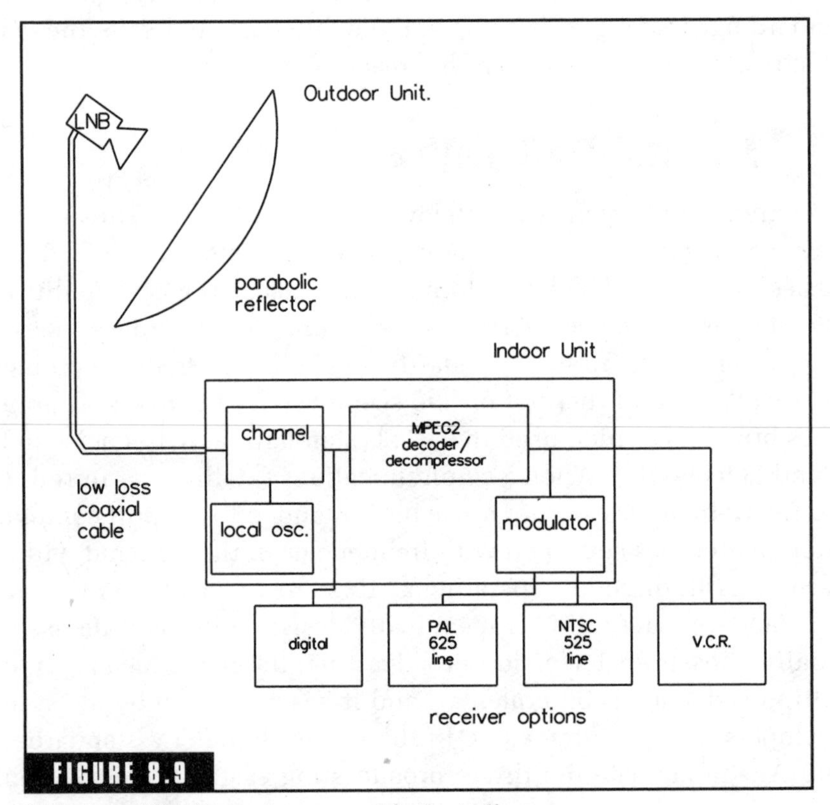

FIGURE 8.9

Typical satellite TV receiver as used in Australia.

if defective. The performance of LNAs and LNBs is specified according to their noise contribution described in terms of illumination from a body of a particular absolute (Kelvin scale) temperature [Figure 8.10 shows the theory behind the figure; to read it, think of the first amplifier as a perfectly noise-free device with three inputs in parallel. One is the genuine signal, another is the noise background to that signal (thermal "blackbody" radiation from an opaque object), and the third is the noise internally generated in the amplifier.]

Outer space is almost completely dark and cold; the actual temperature of the distributed radiation source is about 3 K and for most of the time that figure represents the background noise level of the input signal. If the aerial was looking at an object at room temperature, its noise output would be approximately 300 K. In 1999 a good-quality LNB may have a performance rating of 50 K. Research is continuously underway to make that figure lower; in the 1980s a figure of 70 K was considered good and there will probably be a time not too long into the twenty-first century when even 50 K will be considered poor.

The noise specification of an LNA or LNB may also be described as a "noise figure" in decibels. In reference to Figure 8.10, if the opaque object was at room temperature and the LNB was found to be adding the same amount of noise to the signal it would be making the final signal to noise ratio 3-dB worse. The noise temperature of such a device would be 300 K and the noise figure 3 dB. A better-quality LNB may be found to be adding only 1 dB to the signal resulting from thermal radiation of a 300 K source that is adding about 26% to the output from the theoretically perfect amplifier. If the opaque body is cooled to about 75 K it can be shown that the above LNB is adding 3 dB to that lower level of thermal radiation, and so that device can be specified as having either a 1-dB noise figure or a 75 K noise temperature. In the same way a device that is sensitive enough to be adding 3 dB to the thermal radiation when the opaque body is cooled to only

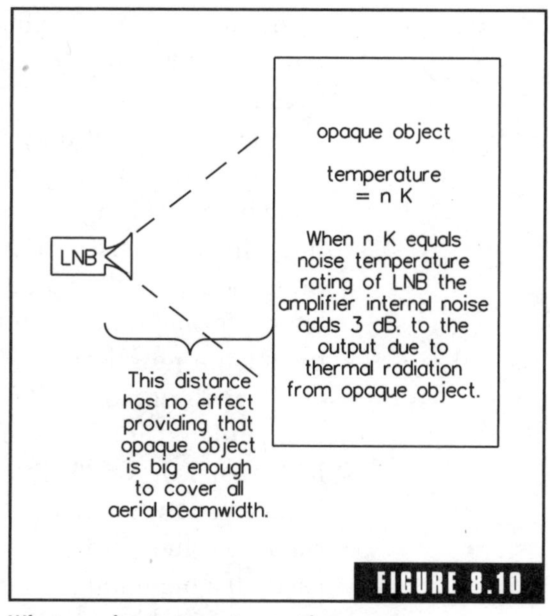

FIGURE 8.10

What a noise temperature rating means.

57 K will be found to have a noise figure in relation to the body at room temperature of about 0.8 dB.

The output of the LNB is a band of signals in a first intermediate-frequency range of 950 to 1450 MHz. For the Ku band, a local oscillator on 11.3 GHz converts 12.25 GHz to 950 MHz and 12.75 to 1.45 GHz. For the C band, the local oscillator is on 5.15 GHz and there is a sideband inversion with 3.7 GHz being converted to 1450 MHz and 4.2 GHz coming out as 950 MHz. There is very little front-end selectivity or image rejection in this conversion; the LNB relies on the directivity of the parabolic dish aerial and the fact that there is only one source of signal in that direction for the bulk of its discrimination against signals from other sources. Output level of an LNB is typically +75 dB μV or about 6 mV, but that is only a nominal figure—there is no gain control in the form of AGC or FM limiting so the actual output-millivolts figure is proportional to the input. The aerial, LNA, and down-converter combination is collectively called the "outdoor unit." The general arrangement of such a station is shown in Figure 8.9.

The outdoor unit is connected to an "indoor unit" by a length of low loss coaxial cable. The maximum length of that cable is limited by losses and the strength of the original signal. The input to the indoor unit must be at least −60 dBm; the nominal output of the LNB theoretically corresponds to −36.4 dBm in 75 Ω, which allows a maximum of 23.6 dB to cover cable losses and allowance for weak signal.

The indoor unit must accept signals in the range 950 to 1450 MHz and select the one required channel from that range. That selection is done by translating the required signal to the passband of an IF amplifier. The unit must also include detector/decoder/decompressor functions to recover the video and audio signals, and if they are required to be in a format other than that received off-air, there must be a modulator for the new format. For instance, in Australia the MPEG2 signal could go direct to a suitable receiver if one was available or it could be demodulated into one of the terrestrial UHF channels as a 625-line PAL signal or a 525-line NTSC signal or as S-VHS for the input to a VCR. In countries with other transmission standards translation of the satellite signal for use with terrestrial receivers can be achieved just as easily if a big-enough market is available to justify the cost of development of suitable circuits.

8.9 Satellite Television Installation and Maintenance

A number of situations are possible. You may be operating in a remote area where the cost of getting people on-site is the major part of the total project budget; you have a good practical mechanical knowledge and some understanding of electrical and electronic principles; and you believe that with some guidance you could do the installation and thereby save considerable transport and accommodation costs. You may be an experimenter or you hold a radio amateur operator's certificate of proficiency and you wish to set up a system in which performance is pushed beyond standard specifications in some directions. You may be someone who knows a little about the subject and you think the next stage in the learning process is to lay the books aside and actually have a go yourself. A number of variations on each of these themes are possible; however, if we can look at each of these three situations in turn it should be possible for an interested person to adapt one of them for his/her own need.

Remote Area with Limited Access to Technical Knowledge and Test Equipment

In this situation you are most strongly advised to buy a complete system and get as much advice and documentation from the technical (not sales) people of the company as you can. It would not be unreasonable to make your purchase conditional on your training in installation procedure; any worthwhile organization will recognize your position and give you as much help as they can.

When you do start practical work, follow the installation instructions to the letter even if they do not seem to be particularly relevant from the mechanical point of view. Some of the factors that are vital to the radio signal could easily be glossed over if you are just considering the project as an electrical or mechanical assembly job. In particular, do not skimp on the rigidity of mounting of the outdoor unit; the aerial beamwidth is less than 1° and wind loading on the finished structure can be a fair proportion of a ton. If there is any flexibility at all in the mounting you may find that you only have a signal when the wind is favorable. You will also need to keep a check on all the factors in the following list; most of them should have been addressed in the installation instructions that came with the equipment.

Experimenter or Amateur with Good Technical Knowledge

The factors to be considered are as follows:

- Are you using magnetic or true bearings for the azimuth angle?
- Do you know the compass deviation for your location?
- Do you have pointing angle information that is correct for your area? Remember, the figures will vary by almost 1° every time you move 100 km across the Earth's surface.
- For the elevation figure, are you sighting to a true horizontal? A moderately sized mountain range even 50 to 100 km away can lift the horizon in that direction by a couple of degrees, and if your installation is on the top of a hill overlooking open sea the horizon can be below the true horizontal by more than the beamwidth of the aerial if you are as little as 1000 m above sea level.
- Final adjustment of the dish pointing must be done using a signal-strength meter on the output of the LNB. Are you sure that the signal you are measuring is a satellite? Is it the correct satellite?
- Does the satellite have a significant drift pattern and if so are you making the final adjustment at a time of day when it is near the center of its pattern?
- Does the beamwidth of the feedpoint agree with the F/D ratio of the parabolic dish? Is the feedpoint aerial mounted exactly at the focus?
- Does the polarization of the LNB match that of the incoming signal? Is the signal linear- or circular-polarized? If circular, is it right-hand or left-hand rotation?
- If the signal is linear-polarized is cross-polarization of adjacent channels used to discriminate between transponders? If so, the adjustment of LNB rotation requires a spectrum analyzer capable of covering up to 1450 MHz to set for maximum rejection of the opposite-polarized channels.

- Are there any significant terrestrial signals in the IF range (950 to 1450 MHz) that are being picked up by the coaxial cable and interfering with the desired signal?
- Does the output of the LNB match the required input to the receiving and decoding equipment?

From the point at which an acceptable signal is delivered to the input of the receiver all the considerations of a terrestrial UHF receiving station of the relevant bandwidth and modulation type can be applied to this station.

As a Training Exercise

All signals from geostationary satellites are very weak and the initial finding of a genuine signal is the hard part. There are several practical steps you can take to break the job down into sections and get each part right as you go.

If a UHF signal generator is available, the receiver (indoor unit) can be proved independently of any satellite receiving equipment by connecting the signal generator in place of the coaxial cable. When the receiver is proved, transfer the signal generator to the input end of the coaxial cable; exactly the same outputs should be seen but at a lower sensitivity due to the loss in the cable.

When the receiver and cable are proved, connect the LNB. There may have been some evidence of noise from the input circuits of the receiver; there should be more thermal noise when the LNB is connected. That extra noise does not prove, however, that the LNB is ready to receive a signal.

A quick rough check of LNB performance is possible by placing a lump of radio-opaque material directly in front of the feedpoint aerial. Material that is an electrical insulator is transparent to radio waves; metals are reflectors and so they do not work either. A damp blanket arranged in a loose pile should give a signal; a plastic bucket full of fairly clean water should also work. A metal bucket or salty water would be seen as conductors which are reflective; distilled water could be sensed as an insulator. The aim is that thermal radiation from the radio-opaque material will give a broadband signal which will be seen as an increase in noise output from the LNB. If it is possible to unbolt the LNB from the dish mounting with the feedpoint aerial and coaxial cable still connected, the signal out of the LNB when it is pointing toward the ground should be significantly higher than that seen when it is pointed toward the sky anywhere well away from the Sun. As the LNB is moved around and pointed to different directions the noise output should change in step with the movement. The beamwidth of the feedpoint aerial is quite broad, perhaps 30° to 40° either side of the center line so no great discrimination will be possible, but any variation at all in step with the movement will show that the LNB is receiving signals of some sort via its input.

When you can prove the system to that extent if you reattach the LNB at the focus of the dish and point it at the satellite you ought to be able to detect a signal of some sort. Once a signal is detected, attach a power output or carrier/noise meter and look for that signal. When you can get a meter reading which you know is due to the desired signal, work through the above list and make adjustments to maximize the signal. Only adjust one factor at a time and make sure that the finished adjustment is locked into its maximum position before you move on to the next adjustment.

There are possibly some interactions between the adjustments so after you have worked through the complete list start from the top and check that each one is still at its maximum signal setting. If any need to be moved it is an indication that there is interaction between that one and some other adjustment.

When you are required to do maintenance or fault finding on a working satellite receiving station the very first question to ask is "Did it ever work properly?" If at any time in the past the system has given satisfactory service your task is to find out and correct whatever has changed since that time.

Before you start making any changes, however, there is one factor that will cause interruptions to the signal but is not a fault. When the Sun is directly behind the satellite the radio noise from it will be strong enough to blank out the signal. That will happen at a particular time of each day for two 2-week periods on the same dates each year. The dates on which the effect occurs will depend on the latitude of your station—at the equator, dates are symmetrical about the equinoxes and at high latitudes they are pushed a few weeks into the winter end of the seasons.

Performance of the parabolic dish is related to its shape and the direction it is pointed at. Any defect large enough to affect the signal will be very obvious to visual inspection. It is possible that changes to the pointing angle could go unnoticed by a casual observer but close inspection of the screw threads would show a change in the weathering pattern of the section that had been adjusted. If the dish is mounted on the roof of a building, any movement sufficient to cause a change to the pointing angles would also cause very obvious cracking of walls or similar structural damage. In most cases, if the station has previously worked well and close visual inspection shows no reason to suspect a defect in the dish, then the fault is probably elsewhere.

A fault observed as loss of sensitivity can be due to a build-up of salt or dust on the surface of the dish in seaside or desert areas. A very readily observed layer is needed to cause significant loss of signal and washing with clean water is normally all that is needed to clear the problem if it occurs.

The LNB can be checked for output on the correct frequency range. To make definitive measurements will require either a spectrum analyzer or tunable UHF receiver. For rough checks the correct level of output signal indicated in a power meter would suggest that you suspect some other part of the system. Note carefully that the "opaque object" test described earlier in this section will not give reliable results with a power meter in the case where the LNB is receiving a satellite signal; the object will block out the satellite signal but will replace it with thermal radiation which may be close to the same total signal level. In remote areas, if a new LNB is available, its cost is probably going to be less than the cost of hiring and transporting a suitable spectrum analyzer so testing by substitution is the best action. If you replace the LNB on a system that uses linear polarization, mark the one you take off in some way and mount the new one in exactly the same alignment to preserve polarization.

The coaxial cable can be checked with a multimeter if it is disconnected at both ends. Check for insulation between conductors and for loop resistance with the far end shorted. If both measurements are normal, the cable can be assumed to be workable unless the fault being traced is due to the type of ghosting described in the latter part of Section 4.7.

For the indoor unit if a receiver (display module) capable of showing a signal from a terrestrial source is being used then that other source can indicate whether the receiver is working. If the equipment offers several different outputs with different transmission standards then checking all of them would be an advantage if suitable receiving equipment is available. Power input to the indoor unit should be checked. If the supply is AC but the source is not from a commercial distribution grid (i.e., from a diesel-driven motor/alternator set or similar), check the supply voltage and frequency at the wall socket with the indoor unit plugged in and running. If the measured figures are not exactly what is specified, make suitable adjustments to the speed and excitation of the alternator so that the power supplied at the actual wall socket is correct then switch off and unplug the satellite receiver, wait about 30 s, plug the equipment back in, and try again.

If all of the above tests indicate that each part of the equipment tested is correct but when all put together does not work the only part left unchecked is the indoor unit itself. Definitive proof-of-performance tests on that equipment require a considerable amount of specialized and expensive test equipment. There will be a main fuse which can be checked with a multimeter, but apart from that, the fact that a fault still exists after you have checked and proved all the other items in the system means that it is the indoor unit itself that is faulty. In most cases that will need to be taken to a centralized depot for repair.

8.10 Go-Anywhere Telephones

Mobile-telephone systems based on terrestrial VHF or UHF links are short-range services. Large systems based on the cellular principle (see Chapter 11) can appear to offer long-range service but only by using many repeaters, each of which only covers a small area. Mobile-telephone systems based on HF links can give wide-area coverage from a single base but for successful use they require that the operator have some familiarity with the principles of radio propagation; there may be some delays due to solar activity and on a worldwide basis there are not nearly enough channels available in the HF spectrum for the total demand for long-range mobile-telephone services. The business community demands and is willing to pay for instant access to untrained operators at any location and communications satellites are able to provide that service.

Satellite-based systems can provide telephone services in several quite radically different ways. Low Earth orbit clusters of satellites can be used in combination with omnidirectional or low-gain aerials to give connection for individual channels to a mobile terminal. High-gain aerials used with geostationary satellites can give broadband links for connection to a switchboard or PABX system in a mining camp, ship at sea, off-shore drilling platform, or other isolated group of people. Land-based terminals must be at a fixed location and aligned for correct azimuth, elevation, and polarization in the same way as is required for the satellite TV service described in Section 8.8. Marine installations require the capability for tracking as described in Section 8.5.

For connection of a single-voice channel from a geostationary satellite to, for instance, a driver of a road vehicle in a remote area somewhat lower aerial gain is acceptable. Aerial directivity can be made broad enough to allow for vehicle movement with either no tracking or a very much reduced demand for the facility. In a typical satellite-based mobile-telephone system there is one transponder on a satellite as the common link with many hundreds or thousands of mobile terminals (portable or mounted in vehicles), each of which requires the service of a single VF channel. In Australia there are two services which offer telephone access via geostationary satellite to individual terminals; each is commercially linked to one of the two major providers of terrestrial telecommunications services. Their technical features are probably indicative of the protocols that apply to most services with a similar function.

One service operates under the name "Mobilesat" and uses a transponder on an Australian-owned satellite. The other geostationary service offers a worldwide range via the Inmarsat service; from Australia communication is via the Pacific Ocean satellite. The similarities between these services are indicative of what features are likely for other satellite-based telephone services. Both operate on frequency ranges in the L band (1500 to 1600 MHz), both use QPSK modulation ("quadrative phase shift keying"), and both use a form of multiple-carrier digital multiplexing scheme.

The differences are also significant and indicate differences in the way the system is used. The Australian satellite covers only the Australian continent and up to 200 nautical miles offshore. Reception requirements are limited to elevation angles of 30° to 50° approximately so the Mobilesat system uses a modified form of colinear aerial which has sensitivity in a cone shape covering that range of elevations all around the vehicle. The Inmarsat system covers all the oceans of the world (except some of the Arctic) and most inhabited continental areas so the telephone that works through it may have any possible combination of elevation and azimuth angles from about 2° above the horizon to directly overhead. That system uses a tracking aerial with a broad beam, engineered to cope with very high rates of turn and the QSB fluttering of mobile signals.

The salespeople for these single-channel systems describe them as "able to work anywhere in Australia." In the broad sense that the satellite footprint covers all of the continent and up to 300 km offshore the statement is justified, but in detail there are some locations where satellite telephones will not work. There must be a true line-of-sight path for the signal so around tall buildings, cliff faces, gullies, and sharply peaked mountains there are shadow areas which are unworkable and in forest areas where the trees are dense enough to form a solid canopy overhead there will be no signal over quite large areas. Satellite telephones work well in deserts, open bushland, and at sea. In sclerophyll forest areas where there are big trees but with spaces in between them it should be possible to get a signal by moving to an open area. The signal path is only a pencil beam so gaps in the forest canopy do not need to be very big and in quite heavily treed areas you may get service if you are lucky enough to find a clear spot. If you are in that fortunate circumstance keep stationary for the period you wish to use the telephone.

The shadows can be plotted by calculation if you have enough information about the local terrain and the satellite position. For example, using the Inmarsat Pacific Ocean satellite in the region of the Stirling Ranges in southernwestern Australia the following calculations would apply: The satellite is geostationary at orbital location 178° E; in the Stirling Ranges area the elevation of the signal path is about 22° above the horizon. The Stirling Ranges rise abruptly from clear farming land generally less than 100 m above sea level. The highest peak, Bluff Knoll at 1073 m, is on the eastern end of the range so casts its shadow over other peaks in the Stirling Range National Park. The second highest point is Toolbrunup Peak at 1052 m on the southern edge of the range and it throws a shadow up to 4.7 km (calculated from the cotangent of 22° × the difference in altitude of ~1000 m) out into the surrounding flatland in a west-southwesterly direction. For all its 4.7-km length, however, the shadow is only as wide as the width of the mountain so if you happened to find yourself in such a position you could find a clear signal by moving sideways a few hundred meters. On the other hand, that particular location is within range of the Indian Ocean satellite at orbital location 64°E at about the same elevation of signal path. In places where satellite footprints overlap you have the option of switching to the other one for emergency communications.

Signals from geostationary satellites are never strong enough to overload the receiver and in clear locations they are about the same strength over very large areas so weak signal problems are all due to obstruction of the signal path, and the cure is to move the mobile terminal to a clear location. If you suspect a fault other than a weak signal problem in a satellite mobile telephone, the paragraphs under the heading "For Mobile Phones" (Section 11.10) in *How Radio Signals Work* have some pointers to fault conditions that may be fixed in the field. If the fault is in the electronics box itself, however, the required test signal is a UHF carrier wave frequency-modulated with a quite complex digital bit stream so the whole unit would normally have to be returned to a service depot for fault finding.

A third satellite mobile telephone system became available in Australia in early 1998 operating under the name "Iridium." It uses six clusters of 11 LEO satellites to provide a worldwide service with characteristics very similar to a cellular telephone service. In Australia the service was marketed as a mobile-telephone service with remote-area capabilities integrated with terrestrial mobile-telephone services. The user's transceiver made connection to a terrestrial cellphone base if one was within range and switched to operate via the satellite when no terrestrial base was available.

Because the Iridium satellites were in low Earth orbit, the high-gain aerials characteristic of geostationary services were not needed and the 220 ms of time delay was reduced to only a few (minimum

~2.5) milliseconds. More detail on LEO mobile-telephone services is included in Section 11.6.

8.11 Global Positioning System

The letters GPS stand for "global positioning system." A cluster of 22 satellites in polar orbits at about 12,000 km altitude radiate continuous signals that give very exact information on the orbital parameters for that satellite and a highly accurate time code. At the receiver, signals from at least 3 and possibly up to 8 satellites are compared. The comparison of signals from 3 satellites well separated in the sky can give the receiver information to compute how far it is from each of the three and the exact location of each at the time of the measurement. That information specifies three spherical surfaces and the receiver location is at the point where all three intersect. Signals from more than 3 satellites are used to improve the reliability of the reported position.

The receiving equipment is a multichannel UHF receiver with all outputs going direct to a very complex dedicated digital processor. The bulk of the circuitry is in the form of VLSI-integrated circuits, so unless the unit was particularly expensive to replace, most GPS units are not economic to repair. Where the GPS facility is built into a larger piece of equipment such as an HF transceiver or telemetry unit it is built on a modular subassembly which can be replaced as a complete unit.

Accuracy of the basic GPS is ±100 m in the horizontal plane. A process called "differential GPS" can be used to give greater accuracy than that over short ranges; it uses a GPS receiver at an accurately known location, a radio link (usually UHF), and a GPS receiver at the spot to be measured with the facility to compare its own reading with the other as reported by the radio link. Differential GPS can offer accuracy to within a small fraction of a meter for all locations within range of the radio link from the reference location.

The GPS was originally developed for marine and aeronautical navigation; these are both functions where a clear view of the whole sky is available at all times. The service is less accurate in mountainous regions and city centers where direct signal paths may not be available and signals may arrive by reflection. A reading taken in such a location in which the receiver only reported signals from three satellites should be discarded unless it was corroborated by another source of information such as a good-quality small-scale map or previous GPS reading in a clear location. With some GPS receivers it is possible to remove the aerial and insert a length of coaxial cable so that the aerial can be placed in a clearer location such as at the top of a mast. Note that in those cases the position being reported is that of the aerial, not the receiver itself; this could become important in some differential measurements.

A second use for GPS has emerged as the system has been used. An essential part of the system is the highly accurate measurement of time, and the system has come into use in a number of scientific disciplines as a transferable time reference. The system is able to deliver time signals to all points on the Earth's surface with accuracy typically in the nanosecond range.

8.12 Orbital Satellites Carrying Amateur Radio

Amateur operators around the world have been close on the heels of the developers of commercial equipment during the whole space age. OSCAR 1 (Orbital Satellites Carrying Amateur Radio) was one of the first nongovernment objects ever to be placed in orbit and at the end of the twentieth century most of the functions able to be provided by satellite, with the exception of the broadband geostationary systems described in Sections 8.8 to 8.10, have been tried in experimental form by amateurs somewhere in the world. For most of the satellite projects so far, the space segment (the bird itself) has been provided by amateurs in the United States, but amateurs form a worldwide nonprofit fraternity so the facilities have been made available to amateurs and others in all countries free of charge. A great many people all over the world have gained a lot of basic practical experience in receiving and tracking amateur satellite signals and for the foreseeable future the amateur service offers a very good way for a would-be satellite experimenter to launch into the discipline of satellite tracking.

You do not need to hold an amateur operator's licence to receive and decode satellite signals. What you will need is information on when the satellite will be above the horizon at your location, expected azimuth and elevation at each time, and the frequency you will need to tune to. The source of that information is sets of figures called "orbital parameters." Orbital parameters give information on the expected position of the satellite in space at each instant of time and some conversion is needed to derive aerial pointing angles and times from them. The calculation is a reasonably involved solid geometry problem to do longhand or a fairly simple exercise for a computer.

Software is available for most of the popular brands of home computers. Programs written in BASIC are available for computers not specifically covered by particular packages. The software is not specific to particular satellites; once installed the same program handles all types providing that the orbital parameters are expressed in a form that it can recognize. There is introductory information about software and orbital parameters on Internet websites and the actual parameters for each satellite are supplied by the group or club which manages that satellite. Most of these are bodies affiliated with the American Radio Relay League so its website, http://www.arrl.org/tis/info/satguide.html, is a good place to start looking.

At the end of the 20th century there are satellite signals of some sort on all amateur bands from 28 MHz to 10 GHz and functions covered include voice signal repeaters, data repeaters, telemetry, data from remote sensing instruments, contact with astronauts, and multiuser operation via packet switching. Signal strength in most cases is adequate for reception using quite small aerial systems with broad directivity patterns and for most two-way uses transmitter power of only 20 to 30 W is adequate for the satellite to receive a good-quality signal.

8.13 Radio Astronomy

In many ways a steerable radio telescope has a lot of similarities to a very big version of the TVRO station described in Section 8.8. There are many other types of radio telescope for other particular purposes; some, for instance, are little more than very long lengths of wire arranged in particular ways. In many cases those special-purpose instruments are built for the purpose of a particular experiment or to collect a particular type of data, and the instrument is designed around the data to be collected.

For the steerable, more general-purpose, telescopes that can be pointed to any spot in the sky and have their configuration changed for a variety of operating frequencies and bandwidths, the description "a large parabolic reflector with a low-noise amplifier mounted at its prime focus" is common to many telescopes around the world. Astronomers using these instruments generally spend most of their time making measurements of one of two types. There are highly directional broad bandwidth measurements of noise power related to the spectrum to indicate blackbody radiation from a source at a particular temperature and there are discrete frequency measurements at the signature frequencies of particular chemicals and their reactions to indicate relative chemical composition and radial velocity of those sources.

A brief comparison with a telescope for optical astronomy points out one of the cutting edges of technology for radio astronomers. Up to a certain point the maximum resolution of a telescope (both types) depends on the diameter of the objective lens or mirror compared to the wavelength of the signal to be received. For optical instruments resolution is specified in terms of how close lines can be to each other and still be displayed as separate lines. For systems such as cameras, projectors, and microscopes where a real image at a well-defined distance is being tested, resolution can be specified as lines per millimeter at the focal plane. For video systems the definition can be related to the total width across the screen with no consideration of how big or small the screen might be. For telescopes there is no particular distance that can be called a focal plane; the display seen at the eyepiece is a virtual image whose apparent distance can be adjusted to suit the observer's eye and the other focus is at an indeterminate distance. Telescope resolution is specified in terms of visual angle; the relevant factor is minimum angular separation between line images. A telescope with an objective 200 mm in diameter will allow definition of separate lines at half the angle of one whose objective lens is only 100 mm. For telescopes operated at the bottom of the atmosphere that relationship only holds for instruments up to a certain size. Due to degradation of the signal in its passage through the atmosphere the 200-inch and bigger optical monsters that have been built are not able to offer any greater resolution than an instrument about 1 m in diameter.

The wavelength of visible light is in the range between 0.4 to 0.7 μm, which is between $1\frac{1}{2}$ and $2\frac{1}{2}$ million times smaller than the 1 m of the telescope of maximum resolution. For a radio telescope receiving signals on, for instance, 30 mm (frequency 10 GHz) to have the same resolution the parabolic reflector would have to be about 75 km in diameter and follow the required curve with accuracy of ±2 mm over that 75-km-diameter area.

For small-diameter radio telescopes, the limitation of sensitivity is determined by the noise temperature of the first amplifying stage in the receiving electronics. Amplifying circuits become quieter as they are cooled so for receiving the really small signals from many light years away running the amplifier at cryogenic temperatures is normal practice. Liquid nitrogen is commonly used; some use is also made of liquid helium if really cold operation is essential regardless of cost. Within certain limits telescopes can be made more sensitive by making the parabolic reflector bigger. The cosmic limit of sensitivity is dictated by the cosmic background radiation which is a thermal signal from a source distributed very evenly over the whole sky with an apparent temperature of just under 3 K. This level of sensitivity can be reached by a dish of about 20 m in diameter feeding signal to a good-quality low-noise amplifier with liquid nitrogen cooling.

For a steerable telescope, the largest dish that is manageable is about 80 to 100 m in diameter; for an 80-m instrument working on a 10-mm wavelength the angle from the center of the beam to the first null is theoretically 25.8 arcsec. If the wavelength were 70 cm instead of 10 mm the width of the beam would be close to 30 min of arc, which corresponds to a spot in the sky about the size of the disk of the Sun or Moon. Those figures indicate the smallest available pixel size if any attempt is made to build up an image of a section of the sky in terms of radio flux.

Radio astronomers use interferometry (described in Section A2 in the Appendix) for resolution of finer detail than that. Signals may be collected by an array of smaller aerials spaced in a line or grid pattern or there may be one aerial that is fixed and one that can be moved on a track. Distance apart of the two aerials determines the angular width of the beam to the first null but for two small aerials widely spaced there are many nulls close together (see Figure 8.11) so there may be uncertainty about which one is the true center of the beam. To resolve that uncertainty, a number of measurements are made of the same source with the aerials separated by different distances and directions. All the data is collected and analyzed by a computer program to indicate the size and shape of the source. Using interferometry pixel size can be reduced to considerably less than 1 arcsec.

The finest possible resolution for Earth-based measurements would be made with an interferometer with aerials on opposite sides of the Earth; that is, with the baseline equal to the ~12,000 km of the Earth's diameter. In the closing stages of the 20th century a research project called the VLBI Project was underway to do just that. Obviously microwave signals from opposite sides of the Earth cannot be compared directly; the technique being used is to sense the phase of the incoming signal and compare it with a highly accurate time code. The time-coded data from each telescope is then sent to a data-processing center where signals are compared and the actual pixel map is built up.

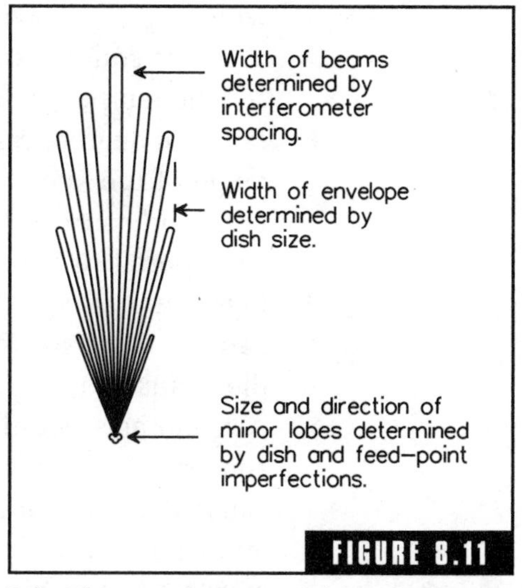

FIGURE 8.11

Polar diagram of a two-aerial interferometer.

Astronomers search for signals that are broad bandwidth thermal radiation (to the rest of the world that is "noise") so their consideration of bandwidth is not related in any way to VF channels. They do have to avoid the frequencies of communications transmissions and there may be times when reception limited to a particular bandwidth may be needed but often the widest possible bandwidth up to several hundred megahertz is aimed for. The center frequency of the measurement may be set in allocated bands anywhere in the range from 1 to 70 cm (corresponding to frequencies from 30 GHz to about 400 MHz), and the aim of the measurement in most cases is to point the telescope at a particular spot in the sky and make meaurements at several different center frequencies. Comparison of radio flux at the different frequencies gives information on the temperature of the source.

The frequency (wavelength) of maximum flux of radiation is related to temperature by Wien's Law, which states that *the wavelength carrying the maximum intensity in a spectrum emitted by a blackbody at a certain absolute temperature is inversely proportional to that temperature*. Mathematically, it can be expressed as follows:

$$\lambda_{\max} \times T = W,$$

where λ_{max} is wavelength, T is Kelvin-scale temperature, and W is a constant equal to 0.289 centimeter-degrees Kelvin.

The temperature range associated with visible light is in the thousands of degrees range, room temperature (300 K) is in the far infrared range of the spectrum, and thermal radio signals are associated with very cold sources. Peak radiation at a 1-cm wavelength corresponds to a temperature of 0.289 K and for a 1-mm wavelength the corresponding temperature is 2.89 K. Most sources are hotter than that so broadband radio measurements are commonly used to investigate the tail of the Gaussian distribution rather than the peak of the spectrum. Astronomers would prefer to make measurements in the infrared spectrum, but that is not possible for Earth-based observations because the atmosphere absorbs the energy in that frequency range. The atmosphere is only transparent in two sections (or "windows") of the spectrum. One window covering visible light and some of the close by infrared and ultraviolet range is called "the optical window." Another covering the frequency range from about 30 MHz to 22 GHz approximately is called "the radio window." Infrared measurements can only be done from satellites or high-altitude sounding rockets and that will probably be one of the directions of great development of astronomy in the twenty-first century.

There are also single-frequency signals in the radio spectrum due to spectral emission lines. The frequencies at which these signals occur is known very accurately for sources which have zero radial velocity relative to the Earth. Movement is normal, however, and radial velocities up to several hundred kilometers per second are observed quite commonly. The Doppler shift (Section 8.5) may amount to several megahertz and for a particular directional position there may be different gas clouds which are moving with different velocities so emission line signals are possible at a considerable range of frequencies either side of the zero velocity frequency.

Frquency shifts similar in effect to Doppler shifts may also result from strong gravitational or magnetic fields and if the gas cloud is rotating the Doppler effect causes the received signal to be smeared over a band of frequencies. Receivers for emission line signals require a selection of very narrow bandwidths and adjustable tuning. The receiver sections of radio telescopes are equipped with bandwidths selectable over a very wide range.

Radio astronomers may need to be able to prove that the signal they are receiving is actually from an extraterrestrial source and one of the surest ways of proving that is by testing its rate of movement across the sky. If the telescope is set to a fixed azimuth and elevation with respect to the Earth's horizon it is, in fact, sweeping across the sky at the rate of the Earth's rotation. Our 24-h clock time is derived from the rate at which the Sun appears to move but the Sun itself has an apparent movement across the sky to the extent that each year the background star field crosses the local meridian one more time than the Sun. A year that has 365.24 solar days has 366.24 "days" in relation to the star field and each of those "days" is about 4 min shorter than a solar day. That time period of 23 h, 56 min, and 3.4 s is called a "sidereal day."

If a source of radiation repeats its passage across the sky in step with the sidereal time period it is almost certainly from a source outside the Solar System. If it repeats every 24 h its source is probably related to the Sun. Each of the planets also has its own rate of movement and that can be used to identify radio sources related to them. Signals that repeat at 24-h intervals but only for 5 days per week are interference from terrestrial industry sources. Other signals have regular repetitions but at rates not related to any known movements in the Solar System and these may signify rotation of the source itself. The repetition period of these variations can range from the months or years of a binary star system down to milliseconds in the case of a pulsar.

The information collected by radio telescopes is of interest to astronomers and is referred to astrophysicists and others for analysis, but modern radio telescopes provide work for many people who are not astronomers and who have only a very distant connection with the centers of academia. In Australia, for instance, there are three radio telescopes operated by the Australia Telescope National Foundation, which is a branch of the CSIRO, a Commonwealth Government subsidiary. There are also solar observatories which use radio observations of the Sun operated by IPS Radio and Space Services, also a Commonwealth Government subsidiary in a different branch from the ATNF. The organization that owns and operates the telescopes is not necessarily the most frequent user of it; telescope observing time is hired out to research groups and individuals in a bit the same way as specialized equipment is hired by construction companies for particular jobs on particular projects. The difference in this case, however, is that observing time is allocated on the basis of scientific value of the research project rather than purely commercial or order-of-booking considerations.

Radio telescopes have employment for radio technicians and engineers, computer experts and programmers of several types, servo system experts, and electrical and mechanical people at all levels from engineers to manual laborers. There are also needs for public relations people, tour guides, and administration and marketing experts; a large and complex structure of that type creates a great deal of public interest. Although radio telescopes are built for the purpose of collecting and disseminating information they are not, in general, technical training institutions; people who seek to apply for technical or engineering jobs on telescope sites should have already established skills that can be adapted to the business of telescope operation and maintenance.

8.14 Moonbounce and Interplanetary and Asteroid Radar

The techniques used in this highly specialized field are all those for the very ultimate maximum sensitivity of reception. Round-trip path loss from transmitter to receiver is often well over 300 dB and the returned signal level may be up to 40 dB below the level of thermal noise in the best available receiver.

Radar which had been developed as a secret weapon during the second World War became less secret and more commercially useful for shipping and aircraft during the decade following that war. The competitive nature of national governments made radar engineers and technicians search for ever longer range until they basically ran out of room on the Earth. In the 1950s they dreamed about the astronomical bodies with the realization that with a powerful-enough transmitter, enough gain in the aerial, and a sensitive-enough narrow-bandwidth receiver radar echoes from the Moon at least should be possible. The first echoes definitely identified as coming from the Moon were being heard in the early part of the 1950s and with rapid advancement of techniques a new more accurate measurement of the Earth–Moon distance was made in 1957. That measurement was the birth of an entirely new branch of science, "radar astronomy."

For the amateur radio fraternity, moonbounce was a difficult but not impossible feat. In the 144-MHz amateur band the path loss, Earth–Moon–Earth is about 240 dB. A 1-kW transmitter and a 100-Hz passband CW receiver can just pass intelligent messages when the path loss between them is 217 dB, which if combined with a total aerial gain of 23 dB would give a just-detectable echo. An extra 10 to 15 dB of gain in the aerial system would improve the chances of success and in the 1950s and 1960s many amateurs with high-power VHF rigs and CW experience began building high-gain aerials for the 2-m band. The major spur for all this activity was the dream of worldwide networks of communication on the VHF bands similar to the conditions being experienced at that time by operators on the HF bands but without the noise and interruptions of ionospheric propagation.

The amateurs at that time were not far behind the heels of the government research workers and by the close of the 1950s there were many people worldwide who were able to hear their own echoes and were searching for the signals of people on other continents. The first contacts were "crossband" with the initial contact being made on the 20-m band using ionospheric propagation to carry a message such as "I'm sending you a signal via the Moon on 144.???, can you hear it?"

Success was delayed for a while because of Doppler shift. The relatively small eccentricity of the Moon's orbit combined with rotation of the Earth on its own axis was sufficient to shift 144-MHz signals out of the passband of the receivers being used (normally less than 100 Hz) and some tuning around at very slow rates was needed to initially find the signals.

The professional researchers soon turned their attention further afield. Venus is the next nearest heavenly body after the Moon in terms of round-trip path loss for a radio signal. Higher-power transmitters were built (which the amateurs were not permitted to do), internal noise level of receivers was made lower by a continuous research effort, and ever bigger parabolic reflectors were being built. A technique of "adaptive filtering" was developed in principle similar to the photographing of test patterns by TV Dx enthusiasts (Section 4.11) which allowed the detection of signals well below the general level of noise due to thermal sources, and the first radar echoes were detected from Venus late in the decade of the 1960s.

This new branch of science has continued to grow from that base with the attainable range being continually expanded. In the closing months of the twentieth century the range at which objects can be detected reached to the Asteroid Belt and the planet Mercury and radiowaves have looked through the thick clouds of Venus to produce a map of surface features sufficiently detailed to show features the size of mountain ranges. Currently, close attention is being paid to asteroids with Earth-crossing orbits, and measurements sufficiently detailed to show size and shape with resolution to within 10 m have been made of those which have passed close to the Earth. To facilitate this detail, radar probes have been flown on spacecraft aimed for passage of the asteroid orbit.

In 1999 the greatest range available for Earth-based radar installations uses the radio telescope at Aricebo, Puerto Rico, operating in the S band in the frequency range 2.33 to 2.43 GHz with a 0.5-MW transmitter, aerial gain of 73 dB, and a receiver with 32 K noise temperature. Actual signal level required for the receiver depends on the bandwidth being received; the full bandwidth is required for the short-duration pulses that are needed for fine resolution of distance, but a tradeoff between resolution and sensitivity is possible if bandwidth is narrowed. Bandwidth narrowed to the width of a 1 VF channel would make this instrument capable of operating with a round-trip path loss of the order of 350 dB, but that degree of limitation of bandwidth would degrade distance resolution to ±50 km.

An important factor in the collection of data is the Doppler shift of each echo. When a planet or asteroid is rotating echoes will be returned with a range of Doppler shifts, and the observed shift can be resolved as a particular zone on the surface of the rotating body. This information is collected by either spectrum analysis or multichannel reception of the returned signal. Receivers with the effect of up to 4000 separate very narrowband channels are currently available.

All interplanetary or asteroid radar experiments are conducted at the largest and most sensitive of radio telescopes so the business implications and job opportunities are identical with those mentioned in the previous section.

8.15 Domestic-Scale Radio Astronomy

In the same way as there are some aspects of visual astronomy in which enlightened amateurs with binoculars or small telescopes can take interest and occasionally make an original scientific contribution, there are a few aspects of the radio astronomy field where very large supersensitive collectors of very weak signals are not needed and many sets of eyes and ears are needed to collect the volume of available information.

Observation of the Sun at all frequencies in the microwave spectrum is a subject of continuing interest and the signals are powerful enough to be detected even by equipment not specifically designed for radio astronomy. The flux of energy from the Sun regularly blanks out signals from geostationary satellites so equipment for satellite TV reception can be modified to use for solar observation. For long-term observation the aerial system must be mounted on a shaft aligned to the celestial poles and equipped with a clock drive to follow the Sun across the sky.

An amateur with the capability to accurately measure the frequencies of signals in the range a megahertz or two either side of the neutral hydrogen spectral line at 1421 MHz would collect information of scientific value about the radial velocity of gas clouds in solar flares and other active regions. There are also spectral lines associated with many other chemicals, and similar measurements of energy flux at the range of frequencies related by Doppler shifts to them will collect scientifically valuable information. There are so many chemicals and such a range of Doppler frequencies to be monitored that there is room for many thousands of people to take part in just that activity alone.

The next strongest radio signals from the sky come from the planet Jupiter and quite low gain directional aerials can be used for their detection. Jupiter emits more energy than it absorbs from the Sun and a large proportion of it is in the radio spectrum. In the same way as for the Sun, observation at many different frequencies is a scientifically worthwhile activity. Similar signals from the other gas-giant planets can be heard with more sensitive receivers and more directional aerials; there is still a great deal of original research which can be done in the field of observation of the Sun and the planets by amateurs and students.

Observations do not need to be tied to particular ranges of frequencies. Observations at a range of bandwidths may be useful, and as a separate factor the spectrum of any modulation present is worth investigating. At the low-frequency end of the modulation spectrum, integration of the signal over quite long time periods; recording peak level and average power over the integration period are all aspects of the signal that are worth noting. At higher modulation frequencies any evidence of periodicity, even if in the form of a periodic variation of a thermal signal, is potentially highly significant as are any observed changes that can be related to observed visual phenomena.

With another step up in sensitivity signals from the major star clusters and objects such as the great nebula in Orion, the center of the Milky Way Galaxy and the Clouds of Magellan can be studied. These are probably beyond the capability of anyone located in a suburban area, a radio-quiet location is the first requirement. A rural valley with few local inhabitants and well away from major power lines should be suitable; an oceanic island or cattle station would be excellent providing it is not in a tropical area where lightning continuously raises the background noise level. Actual observations to be done on stars are the same as are done on the Sun with due notice taken of the higher sensitivity and greater directivity needed for the receiving system.

CHAPTER

9

Radio Communications

9.1 The Range of Services Offered

As explained in Chapter 1 the uses of radio may be divided into three broad classifications. A radio signal may be used for broadcasting, communications, or power transfer. This chapter introduces those radio systems in which one person speaks into a microphone and transmits a signal expressly intended for another particular person to hear and reply to. It may also be used to describe the sending of messages for a small group of people to hear and acknowledge in turn. In modern times "communications" may also include links where a computer originates a message and transmits it to another particular computer and the data transmitted could have the significance of any of the whole range of purposes for which computers use electronic data (transmitting text or a picture or switching a machine on or off, for instance).

Almost all radio services may be operated from a moving object (motor vehicle, ship, aircraft, etc.) and there is an incredible variety of communication systems which may come under the general banner of "mobile." The next chapter is devoted to mobile operation and an aspect of the subject is also dealt with in Chapter 11.

When mobile communications are being considered the link is generally assumed to be two-way but even within that boundary the list includes the following:

- mobile phones
- taxi and delivery dispatch systems
- tradespeople's private channels
- emergency services
- navigation systems for ships and aircraft
- air traffic control
- aircraft flight services
- shipping in-port movement control
- ship and aircraft general communications (including messages for passengers)
- amateur radio services
- mining and industrial personnel location and safety systems
- citizen's band radio

The essential difference between communications and broadcasting is that communication systems have a particular receiver who is known to the sender and usually has the opportunity to return a message; the broadcaster does not know who the listeners are but hopes there are lots of them. The broadcasting services are normally one-way only and include the following functions:

- reception of broadcast transmissions
- directional and distance-measuring beacons
- the Global Positioning System
- reception of time signals

For these classes of service a receiving system using "communications" principles can be used. In almost all cases when the link or broadcast is unidirectional the transmitter is at a fixed site (except for satellite services) and the mobile station has the receiver.

For most of these services the requirements for high-quality reception of signals while mobile are closely related to the same class of service for a fixed transmitter and receiver with some special considerations for the mobile mode of operation. The particular features of mobile communications systems will be dealt with in the next two chapters.

The types of services in which radio communication may be used in a fixed link are as follows:

- radio telephones in remote areas or over difficult terrain
- broadband telephone bearers
- telemetry as, for instance, unattended weather or stream-gauging stations
- amateur radio operations
- citizen's radio services
- emergency telecommunications after natural disasters

Communication systems may use a single channel or a broader band subdivided into a group of channels (multiplexed), and the channel, which in many cases will be just wide enough in the spectrum to pass the intelligence of a human voice (about 3 kHz bandwidth), can in fact be as narrow as less than 100 Hz or as wide as a high-speed datalink with up to 20 MHz bandwidth. The next seven sections of this chapter deal with communications using frequencies higher than 30 MHz. For frequencies below 30 MHz the aerial for mobile operation is a critical factor so those systems are dealt with in the next chapter, particularly in Sections 10.4 to 10.7. There is also related information in Chapters 3 and 6.

9.2 Single-Channel VHF/UHF Systems

This section deals with radio communication using voice channels in the frequency range from 30 MHz to slightly higher than 500 MHz with power output usually between 1 and 50 W and usually but not exclusively using narrowband angle modulation (either FM or PM). The band includes a few services which use amplitude modulation and some digital services which use pulse-coded modulation. Within that frequency range channels are allocated assuming in most cases a 1VF channel (maximum modulating frequency 3 kHz; see Section 4.10 in *How Radio Signals Work* for a definition of the 1VF channel) with maximum deviation in the range 3 to 5 kHz. This requires channel spacing of between 15 and 30 kHz.

The earliest "two-way radios" were simply a transmitter and receiver installed in a vehicle and tuned so that the receiver could be used to listen to a similar transmitter installed in a base or another vehicle and the transmitter could be used to send a signal to a receiver installed at the same location as that other transmitter. Almost as soon as that class of service became established the manufacturers began to develop units in which a compatible transmitter and receiver were installed in a single box and the circuits used some components in common for both transmitting and receiving as shown in Figure 9.1. These pieces of equipment were known as "transceivers" and were from the start made with the intention that they should be used for mobile operation. The present-day 27-MHz citizen's band services are very little different in function (in countries where their use is permitted) from the first commercial two-way radio services. Note that the equipment used now is vastly smaller, neater, and more efficient, but the essentials of the service it provides is the same.

In the frequency range of 30 to about 500 MHz systems can be assembled with all transceivers mounted in motor vehicles or boats or airplanes as in the example of the 27-MHz CB service, but the bulk of commercial communications traffic is carried by base-to-mobile rather than by mobile-to-mobile services. One of the transceivers is placed at an office, workshop, or depot in a fixed location and works through a more efficient (usually elevated on a sizeable tower) aerial. The "base"

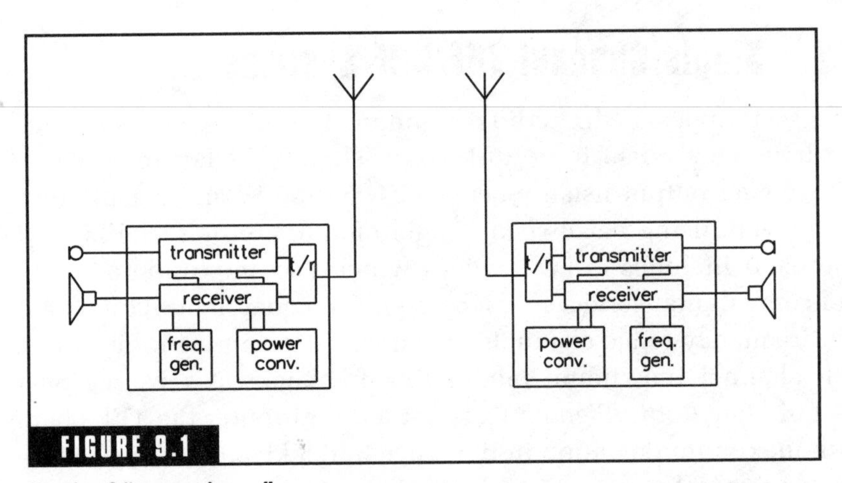

FIGURE 9.1

A pair of "transceivers."

can also be a repeater in a position chosen so as to give particularly good coverage of the area in which the business operates. In most of the larger cities of the world large numbers of small to medium-sized businesses' mobile communication needs are completely met by a service of that type.

Modern systems using these classes of service use tone signaling to provide "quietline" operation. The transceiver is operated continuously (in vehicles power is connected through the ignition switch) in the receiving mode but with the audio stages muted. When a coded tone detector senses receipt of a call intended for that particular station it switches the audio output on and the message is received. The two most common quietline systems are known by the names "CTCSS" and "selcall."

CTCSS: "Continuous Tone Coded Squelch System"

The system uses a low-volume tone in the frequency range below 300 Hz and a narrowband tone detector at the receiver. This tone is added to the modulation of all transmissions for as long as the transmitter is operating; hence the first two words of the title. Tone frequencies can be specified to within 0.1 Hz and there is room in the range below 300 Hz for up to about 35 separate tones to be used; thus services can be coded by the frequency of the tone used into 35 separate groups. When you first install a CTCSS you must consult with all other users of the channel that may be within range of your base to ensure that your tone is different from theirs. Figure 9.2 shows a block schematic diagram of a transceiver with a CTCSS facility added; you should be able to compare it with half of Figure 9.1 to see the changes.

Once the system is properly set up it has no effect on actual operation; the operators may be vaguely aware that there are other users of the channel that they do not hear but the users do not need to do anything to use the CTCSS—it just happens.

The term "out-of-band signaling" may also be used to describe CTCSS, but CTCSS is not the only type of out-of-band signaling in use.

Selcall

To the transceiver operator use of a selcall system may be very similar to dialing up a telephone number, and in fact there are some new developments on the market which can allow a telephone number to be

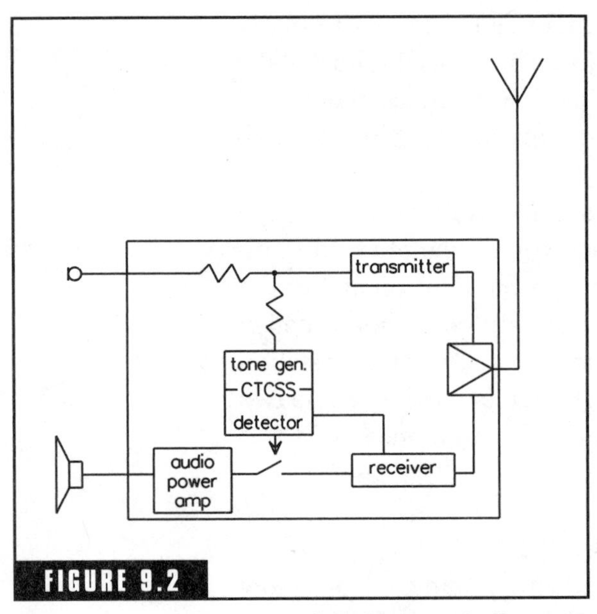

Transceiver with CTCSS added (block schematic diagram).

direct-dialed from a radio transceiver by an extension of the principle of the digital selcall facility.

When you commence a call on a radio system with selcall you may press several buttons to define a code for a particular transceiver or group of transceivers then press a send button. Your transceiver transmits a short burst of signal that contains a data message, which is sometimes described by non-technical people as "a short bit of electronic music." If that message is decoded by a transceiver anywhere that station will send back a signal called a "revertive tone" to advise you that the call has been decoded. At the same time the receiving unit will open the mute on its own receiver and generate an alarm signal to alert the person you are calling. The system can also send information to identify the calling station which is stored and displayed by the receiving station. Figure 9.3 shows how selcall is added to a transceiver; comparison with Figures 9.1 and 9.2 will show the technical similarities between the two systems.

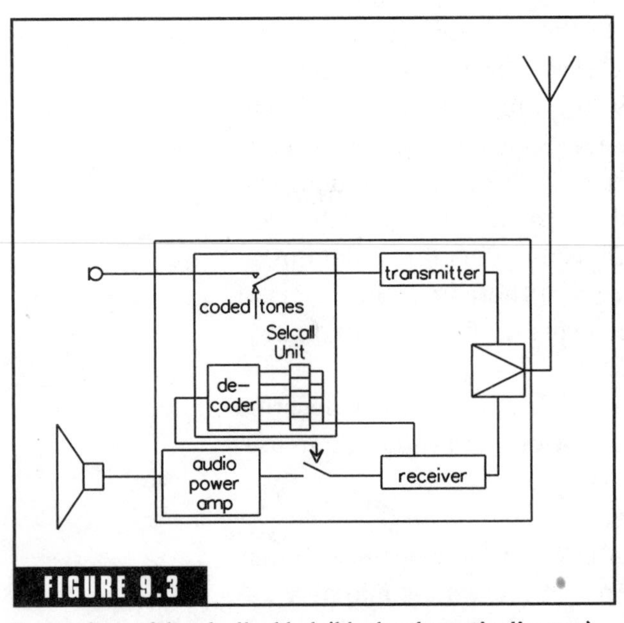

Transceiver with selcall added (block schematic diagram).

Whereas the CTCSS can only cope with a maximum of about 35 stations (or groups) on the channel, a digital Selcall facility can be made much more versatile. A code allowing for four decimal (or BCD) digits may be used to address up to 10,000 stations. The system can also cope with division of these 10,000 into 1,000 groups of 10; 100 groups of 100 or even mixed-sized batches where some groups are thousands, some are hundreds, and some are tens. Selcall systems are available which specify two digits (100 stations) or up to about 9 or 10 digits (large groups of stations). There is no technical limit to how many digits may be used, but the extra time required to transmit unnecessary digits may be noticed.

Along with the extra versatility comes extra cost. The selcall facility in a mobile may be two or three times more expensive than CTCSS; however, that is 10 to 30% of the total cost of the mobile so selcall is a noticeable but not prohibitive addition to the capital investment. Apart from depreciation/maintenance there is no running cost associated with either CTCSS or selcall. There is more information on signaling systems in Chapter 12 ("Data, Codes, and Selcalls").

A radio system using selcall can be made marginally more sensitive than one using CTCSS. The CTCSS tone is transmitted at a low level and must be decoded all the time otherwise the receiver is muted. Selcall tones are transmitted at full modulation and once the channel is opened up the only limitation is the intelligibility of the received voice.

Before leaving the subject of quietline operation it is worth making a note about privacy. The muting of your receiver does not give you any privacy from other people; it only gives them privacy from you. Your privacy comes from the muting of other people's receivers. If you require to transmit any information that is commercially sensitive or otherwise potentially embarrassing you should always assume that your most active competitor is listening on the channel with the receiver mute open.

The VHF band can be used for all of the classes of service listed in the previous section for distances up to 40 km. In some circumstances distance up to about 160 km can be covered in one hop using a VHF signal, but particular attention must be paid to siting of transmitter and receiver aerials and the link becomes much more susceptible to disturbance due to atmospheric effects on signal propagation (see Sections 5.5–5.7 and 6.6 in *How Radio Signals Work*). For most VHF communication services one of the stations is mobile, the next chapter deals specifically with mobile operation. The next section in this chapter gives some more detailed information on one way in which VHF services may be used for communication with both stations fixed.

9.3 A VHF Radio Telephone

Over the years there have been many telephone systems which have offered a remote area or mobile capacity by using a radio link. In the simplest of them a single voice-channel duplex link is combined with sufficient tone signaling equipment to transfer information related to on-/off-hook, ringing, and dialing-out conditions. Even in this simple single-channel form there is considerable equipment required to interface between the physical pair of wires of the telephone line and the basically unidirectional voice frequencies only capabilities of a radio link. The block schematic diagram of Figure 9.4 shows the equipment required for a single-channel radio-linked subscriber's telephone extension.

Figure 9.4 is related specifically to the Australian telephone network. There are a large number of sets of standards for telephone lines around the world so a subscriber's line radio telephone system would need considerable adaption if it were to be made to cope with all possible standards; however, it is only the exchange-end relay set which must be changed. That is the interface between the telephone and radio systems.

DC conditions cannot be signaled over the radio link so all signaling must be transmitted as tones. The total modulation bandwidth up to 4 kHz is used, but only the section from 300 Hz to 3.4 kHz is used for speech. The section from 3.4 to 4 kHz is used for a frequency shift keyed telegraphy channel to signal a ring-tone in one direction and dial impulses in the other. This FSK telegraphy is called "E and M signaling" and may also be described as "out-of-band signaling." Figure 9.5 shows how all the functions are interconnected; this diagram is specific to the Australian telephone system, and there may be differences in other countries. If a particular service uses tone dialing the tones are generated at the keypad of the handset and are transmitted in the voice channel (in Australia using DTMF format), and in that case the FSK signal is only used for the "off-hook" condition. Dial, engaged, and number-unobtainable tones from the exchange are audible signals in the voice channel and no equipment is needed in the relay set to deal with them.

When a relay set of the type illustrated in Figure 9.5 senses a ring tone (a 17-Hz sawtooth wave at up to 90-V peak) it switches on the

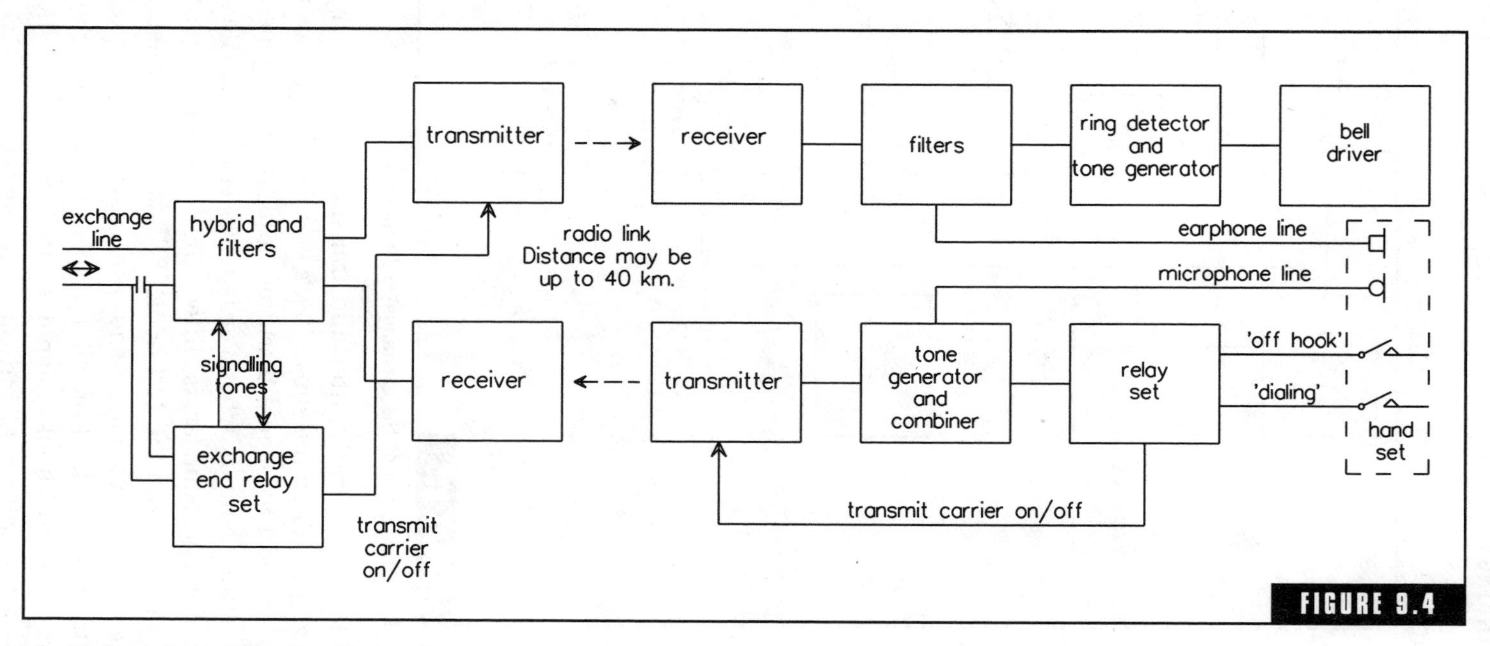

Single-channel subscriber's radio telephone.

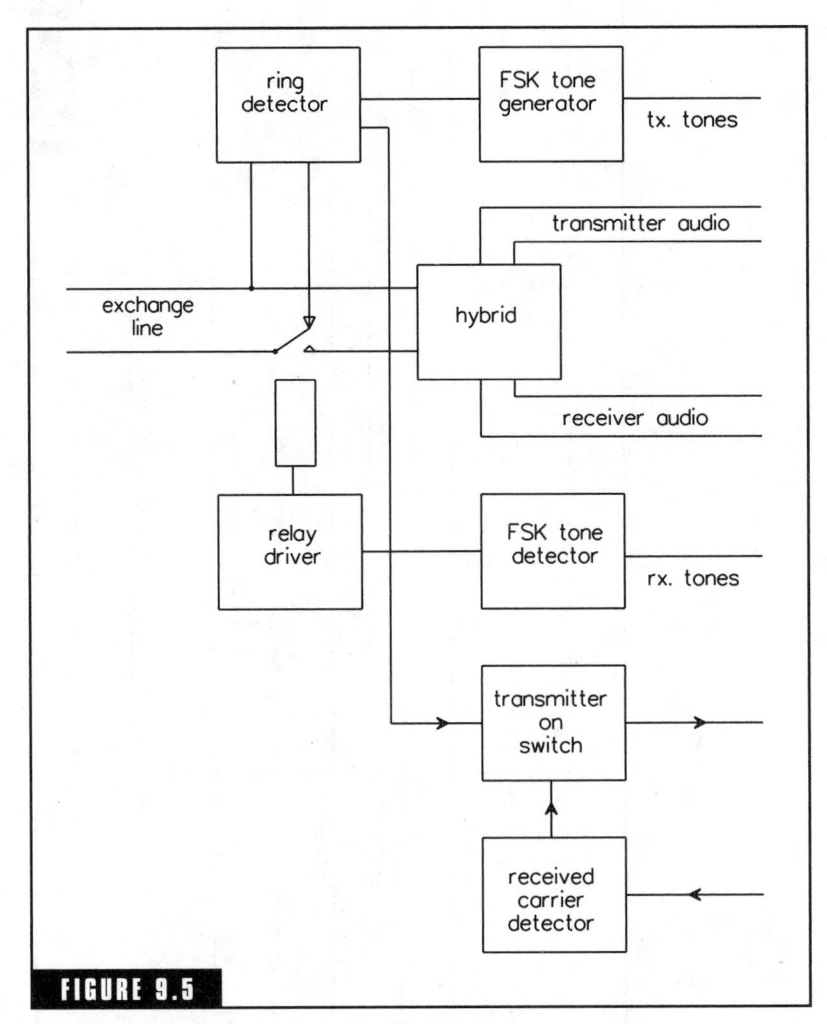

FIGURE 9.5

Basic functions of the exchange-end relay set.

exchange-end transmitter and transmits the ring signal via the FSK telegraph. When the subscriber's end receiver senses the presence of a carrier and FSK signal it triggers a local ringer generator which rings the bell or sounds whatever is used as the audible alarm. When the subscriber's handset is lifted the subscriber's end transmitter is switched on and the "off-hook" FSK signal is transmitted. The carrier detector in the exchange-end receiver latches the exchange-end transmitter on and the FSK signal operates the DC relay contact, which refers the "off-hook" signal to the exchange. When the exchange senses the "off-hook" indication it removes the ringer signal and connects the line for voice communication. When ending of the ringer is sensed by the radio telephone exchange-end relay set it shifts the frequency of its FSK telegraph signal to the busy state, and that has the effect of switching off the ring generator at the subscriber's end. Each of these processes only takes a few milliseconds so the total time for setting up a call is usually less than the time needed by a person to put the handset to his/her ear.

For a call in the other direction the initial lifting of the subscriber's handset switches on the subscriber's end transmitter; sensing of a carrier at the exchange end switches on the exchange-end transmitter; and sensing of the FSK signal switches on the DC relay connection, which seizes the line at the exchange. When the exchange has prepared itself to receive dial information it sends a dial tone, which is an audible signal transmitted via the radio link to the earphone. If impulse dialing is being used the dial mechanism in the subscriber's handset generates a corresponding FSK telegraphy signal which pulses the DC relay in the exchange-end relay set. The exchange senses those pulses as dial impulses exactly the same as those from a wire-connected handset. DTMF tone dialing is transmitted via the voice channel and FSK signaling is not involved.

The purpose of the "hybrid" shown in Figure 9.5 is to separate directions of the audio signals. The receiver output must be prevented from reaching the input of the transmitter. Hybrids use a principle similar to that of the Wheatstone bridge to provide separation between the ports as shown in Figure 9.6. The "net" is a passive impedance which combines resistance with a reactive component; for maximum separation (named "hybrid balance") its impedance must exactly match the line impedance over the whole range of audio frequencies. In ideal test conditions figures for hybrid balance of 30 to 35 dB are easily obtainable but for a real telephone system the line impedance can vary and may not be constant from call to call so for normally installed hybrids in telephone exchanges a figure for hybrid balance of 10 dB may be more usual. (This means that in a two- to four-wire transition the signal leaking across the hybrid from the receive side to the send side of the four-wire line may only be reduced in level by 10 dB.)

Although the purpose of a hybrid is to provide separation between directions of signals, its action is to give isolation between the two ports on the four-wire side; energy from the two-wire side is coupled

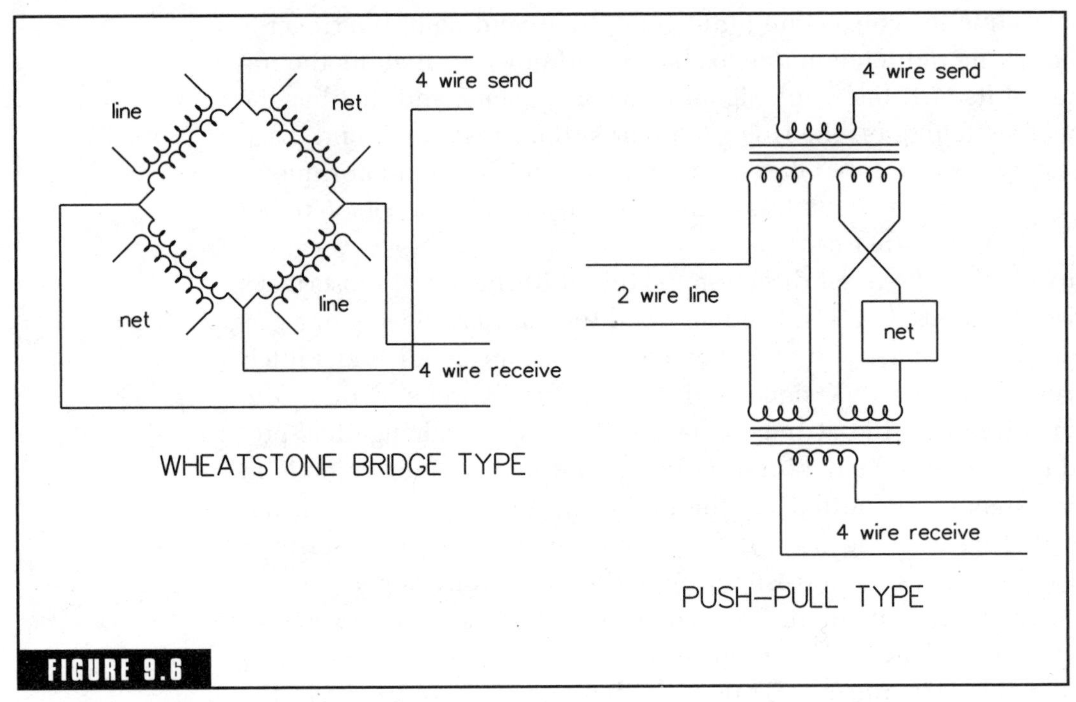

FIGURE 9.6

Schematic diagrams of hybrids.

equally to both of the four-wire ports, but each four-wire port is coupled to the other via two paths which are arranged to be exactly equal in magnitude and out of phase so that coupling between them is canceled out. Separation of directions of signals occurs because the receiver output is not listening to signals incoming from the two-wire line and, providing the transmitter input is impedance-matched to the line, it absorbs all the energy sent to it. In Figure 9.6 the two coils labeled "line" in the "bridge" diagram can be connected together either in series or in parallel but, whatever is done to them the same must be done to the "net" coils. In the "push–pull" version when the "net" impedance exactly matches the "line" impedance the two links carry equal currents which are made out of phase by reversal of a link.

When the hybrid is balanced it introduces a loss of 3 dB to the signal in each direction. This is because its action splits the power of the signal into two equal components, and the power of one component is delivered to the line or load while the power of the other is dissipated in the net.

If the radio system must use hybrids at both ends of the link so that the connection to the subscriber's handset is a two-wire line a feedback loop is possible by the mechanism illustrated in Figure 9.7, and hybrid balance figures of 10 dB at each end may not be enough to prevent oscillation around the loop. Because of that limitation, VHF radio telephones work best if the subscriber's end receiver output is kept isolated from the transmitter. If a two-wire connection of the subscriber's end is an essential part of the design of the system then the installers will need to pay close attention to hybrid balancing and level and gain control settings to avoid feedback.

The earliest mobile telephone systems were all of the type shown in Figure 9.5 and the principle is still used whenever a single-channel link (these days usually between two fixed points) is all that is needed. Services of this type are commonly used to connect between offshore islands and in places where terrain is difficult for laying cable and only one or a small group of homes require the service. Most of these services are subscriber's lines but they can also be connected to a line concentrator or PABX to serve a group of handsets within the limitation that switching must be done to a total of six lines as a group or a very close control must be kept on line impedance and levels.

When system maintenance or proof-of-performance tests are being

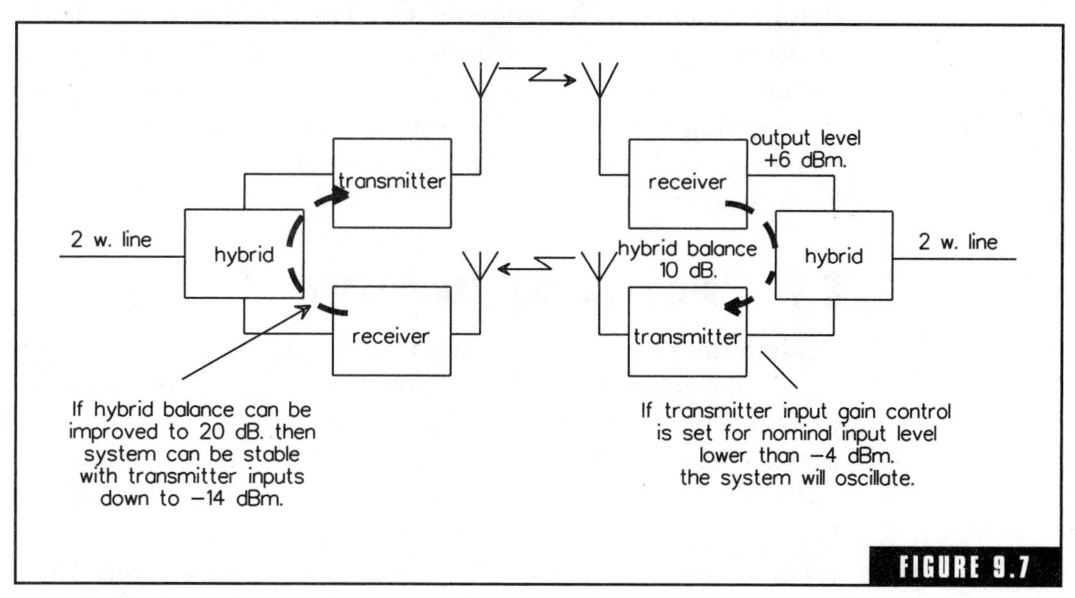

FIGURE 9.7

How a feedback loop is formed.

done the overall sensitivity of the complete system is only marginally affected by the normal sensitivity limitations of a radio system (such as transmitter power output and frequency alignment). The setting up of calling conditions is controlled by carrier and tone detection so the critical factors are tone frequency and level, detector sensitivity, and tuning center frequency. Tone level is measured as a particular level of deviation of an FM signal, and frequency can be measured at the tone generator output point. The corresponding measurements of sensitivity of the detector are more tedious but equally as critical for correct operation.

Radio telephones operating in the VHF band use frequency modulation so, providing the signal is above a certain minimum strength, the level of signaling tones out of the receiver depends very little on signal strength. The level of received tone only needs to be about 6 dB higher than the detector threshold for proper operation so a relatively slight change in tone level or frequency (or the corresponding detector settings) can make a gross change in operation. The dynamic range of the detector may be quite small in decibel terms; if the receiver output level is too high, distortion and intermod components of the speech can cause false operation of the tone detectors. Output power of the transmitters and propagation disturbances have a minor effect on system performance, which shows up in marginal conditions as a reduction in reliability. The equipment designer tries to ensure that in the presence of gradually decreasing signal strength the ability to call will be the first function to fail with the aim that a signal which is initially strong enough to allow establishment of the call will support conversation down to a significantly lower signal strength before contact is lost.

9.4 Path Loss and Link Budgets

The minimum workable signal at the input of an FM receiver for a 3-kHz voice channel with maximum deviation of 5 kHz must be close to 0.3 μV in 50 Ω. Power level of that signal is 0.0018 pW. Compared to the output of a 25-W transmitter, that figure is 14 thousand million million times smaller. The total loss due to all causes over the path from transmitter to receiver must be no greater than that figure. Other ways in which that power ratio can also be expressed are 1.4×10^{16} or in decibels—161 dB maximum total loss.

The total path loss from transmitter station to receiver station can be a few decibels higher than that figure if directional aerials can be used. For instance, if colinear aerials with gain of 6 dB each are used at both stations and feeder loss of 3 dB must be allowed for, the extra 9 dB can be added to the above figure, making the maximum permissible path loss 170 dB. Very high gain arrays (figures from 25 to 40 dB are possible for fixed installations on the UHF band) may allow a total path loss of over 200 dB to be tolerated.

For systems with a fixed base, the expected signal strength at a particular distance from that base can be calculated to within a tolerance of about 8 to 10 dB. The calculation can be fairly involved and the derivation of the equation for it is one of the major subjects of a university course with postgraduate studies in radiophysics. There is an outline explanation of the principles in Sections 3.6, 5.5, 6.6, and 6.7 in *How Radio Signals Work.* Calculations of that type are called "link budgets." That book will not give you enough detail to calculate a link budget but hopefully will give you an understanding of the principles.

Although the calculation of an expected signal strength from scratch is fairly involved, the estimation of the effect of changes to an existing system is a lot easier with less loose ends to tie in. In the case of a VHF/UHF base-and-mobiles system operating over smooth Earth (for VHF radio signals, that means no sharp cliffs or gullies and local rises up to no more than about 5 to 10 m high) the factors that will affect distance over which a particular signal-to-noise ratio at the receiver can be achieved are as follows:

Aerial height: The distance to the horizon is proportional to the square root of the height.

Aerial gain if the other aerial is in the direction of the main beam: The relationship between aerial gain and operating distance may depend on several other factors (see next item and Figure 9.3).

Transmitter power output and feedline losses: For both these factors and aerial gains if the aerials are within radio line of sight of each other a 6-dB increase in effective radiated power will double the distance to a particular signal strength (inverse-square law of optics). For paths which go into the diffraction zone (refer to Figure 5.4 in *How Radio Signals Work*) the loss rate per distance is greater than the inverse-square law so the practical effect of increases in distance due to marginal improvements of 1 or 2 dB in equipment technical performance may be too small to be noticed by a casual user.

Operating frequency: Lower frequencies travel further due to both the greater degree of refraction and also a capability of greater distance beyond the radio horizon due to the effects of ground-wave/diffraction.

Receiver bandwidth: Assuming that the receiver is sensitive enough so that when the signal disappears it is replaced by noise (so audio output stays about the same) the limiting sensitivity is inversely proportional to the square root of the bandwidth.

Any RF noise sources in the local vicinity of the receiver which mask the incoming signal.

Weather conditions if the path length is over about 20 km (see section 9.7).

Note concerning the relationship between noise and bandwidth that most VHF and UHF single-channel services use narrowband frequency modulation. In Chapter 1 there is a statement to the effect that signal-to-noise ratio is improved by reducing audio bandwidth. For frequency modulation, some qualification of that may be needed stressing that it is only the ***audio*** bandwidth (rate of data transfer in a digital link) that obeys that relationship; if the deviation of the FM signal is reduced to allow for a narrower bandwidth of the receiver RF and IF stages the signal-to-noise ratio is actually degraded by the change. Transmission of wider deviation improves signal-to-noise ratio; the receiver gives best results when its RF/IF bandwidth is matched to the transmitted deviation. Wider-than-required receiver bandwidth will still make the signal-to-noise ratio worse despite the relationship stated above, and if the receiver bandwidth is too narrow the recovered signal will be distorted. All this only applies to frequency-modulated signals; if the service uses amplitude modulation or a variation of it, the bandwidth/noise relationship described in Chapter 1 and explained in more detail in Section 12.2 applies to all stages of the receiver.

If the Earth is not "smooth" in the radio sense features of the terrain will affect signal strength at particular locations. In most cases obstructions and objects causing absorption (such as areas of thick vegetation) will make the signal weaker, but there are a few occasions when an obstruction can be placed so that it gives an aiding reflection, making the received signal a few decibels stronger (see Sections A1 and A2 in the Appendix). Path loss can be calculated or estimated by the following formula:

$$\text{total loss (decibels)} = \text{free} + \text{smooth} + \text{obstruction} - \text{gains}.$$

Taking each of these terms in turn:

Free space loss is a constant figure which depends only on the distance apart of the transmitter and receiver aerials and the frequency of operation. If two half-wave dipoles are aligned for maximum coupling and placed 1 wavelength apart the coupling loss between them is 22 dB and the loss figure increases by 6 dB each time the distance apart is doubled. Two wavelengths of separation gives a 28-dB loss, four wavelengths gives 34 dB, eight wavelengths gives 40 dB, and so on. The frequency factor comes into play when the free space path loss is expressed as a loss over a path of a particular number of meters or kilometers. For higher frequencies a set distance is a greater number of wavelengths so the loss figure is greater. There is an explanation of the principles of free space propagation in Sections 3.4–3.6 and 5.1 in *How Radio Signals Work*. The extra loss due to the signal traveling over a smooth earth surface can be calculated with very little more complication than is required for the calculation of free space path loss. This factor applies mainly to that section of the path in the gap between the radio horizon for the transmitting aerial and that for the receiving aerial as illustrated in Figure 9.8. Operating frequency is a factor in this calculation and the figure is also affected by ground conductivity in the gap section.

The "obstruction" factor is less easy to calculate. When the obstruction is a single sharp ridge whose top is within radio line of sight of both transmitter and receiver aerials the path can be treated as two sections of free space path and extra loss due to the ridge can be calculated with fair accuracy. If the ridge is not sharp then the ra-

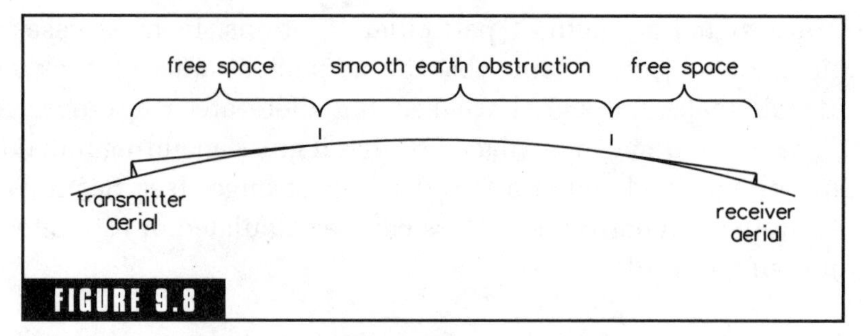

FIGURE 9.8

Where the "smooth earth" factor is relevant.

dius of curvature and the angle through which the signal must be turned become factors in the calculation. When there is more than one obstruction the calculation becomes more complicated and less certain and when the country is generally rough with many small hills in the radio path the tolerance on calculations will become so wide that the best estimate will probably be that which would be made by an experienced practical operator looking over the country from a high point and comparing it with memories of previous similar situations.

In reference to the "gain" factor in the equation, it may seem strange at first to be subtracting it; the quantity being calculated is a path loss so there is a double-negative situation. Gains usually are derived from reflections which produce multipathing and do not usually amount to more than a couple of decibels. For broadband signals, the "gain" due to multipathing may be observed as a ghost on a television picture or smearing of the transitions of a data stream so a high figure for "gain" may still be an indicator of a degraded signal.

There is one case where propagation "gains" of up to about 90 dB have been quoted. The effect of knife-edge refraction is sometimes described as "obstruction gain." Be warned, however, if you see a figure quoted for that factor; the obstruction always causes a net reduction in signal strength. The factor arises when a signal over a very long path would have substantial loss due to the smooth earth factor, and that loss is reduced by the presence of a ridge in radio line of sight of both aerials. The signal is always weaker than that which would be received over that distance if the aerials were in radio line of sight of each other and free space path loss was the only relevant factor.

For radio paths of the type shown in Figure 9.8 weather conditions have a substantial effect on the actual signal received at any given time. Atmospheric conditions can cause the apparent distance to the radio horizon to vary, which will have a pronounced effect on the length of the gap section of the path.

On the HF band, signal strength for skywave signals is mainly determined by the relationship between the operating frequency and the actual MUF that would apply to that path at that time. The MUF depends on the factors listed in Section 5.8 of *How Radio Signals Work.* Calculations of the link budget for an HF communication system are not normally done to specify a figure for expected path loss but to test the path loss due to using a selected frequency and compare the results for several selected frequencies.

Figure 9.9 is a graphical presentation of the effects of all the factors mentioned in relation to a link budget. This diagram is not to scale; the small factors are enlarged to highlight their effect.

The next three sections give some practical examples of expected range for terrestrial signals in the VHF/UHF spectrum and the effect on them of various types of weather conditions. They are intended to be stand-alone practical examples so some of the statements may duplicate information in this and other sections.

9.5 477-MHz UHF CB Maximum Ranges

In Australia there is a group of 40 channels in the frequency band close to 477 MHz which is allocated for casual use for voice communication, commonly from mobile to mobile, by anyone who has equipment approved for that band. The Australian Communications Authority holds a group license and individual station licenses are not required. Transmitter power output is limited to 5 W and frequency modulation is used with maximum deviation of 5 kHz. The following notes on maximum expected ranges can be adapted to most other mobile-to-mobile operations in the VHF and low-frequency end of the UHF bands if appropriate scaling factors are applied.

Some of the early lunar probes used 5-W transmitters on a fre-

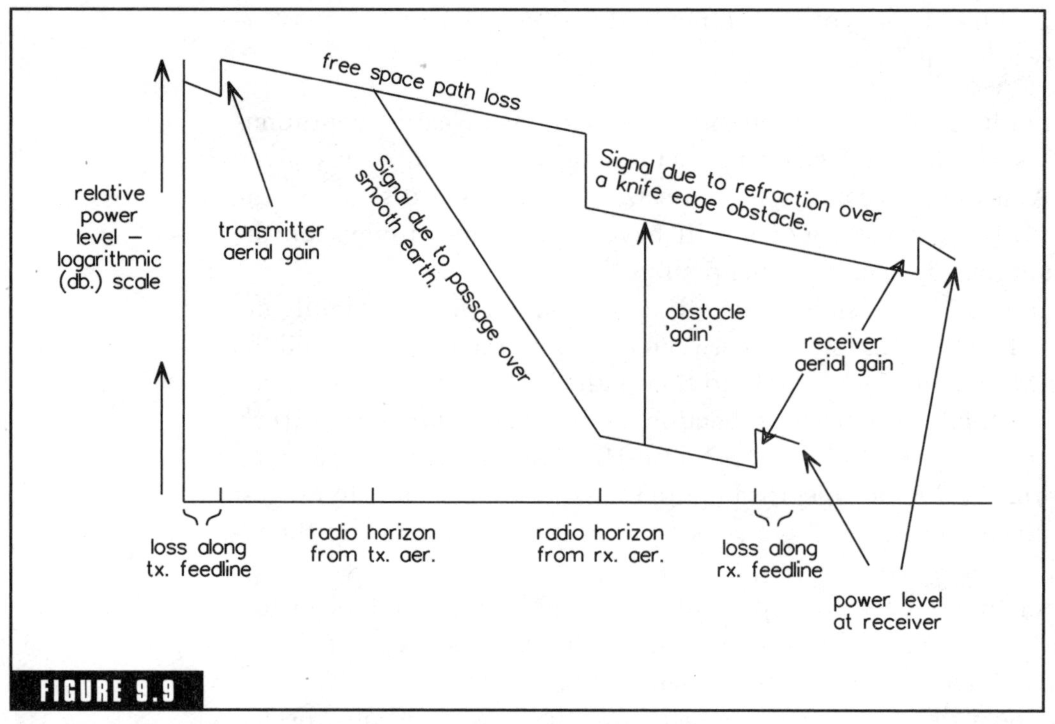

A typical link budget presented in graphic form.

quency close to 470 MHz for communication from the Earth to the Moon. On the other hand, if you have two mobiles in dense scrub with a ridge between them, maximum range could be as little as a couple of hundred meters. The distance over which a 5-W signal could be used would be maximum if there were no obstructions such as the Earth. That distance is called the "free space path" and for a pair of typical UHF CB sets working into quarter-wave whip aerials would correspond to about 2,500 km. In free space the total loss is reduced by 6 dB if the distance is halved and increased by 6 dB each time the distance is doubled. Directional aerials can be used to concentrate the signal, which increases the allowable loss by maybe up to 10 dB for each aerial (for mobiles used in motor vehicles). Replacement of either aerial with a 6-dB gain whip would allow for a maximum free space path of 5,000 km and changing aerials at both ends would allow that distance to be increased to 10,000 km.

If an aerial gain of 36 dB was used, the maximum free-space path would become 160,000 km. For Earth–Moon communication the combined gain of the aerials at both ends must be about 45 to 50 dB. Total aerial gain of 50 dB could be achieved in practice with a large horn or parabolic reflector at the Earth station giving a 36-dB gain and a Yagi or screen-backed array at the Moon giving 14 dB of gain.

The expected range of a 477-MHz transmission over a terrestrial path is usually expressed as "radio line-of-sight," which is a complex concept involving consideration of several factors such as:

- free-space path
- size and shape of the obstructions
- possible reflections
- buildings and vegetation in the path
- weather conditions

The effects of most of the factors listed above can be calculated in terms of extra loss so it is possible to calculate the signal strength expected of a particular transmitter and aerial system over a particular path. On a practical path of 20 to 40 km the free-space loss may be only about 110 dB; the other 40 to 50 dB is accounted for by obstructions and the curvature of the Earth as shown graphically in Figure 9.9.

On UHF CB the price of the engineer's time for link budget calculations will often be close to the cost of buying a couple of transceivers and simply going out and doing a few practical tests to see how far you get. For the local delivery or home-maintenance service type of use in a suburban or light industrial area with reasonably flat terrain you could probably rely on UHF CB on a direct channel to give about 10 to 15 km of coverage for reliable signals from a base with an aerial system at 10 m height and about double that if the aerial height is increased to 40 m. In open country, if the base was on top of a high mountain and the mobile on a similar peak within sight of the base, they could get quite workable signals even if they were a couple of hundred kilometers apart. The line of sight is essential though and the signal would soon be lost if the mobile moved behind even the slightest obstruction.

Commercial services using 25- instead of 5-W transmitters can have total path loss 7 dB greater than for a UHF CB service. That will result in noticeably greater freedom from noise but in some directions only a marginal increase in maximum range. When the UHF CB receiver is compared to that of an equivalent commercial transceiver the CB receiver has a lower specification for adjacent channel rejection and skirt selectivity; both these result in higher noise level and therefore reduced sensitivity when compared to the commercial receiver. The combination of higher transmitter power and greater receiver sensitivity means that the working range of a commercial system with direct transmission from a 10-m high aerial can in many cases be in the range 15 to 30 km in moderately rolling suburban areas.

9.6 Scaling for Operation at Other Frequencies

The notes in the previous section are specified for the part of the band close to 477 MHz but can be adapted to other VHF and UHF channels in the range from 30 to 500 MHz by taking into account the following considerations:

Frequencies above 500 MHz will be of restricted usefulness for mobile systems due to limitations of capture area of an omnidirectional aerial, but they can offer very low-noise, wideband communications over a very short range (mobile phone cells) or strictly line-of-sight (satellite) paths.

For lower frequencies, free-space path loss for a given distance is lower by a few decibels because aerial capture area for a quarter-wave whip is bigger (6 dB for each time the frequency is halved), but the most noticeable difference due to frequency is the "over-the-horizon" performance. In exactly the same way as red light is bent through a prism more than blue light lower-frequency signals are refracted by a greater angle for a given set of physical and meteorological conditions. That has two effects; the radio horizon itself is further away and signals a certain distance into the diffraction zone are less attenuated by the obstruction.

Noise is much more liable to be a problem for lower frequencies. Electrical noise due to switching of power circuits is spread over the whole spectrum but with energy concentrated at lower frequencies. Noise from data switching and computer and television deflection circuits occurs at harmonics of the switching or deflection frequencies also with energy concentrated in the lower-order harmonics.

At the low-frequency (30 MHz) end of the range there will be a few seasons around the peak of each sunspot cycle when local services will suffer noise and interference from a very long way away due to skywave reflections. The F layer MUF can go as high as 35 MHz and in the band from there to about 70 MHz there is a possibility of interference due to sporadic clouds of strong ionization in the E layer mainly in the mid- to late afternoon of summer days.

In low-noise locations the conditions for a particular signal strength at 50 km when the frequency is 500 MHz may give the same strength at 150 km on 30 MHz. However, if the receiver must be operated in the CBD of a large city or at any other place close to where large quantities of electrical power are being consumed, the best signal-to-noise ratio will be achieved at the highest available frequencies. In areas of moderately intensive land use the selection of frequency for maximum distance to a particular signal-to-noise ratio will represent a tradeoff between these two factors. Chapter 5, titled "TVI," gives information on noise and interference for VHF and UHF services which can be applied with very little adaption to single-channel voice- and data-communication services.

9.7 Effects of Weather on UHF and VHF Signals

What follows describes the effect on all terrestrial signals: CB, commercial, governmental, aeronautical, and so on. Signals from satellites will be affected by the same mechanism but the effect will show in ways that are different.

The atmosphere will noticeably affect the reliability of UHF signals whenever the range of transmission is more than about 20 km, and effects are quite marked for ranges over 50 km. The natural tendency of the atmosphere to have reduced density, lower temperature, and lower humidity as height increases causes a refraction effect on all VHF and UHF signals tending to bow them back toward the surface of the Earth. This means that for radio signals the horizon is a bit further away than the visual horizon (see Section 5.5 in *How Radio Signals*

Work). At times when the atmosphere is well mixed these normal tendencies are reduced and the radio horizon moves closer. This is typical of conditions that occur on a windy night with some showery rain.

At the other end of the scale there can be times when the atmosphere is so strongly layered that refraction may cause the signals to bend, as shown in Figure 9.10, with sharper radius than the curve of the Earth so that the radio horizon is theoretically an infinite distance away. If we could see by radio waves as we do by light, the Earth at these times would appear as an immense bowl; our observing point would be at the bottom of the bowl and the rim of the bowl would be at the edge of the weather cell which may be up to about 500 km away. The weather conditions that most commonly cause this effect occur in the last few hours before an approaching thunderstorm, and extensions of the radio horizon are also likely whenever cumulus clouds are forming.

There are a host of other meteorological factors which affect the maximum range of radio signals ranging from the not uncommon to the very rare; a lot of them are the result of interaction between weather and topography so are specific to particular locations. Local knowledge can be important.

In practice there is a tradeoff between distance and reliability. If you require a totally reliable service you will need to make coverage

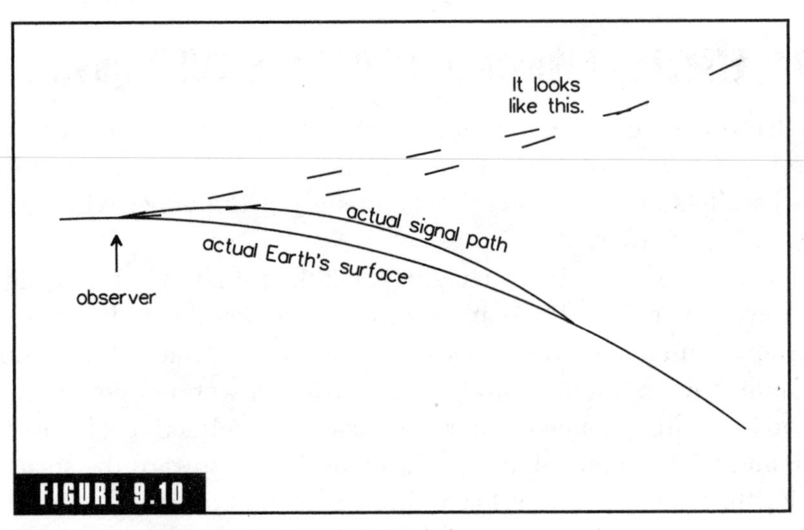

FIGURE 9.10

Signal path when radio horizon is extended.

area estimates based on tests done on wet and windy nights; on the other hand, if you want longer range you may at these times be able to say, "Well I know it won't work now but if I wait 'til morning I should get through." Obviously if you are operating any sort of rescue or emergency service you will find that just the conditions which produce minimum radio range are those most likely to produce clients for your service! You will just have to live with that.

9.8 Broadband Microwave Links

In many countries of the world a considerable portion of the trunk telephone system is carried as many parallel channels (usually several hundred) multiplexed onto a radio bearer. Modulation bandwidth of the multiplexed baseband is usually in the range from 5 to 20 MHz. In order to accommodate that bandwidth the carrier frequency must be in the high-frequency end of the UHF range or in the SHF band.

In a typical broadband microwave link the carrier frequency in one direction may be somewhere in the range of 2,000 to 10,000 MHz and the carrier in the other direction several hundred up to 1,000 MHz different from it. Full duplex operation is normal. Transmitter power output up to a maximum of 5 W is possible, and for shorter-range links power down to a few milliwatts may be all that is needed. Aerial gain is usually over 40 dB so ERP is in the kilowatts range. The aerials for these systems are mounted on high towers with radio-line-of-sight contact with the aerial at the other end of the link. With careful design for high reliability a link of that type can be used in a repeating mode and long chains of microwave links have been built to carry trunk telephone traffic over thousands of kilometers.

Long-distance broadband microwave bearers using chains of repeaters were originally developed in the years immediately following the second World War. The multiplexed modulating signal used a complex frequency-division multiplexing scheme to carry several hundred analog voice channels in a format similar to the following:

Twelve individual VF channels were combined into a group. Channel spacing within the group was 4 kHz with transmission by four-wire with E and M signaling in a similar fashion to that used in the single-channel system described in Section 9.3. Modulation of these channels was by a SSB process with a pilot tone transmitted to lock the frequency of the carrier reinsertion oscillator. Quite close control of signal levels was needed to ensure that the sideband power used was equally divided between the channels without intermodulation due to overloading of any stage.

Five of these 12-channel groups were combined into a "supergroup" of 60 channels also by SSB with a pilot tone with close control of signal levels.

Finally, a number of these supergroups were combined to form the broadband baseband signal which was modulated onto the microwave carrier by frequency modulation. The capacity of the system depended on the number of supergroups used in the final baseband; systems with capacity in the range from 300 to 1,800 channels have been used in Australia. One scheme that was common for many years was a capacity for 16 supergroups or 960 channels which used a bearer of the same bandwidth as a 625-line PAL television signal. In systems where both telephone and television relay traffic must be carried, a single standby bearer could be used for fault protection on both if both used the same bandwidth of modulation.

Later digital transmission systems for voice channels were developed and broadband multichannel bearers were assembled using time-division multiplexing. One advantage of TDM was that it reduced the damaging effects of intermodulation and removed the critical dependence on fine control of signal levels that was necessary with FDM analog signals.

The next development in the story of microwave links was the introduction of digital minilinks. These were designed to transfer a broadband TDM signal over a few kilometers using a low-power transmitter with the RF section of the transmitter and receiver mounted either at the focus of the dish aerial or bolted to the back of the dish with a short section of waveguide feeder and a small horn aerial to illuminate the dish. The cables that run down the tower to give connection to ground-based users of the system carry power and baseband data. In the earlier systems the transmitter and receiver electronics were mounted at ground level with a long very low loss waveguide to carry the RF energy to and from the aerial at the top of the tower.

The historical wheel has turned almost full-circle with regard to microwave links. The original broadband telephony bearers were very expensive, very carefully engineered to give maximum reliability, and were only used across rural and remote areas where there was not much traffic to intermediate stations along the way. An alternative broadband bearer system using coaxial cable was available and more suited to short runs in city and suburban areas where the long-range trunk network could share cable ducts with subscriber's lines and shorter-exchange interconnections using very large multipair copper cables.

At the end of the 20th century the intercity links are being superseded by cables using optical fiber buried deep underground, below the working depth of the fiercest of agricultural machinery, and the major use of microwaves is in short links between city buildings where the major advantage of the link is that it does not require commitment of space in possibly overcrowded cable ducts and is not at risk of being ploughed up by earthmoving machinery.

The price of equipment for microwave minilinks has become so low that a great investment in careful engineering planning for maximum reliability has become uneconomic in many cases; links are being installed on a "stick it up, line it up, and see if it works" basis with the preferred cure for cases of marginal performance being to convert a long link into two shorter links with a repeater. Repeaters do not need to demodulate the multiplexed signal; the IF output of the receiver can be plugged straight into a compatible input to the transmitter and sent on its way.

For fixed links, marginal signal conditions are seen as a reduction in reliability of the transmission, which for digital signals is seen as a change in the bit error rate. The minilinks use digital signals with frequency modulation. The capture effect of the FM process ensures that if a signal is present at all it will be strong and clear enough to transmit data; however, in order to achieve a bit error rate of 1 in 10^9, which is a figure that the data industry likes to work to, the important factor related to signal propagation is the presence or not of occasional deep fades. For a fixed link, there is a variability of signal strength which for a large number of small samples at random times has the statistical properties of a skewed Gaussian distribution. When the link is short the signal is strong and clear and the variability component of the signal strength is small.

For longer links all the factors conspire to reduce reliability. Signal becomes weaker, the variability component rises, and the chance of a deep fade is increased. A fading margin can be designed for; the actual figure needed will depend on the length of the link and the raw bit error rate that must be achieved. When the link is close to its maximum workable length and BER must be 1 in 10^9, the fading margin may need to be as high as 40 dB. A good error correction algorithm can make a lower-BER figure acceptable, which could mean the fading margin could be perhaps 10 to 15 dB lower than that figure. Increasing the energy allocated to each bit will also improve the overall reliability of transmission; this can be done either by increasing the output power of the transmitter or by sending the data at a slower rate. There is more on data rates, error correction, and so on in Chapter 12 ("Data, Codes, and Selcalls").

9.9 Telephone Connection to HF Services

The system described in Section 9.3 only applies to services using frequency modulation on the VHF or UHF bands. The design of a service to do the same function using single-sideband modulation on the HF band is considerably different in technical function. An HF radio telephone link may have approximately the same functional block diagram as the VHF-linked service illustrated in Figure 9.4, but the functions of the blocks are considerably different from the description given in Section 9.3.

HF radio-telephone services in Australia are usually configured as a base that provides connection to the telephone service to a group of outstations or mobiles. In some other countries a pair of stations operating as peers may be used to temporarily extend a telephone service; for example, in the United States amateur operators are authorized to operate "phone patch" services on a noncommercial basis even if a commercial trunk circuit is available over the same path. HF SSB links cannot be operated duplex so transmit/receive switching is required. If the service operates on a single-frequency (t/r) switching is required at both ends of the link. The equipment required for a service in which a human operator is available to attend to t/r switching can be significantly less complex than one in which the switching must be achieved by operation of a VOX circuit. With human operators the radio link can be achieved by little more than extending the audio circuits of a standard telephone handset via relays to the speaker output and microphone input of an HF transceiver. A line interface unit is needed to guard against the possibility of stray voltages on the telephone line and in the initial setting up, attention must be paid to audio level and impedance but no special hybrid balance adjustment is needed and ringing and dialing can be extended by voice calling.

In Australia the major providers of HF base-station services are Telstra and the Royal Flying Doctor Service (RFDS). Both organizations provide services to fixed outpost stations in remote areas and to land-mobile stations. Telstra also provides a service for marine users. The RFDS version is designed to work with the most basic of transceivers in the type of service described in Section 10.5 with no requirement for selcall or any other special facilities at the outstation. Operation is on a single frequency, and a human operator is available at the base so the interface equipment used has a lot of similarities in function to the phone-patch equipment used by amateurs. Telstra provides service at three functional levels; there are some frequencies which are used for a voice-calling-to-manual-operator service substantially identical with that provided by the RFDS, there are other channels which can accept calls via selcall to a manual operator and there are other channels which can offer a fully automatic direct-dialing service to processor-controlled transceivers equipped with software to match the ADD program.

The base stations of the Selcall and ADD channels operate in the two-frequency simplex mode with receivers at remote locations several kilometers from the transmitter. The receivers run continuously and when no signal is being received the AGC circuit adjusts the receiver gain to almost full volume output on channel noise. Hybrids are used to combine the tx input and rx output, and balance of that hybrid must be kept accurate to avoid the speech input to the transmitter being swamped by noise from the receiver. In the ADD services adaptive hybrids must be used which are quite complex electronic circuits capable of increasing the hybrid balance figure to a minimum of 25 dB. There is also an advantage if the connection from the radio base to the local telephone exchange can be made four-wire or six-wire so that the nearest hybrid is moved further away from the radio and nearer to the wired telephone subscriber.

Whatever the configuration of the base and whatever level of service it provides the transmit and receive functions of the outstation transceiver are as described in Section 10.5. The equipment operates simplex, transmit/receive switching is required, and the outstation receiver is muted as soon as the transmitter is operated. This implies that the wired telephone subscriber cannot interrupt the radio subscriber. The telephone subscriber may not be aware of that so it is the duty of the radio subscriber to advise the other of that circumstance and keep his/her talking to short overs.

9.10 The International Amateur Radio Bands

In most countries of the world there are narrow frequency allocations just above the multiples and submultiples of 7 MHz which are allocated for use by amateur transmitter operators. In all cases the lower edge of the band is exactly related to 7 MHz but the width of each band is liable to be different in different countries. Actual lower-band limit frequencies are 1.75, 3.5, 7, 14, 21, and 28 MHz. Amateurs know these bands colloquially by their wavelength respectively as 160, 80, 40, 20, 15, and 10 m. These allocations cover the HF band that supports skywave propagation and give access to worldwide informal communication between friends. There are also amateur allocations in the VHF, UHF, and SHF parts of the spectrum, but the long distance potential of those is more limited so the actual allocations are more diverse within different countries.

Amateur transmitter operators must pass a written examination which tests their knowledge of radio and electronic principles and the legal requirements of operating a transmitter. Once that licensing examination is passed and a transmitter station license issued the operator is authorized to assemble the equipment for a transmitting station, put it on-air, and transmit signals. Equipment may be home-designed and -constructed providing that the operator takes adequate steps to ensure that output power is less than the relevant authorized limit and that all signal emissions fall within the frequency range of an amateur band.

There are restrictions on the type of messages that may be passed using amateur transmitting equipment; the message must be noncommercial to the point that if a telephone service is available at that location and the message content is important enough that it would normally be transmitted by telephone the amateur service should not be used to divert traffic from the telephone. There is no privacy on the amateur bands so confidential information should not be transmitted and in most countries music or other programs designed specifically for entertainment of a broadcast audience are not permitted. Information that has a secrecy classification or which denigrates the reputation of the country where the transmitter is located is banned.

Amateur or noncommercial reception of signals is permitted with much less restriction than applies to amateur operation of a transmitter. In most countries no special qualifications or license are needed to set up a receiving station to listen to signals on the amateur bands.

Amateur operation (of a transmitting and receiving station) has grown over the years to a specialist hobby with support groups or clubs in almost every country where amateurs reside. The groups are in communication with each other and many are devoted to promotion of amateur radio worldwide. The two organizations most active in the international scene are The American Radio Relay League and The Radio Society of Great Britain. The Australian equivalent organization is The Wireless Institute of Australia. All these and several related organizations host sites on the WorldWide Web. These organizations are heavily involved in education and training and are particularly good at the entry level of the subject at which school-age children are deciding for themselves whether they are interested in pursuing knowledge or whether it is all too hard to get involved in. One activity of this type that has grown into an annual event is the Jamboree Of The Air, where amateur operators worldwide make their stations available to members of the Boy Scout/Girl Guide movement for informal international contact.

The amateur service is also well-adapted to emergency response to major natural disasters like earthquakes and floods. Because there are a large number of transmitting stations scattered over most inhabited areas with the capability of being switched on at short notice and passing messages to a point well outside the danger area without needing an infrastructure such as roads, bridges, and telephone lines it is common for the first reports of a major disaster to be linked via an amateur radio operator. Some operators make special provision for disasters by setting up self-contained emergency power supplies and demountable aerials so that their stations can be very rapidly rerigged as soon as the worst of the event has passed.

The amateur radio licensing examination is kept as simple as possible to allow as many people as possible to have a chance to enjoy the hobby, but the learning required to pass that test is only the start. An active amateur operator will keep learning and what he or she learns will be of practical value. Some of the most technically advanced people in the radio discipline are active on the amateur bands and are willing to share their knowledge free of charge with other operators. If you take up the hobby as a path of learning there is almost no limit to the depth of knowledge you can gain from it.

Because amateurs are only licensed for operation of relatively low-power transmitters and many of them are interested in signals from as far distant as possible they are commonly dealing with weak signals and trimming and adjusting for maximum receiver sensitivity. That expertise can be transferred to other classes of service so the amateur fraternity represents a considerable fund of practical experience in maximizing receiver performance.

CHAPTER 10

Going Mobile

10.1 The Mobile Environment

When radio equipment must be installed in a vehicle, ship, or aircraft, certain constraints are intrinsic to the installation:

- the metalwork of the vehicle forms part of the aerial system
- size of the vehicle is limited and in many cases it may be much smaller that a quarter-wavelength of the signal frequency
- the aerial must be omnidirectional in the horizontal plane or if directional must be capable of tracking very rapid changes of direction
- aerial height will be limited
- power supply must come from the vehicle's own electrical system
- the equipment must be built to withstand high vibration, changeable temperature and humidity, and variations of power supply voltage

The range of frequencies/wavelengths that is usable for mobile operations is more limited than the general radio spectrum. At the LF end, frequencies lower than 0.4 MHz correspond to wavelengths longer than 750 m and even the largest of ships are not big enough to carry the aerial structure required for efficient operation of a transmitter. For the reason detailed in Section 8.11 in *How Radio Signals Work* signals which are transmitted in the frequency range below 0.5 MHz by fixed stations with the necessarily large aerials can be efficiently received by mobile stations using electrically short aerials. At the high-frequency end the need for possibly omnidirectional response puts limits on the capture area that can be built into a UHF/SHF mobile aerial so frequencies above about 2 GHz are not commonly used. In between those two extremes there is a boundary at about the frequency for which the vehicle is approximately a half-wavelength. Aerial and earth designs are radically different for frequencies below that boundary than for frequencies above it.

This chapter will concentrate on those factors which are special to the mobile or portable aspect of the signals so it is intended to be read in conjunction with another chapter that deals with similar types of signals in the case of fixed stations. For example:

For operation of car radios on the MF band refer to Chapter 2; Sections 3.2, 3.4, 3.5, and 3.8 of Chapter 3; and Sections 10.2, 10.4, and 10.12 in this chapter.

For operation of ship stations on the long-wave bands for which the only effective propagation mechanism is groundwave refer to Sections 2.1, 2.2, and 2.3; all the sections of Chapter 3 which deal with daytime operations; Section 10.4 of this chapter; and those sections of Chapter 12 which deal with very narrow band systems (12.2, 12.3, 12.4, and some of 12.7).

For operation of mobile phones and trunking systems refer to Chapter 11 in addition to the VHF/UHF sections of this chapter (Sections 10.3 and 10.8–10.10).

The next six sections give some examples of the types of services which may be included under the umbrella of "mobile radio-communications equipment."

10.2 Land and Maritime Mobile below 30 MHz

The significance of dividing the spectrum at 30 MHz is that above that frequency even the smallest vehicles are big enough to be comparable in size to the quarter-wavelength required to form a ground-plane structure at the base of the aerial. This section is really about dealing with installations in which the vehicle is electrically short compared to what is required to form an efficient earth reference. Figure 10.1 gives some size comparisons of typical vehicles and vessels. If the mobile stations are big enough (i.e., the size of ships) so that efficient low-frequency aerials can be mounted on them then operating frequencies in the range below 2 MHz can be used to give groundwave communication over long distances. See Section 5.4 in *How Radio Signals Work* for information on groundwave operation.

HF transceivers installed in motor vehicles, small boats, or light aircraft operating in the frequency range 4 to 30 MHz can offer up to worldwide range if the operating frequency is correct for skywave propagation at that distance. Section 5.8 in *How Radio Signals Work* has information on skywave signals. There is a frequency band between 2 and 4 MHz in which results of land-mobile operations are liable to be disappointing except at night in the years near the sunspot minimum; motor vehicles are not really big enough to provide a base for an efficient aerial system for frequencies in the range below 4 MHz.

For all operations in this part of the spectrum the limitation on sensitivity is signal-to-noise ratio that can be achieved so the absolute signal strength may be less important than the noise and interference on the channel. All of the electronic measures that can be applied to improving received signal-to-noise ratio have already been covered in Section 4.2 ("Features of a Good Short-Wave Receiver"). HF mobile communication systems use single sideband as the method of modulation almost exclusively so the notes on ISB reception and diversity are irrelevant.

There are many sources of noise in a moving vehicle and for most of them noise energy is concentrated at lower frequencies; the details will be outlined in a later section of this chapter. One question that the operator should ask is "Is it really necessary for the vehicle to

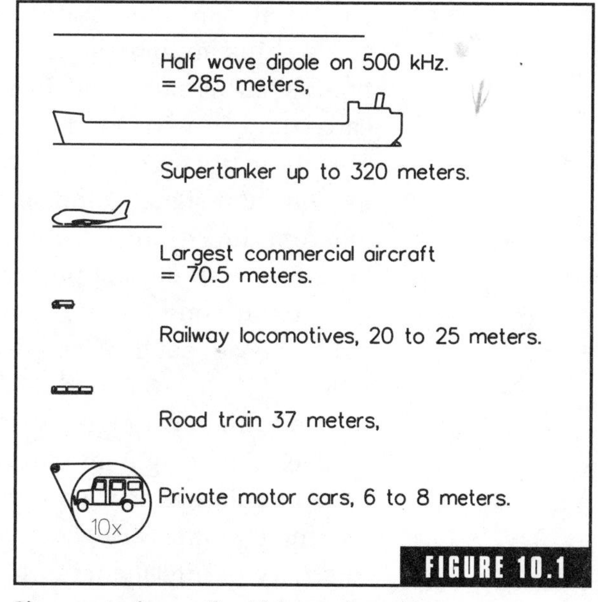

FIGURE 10.1

Size comparisons of vehicles and vessels.

be moving while communication is in progress?" Obviously ships and aircraft do not have the option of stopping for a chat but if the vehicle can be stopped and the engine switched off while talk is going on most of the sources of noise can be sidestepped and there are some measures available to improve signal strength above that which is possible for a mobile installation. See Section 10.6 for one suggestion.

Aerial directional response can improve signal-to-noise ratio but, apart from the factors noted in Sections 10.6 and 10.7 relating to quarter-wavelengths of wire, practical opportunities for it at the mobile station are severely limited for operating frequencies below 30 MHz. For road motor vehicles of passenger-car size, the physical size of the vehicle is so small compared to the wavelength of the signal that (apart from the couple of exceptions mentioned in Section 10.7) it will always behave as a point source so directional effects are not generally possible. A railway locomotive may be big enough for a certain amount of directivity to be built into an aerial for the HF band but that would only be useful if the track was sited in a fairly straight line from a main station to a terminus such as would apply to a private railway between a mine and a port; otherwise directivity could actually reduce the received signal over certain sections of the track. Some of the larger sailing yachts may be tall enough to carry a directional HF aerial which is omnidirectional in the horizontal plane. Ocean-going ships are big enough to carry directional HF aerials of the Yagi type; such aerials need to be particularly waterproof and mechanically rigid to withstand the continual assault of saltwater and rough seas. In many cases the factor about the aerial that may have most effect on final signal-to-noise ratio is the selection of a mounting point where induced noise from the vehicle itself is minimized; there will be more detail on that point in the last section of this chapter.

Ground conductivity will have some effect on strength of HF signals. The subject has been dealt with in some detail in Section 3.3 and Figures 3.3, 3.5, 3.6, and 3.7. The particular points of relevance are in relation to short-range, low-frequency (i.e., groundwave) signals and the electrical height and loss characteristics of the soil close to the aerial. As explained in Chapter 3 earthing is good when soil is either very conductive or a very good insulator and the earth is lossy when conductivity is intermediate. If the ground is highly insulating, such as clean, very dry sand or hard frozen snow, the electrical ground level will be several meters below the physical surface and the aerial height may be good enough (check Figure 10.4) even if the aerial wire is laid straight on the ground surface. On the other hand, for lossy soils such as clay with ironstone nodules or salty sand which is slightly damp, the loss will be minimized with the aerial rigged as high as possible (ideally vertical), and the tests shown in Figure 10.5 will give no stronger signals than those obtained with the aerial vertical.

10.3 Mobiles at the Shortest Wavelengths

At 500 MHz a half-wave dipole is 285-mm long and capture area is 7310 mm^2 or 0.0073 m^2. To produce an adequate signal at the input of the receiver with an aerial that small the field strength in terms of power density per square meter must be quite high. As the frequency is raised above 500 MHz that factor becomes more restrictive at a rate equivalent to the square of the frequency. Capture area can be increased by combining the effect of several dipoles into a directional array or by using a parabolic dish or horn antenna as the aerial element. There is a practical limit, however, to how much directivity can be used with an aerial fastened to a mobile vehicle; the beam of maximum sensitivity swings like a searchlight beam as the vehicle changes direction, negotiates hills and valleys or rough spots on the road, or is buffeted by waves if at sea. For directional aerials if the beam does not point in the direction the signal is coming from the sensitivity of the system is actually less than would be achieved with a nondirectional aerial. Figure 10.2 shows a specific example of one effect of that principle.

There are bands in the spectrum at 900 and 1500 MHz which are used for mobile services; these are broadband high-capacity services used as demand-assigned multiple-access systems to provide services such as cellphone and trunking radio to a very large number of people. They are arranged with multiple bases so that each mobile station works through a base that is close at hand as described in more detail in Chapter 11.

Satellite mobile communication systems can use frequency allocations in the 1500-MHz range and directional aerials are used. Section 8.8 gives information on that application in relation to single-channel mobile telephones. High-gain highly directional aerials can be used on

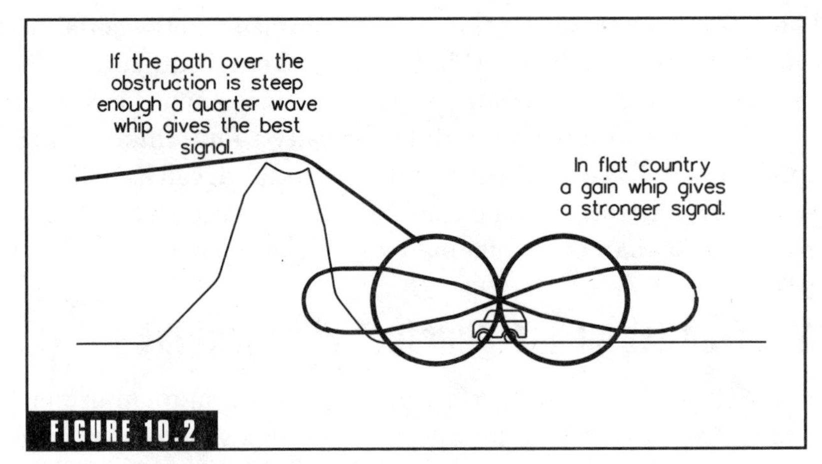

FIGURE 10.2

Directional aerials on mobiles.

mobile stations if the aerial can be made tracking. A parabolic dish or horn assembly, for instance, can be mounted on a gimbal system and powered by stepping motors to follow the direction of the source of the signal. Systems of that type are really only practical for marine installations using satellite links. The gimbal and stepping motor mechanism is limited in its rate of movement and for land-based uses the multiplicity of reflected signals causes the apparent direction of the source to flutter rapidly as the vehicle moves. Portable stations in which the equipment is taken to a location but is stationary while actually being used are feasible with high-gain tracking aerials. Stations of that type can use the highest frequencies in the radio spectrum including all the microwave range to provide broadband or multichannel telecommunications links via satellite to isolated communities such as islands and mining camps.

10.4 Aeronautical and Maritime on the Long-Wave Bands

The wavelength for a frequency of 530 kHz is 565 m. In all the frequency range below the MF broadcasting band the wavelength is longer than that and most land-based vehicles are too small for aerials to work efficiently. The only "vehicles" which are large enough to support an efficient long-wave aerial are ocean-going ships. The relative sizes of aerials and vehicles are illustrated in Figure 10.1. In the earliest days of radio Morse code transmissions from ships were the life support system that brought radio into the focus of worldwide attention. These early uses of radio were all on the long-wave band (30 to 300 kHz) for best use of the long-range aspect of groundwave propagation.

When continuous-wave transmitters began to replace sparks, an international distress and calling frequency was established on 500 kHz for CW transmissions using Morse or similar telegraphic code. The spectrum also included a group of channels either side of 500 kHz for passing of messages; every station listens for calls on 500 kHz, but once contact is established both stations switch to one of the working channels to exchange information. The CW band around 500 kHz has carried a great deal of international shipping traffic over many years. However, as microphones have replaced morse keys and in recent years, even more advanced techniques for automatic reporting of distress situations have become available, frequency allocations for spark transmissions on the long-wave band have been withdrawn and the importance of the CW allocation around 500 kHz has diminished.

Almost all other use of frequencies below 500 kHz is built around one-way flow of information from fixed transmitters with very large aerials to receivers which may be either fixed or mobile. The following are the major forms of use of these frequencies:

Nondirectional beacons primarily intended for use with aircraft ADF receivers. There are at least two or three beacons associated with each international and major commercial airport worldwide and each must be the only signal present on that channel over a radius of several hundred kilometers. That use takes up the majority of the spectrum between approximately 150 and 480 kHz.

Time-signal and weather-report broadcasting which uses a frequency low enough for propagation over the required distance by groundwave.

Military uses specifically naval communications to submarines.

The atmospheric noise level on all frequencies below 500 kHz is quite high. For efficient radiation of power, the transmitter aerial must be very large but for receiving the aerial only needs to be big enough to collect sufficient signal to overcome the internal noise of the receiver input so quite gross inefficiencies of receiving aerials can be tolerated. Section 8.11 in *How Radio Signals Work* gives an explanation of the factors at work in receiving aerials for low frequencies specifically related to a mobile station.

The active aerial described in Section 3.9 and illustrated in Figure 3.16 can also be used by mobile receiving stations working on the long-wave bands and over a more limited range of frequencies could offer higher signal level at the receiver input than would be provided by a passive aerial of the car-radio type. A tuned loop with a ferrite rod in the magnetic circuit can produce an even higher signal level but only at one frequency; it must be retuned each time the frequency is changed.

The fact that the whole LF band is only a few-hundred-kilohertz wide means that channels must be narrow and only very low speed data can be accommodated. Amplitude-modulated signals with 3 kHz of maximum treble response indicates about the widest channels that can be allowed in modern times. Morse code signals suit these bands very well; they can be accommodated within a bandwidth of about 100 Hz, and the reduction of receiver bandwidth to that degree offers a very useful reduction of noise level, which has the effect of significantly increasing the usable sensitivity of the receiver.

The limit of sensitivity for reception of all signals of frequencies lower than 500 kHz is noise external to the aerial. Increased ERP of the transmitter station (increased power output, improved aerial efficiency, or directional aerial) will improve the received signal-to-noise ratio. If receiver-internal noise is detected, a fault in the aerial is indicated (it may have been designed too small). If a directional aerial is feasible (see Chapter 2 for suggestions) it will improve the overall sensitivity of the complete system. The only other way to make a receiving system on these frequencies more sensitive is to reduce the bandwidth of reception; Section 12.2 treats this subject in some depth.

10.5 HF Single-Sideband Land-Mobile Systems

Services using HF single-sideband transmission can be used in conjunction with the VHF/UHF narrowband FM services defined in Sections 9.2 and 10.8 or they can be treated as a longer range alternative to them. Operators of HF services are required to have some familiarity with the technicalities of propagation via skywaves, and there is benefit to them in keeping up to date with propagation conditions relevant to their service over the previous few days. There are some times when propagation may be suspended over particular circuits for a few hours at a time.

In recent years (1980s to 1990s) HF services have not been favored by business users because many sections of the business community demand that the service they use has capability for instant communications with untrained operators. That capability is expensive to provide over long distances. Systems using HF single-sideband transceivers offer the cheapest way of passing information over distances up to several thousand kilometers, but to use them properly, some training of operators is required and lead time for messages may be up to a few hours on some occasions.

In comparison to transceivers for VHF/UHF communications those designed for HF land-mobile use are more powerful, physically bigger, and somewhat more expensive. General specifications for equipment in this class of service are as follows:

Transmitter Functions

Power output

- in the range 100 to 400 W PEP for mobiles
- up to 10 kilowatts PEP for base stations.

Modulation bandwidth

- nominally 1VF channel single sideband
- audio frequency response 300 Hz to 2.4 kHz.

Channel frequencies

- as allocated in range 1.6 to 30 MHz
- crystal or synthesiser controlled
- wider range may be possible in receive only mode
- operator cannot change allocated frequency of a channel from transceiver front panel.

Digital selcall

- permitted using data protocols based on or similar to CCIR 493-2
- multifrequency tone calling may also be permitted
- tone duration of 3 to 8 s is normally provided for noise immunity.

Receiver functions

Clarifier

- up to ±150-Hz shift of carrier insertion oscillator.

Scanning

- programmable channels
- may be set to stop on AGC or to stop on selcall preamble.

AGC

- peak audio with hang function
- hang time-constant may be switchable.
- impulse noise limiter.

Mute

- AGC-operated or channel-noise-operated

The measures described in Chapter 6 ("International Short-Wave Bands") for maximizing HF receiver sensitivity are equally applicable to this class of service; there is some general information on aerials for the HF band in Section 3.10 and the requirements for using skywave propagation are outlined in, Sections 5.8 and 6.5 in *How Radio Signals Work*. The multitapped helical whip illustrated in Figure 8.7 of that book is typical of one type of multifrequency aerial commonly used with HF transceivers in motor vehicles. There is also some instructions for in-the-field localization of faults in Section 11.10 of that book.

Power-supply impedance and/or regulation is one of the critical factors for good performance of HF transceivers installed in motor vehicles. A 100-W transceiver operating from a 12-V battery will draw current when transmitting, which is about 3 amps with no microphone input and pulses up to 20 amps in step with the voice. If the sudden drawing of 20 amps causes the voltage at the power input to the transceiver to drop by even a volt or so the quality of the output signal is degraded. HF transceivers operating on a 12-V battery supply are best installed with their own power cable direct to the battery terminals with no connection to the rest of the vehicle wiring. That power cable should consist of two wires with sufficient current capacity to carry 20 amps over the required distance with ideally only about 0.1 to 0.2 Volts of drop.

The other significant factor about installation of an HF transceiver in a motor vehicle or small ship is that the whole vehicle or vessel is included in the aerial circuit to the extent that aerial currents will flow in all external surfaces of the bodywork. At the same time as that is going on the internal surfaces of the body metalwork are the earth reference for the RF circuits of the transceiver. The distinction between aerial, earth, and power leads can become blurred.

HF mobile aerials usually work as "electrically short antennas" (refer to Section 8.4 in *How Radio Signals Work*) and one of the implications of that is that they are sharply tuned. For a helically wound whip 2 m long which is required to operate on a frequency in the 2-MHz range the bandwidth may only be a few kilohertz so basically a separate tap or tuning point is needed for each frequency used in that range. For higher frequencies the tuning in terms of percentage change is less critical and also that percentage corresponds to a greater number of kilohertz. The same 2-m whip operating at 8 MHz may be able to offer a bandwidth of 50 to 100 kHz rising to several hundred kilohertz at 15 MHz. Longer whips offer better bandwidth; a 4-m whip operating at about 17 MHz is a full quarter-wavelength so bandwidth can be as high as 5% or about 850 kHz.

When considering faults and indications of poor performance in relation to HF SSB communication links it cannot be stressed too strongly that correct choice of operating frequency is the most important key to success. If the station is properly installed, there are no electronic faults in the equipment, and you still experience poor communication then almost certainly the chosen frequency does not match the requirements of the ionosphere for that path. Sections 5.8 and 6.5 of *How Radio Signals Work* and Sections 6.13 and 6.15 of this book give relevant information. If you need up-to-date information specific to a particular path you will need to contact the prediction service which covers your country. In Australia that is IPS Radio and Space Services.

10.6 Wire Aerials for HF Land-Mobile Systems

In the case of a road vehicle using a multichannel transceiver with a tunable or multitapped whip aerial, if the vehicle can be stopped for the period of the radio contact, a significant improvement can be made by changing the aerial. The mobile aerial can be removed as shown in Figure 10.3 and replaced with a length of wire which is cut to be an electrical quarter-wavelength at the chosen operating frequency.

If you are planning a trip and may need to use wire aerials they should be prepared before the start of the trip with a separate piece of wire cut to each frequency you expect to use for transmitting. (Unless

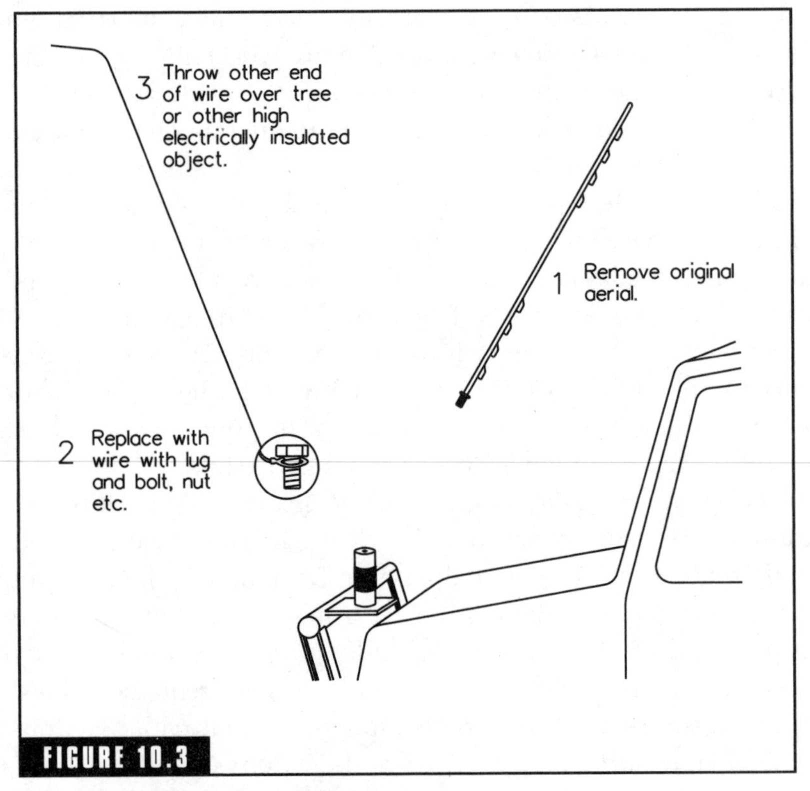

FIGURE 10.3

Using quarter-wave wire aerials.

several channels are within a 5% bandwidth in which case a wire matched to a frequency near the center of the band will cover all of them.) The wire can be almost any flexible, insulated electrical type; radio hook-up wire is a bit flimsy for repeated use but would work well for occasional use on a single expedition; electrician's single-conductor building wire is ideal if it is multistranded—at least seven strands or more if you can get it. One end of each piece of wire should have a lug and bolt or nut suitable for making a firmly bolted connection to the socket or base of the mobile aerial and it should initially be cut to a measurement slightly longer than the calculated quarter-wavelength for the working frequency then finally trimmed using a directional power meter with the transceiver itself as the power source (making sure you are not causing interference) in a manner similar to the tuning of a CB aerial.

The open circuit end will need a weight of some sort; for single use a clove hitch around a stick will work but for repeated uses a fishing weight firmly fixed near the end of the wire and sheathed with either tape or heat-shrink tubing will do better. The aim is that you should be able to sling the weight over a high branch and then pull it back clear again without snagging.

As an alternative to using a separate wire for each channel you may use one piece of wire (which does not need to be exactly tuned) and an aerial tuner. The length of wire is chosen to be approximately a quarter-wavelength at the highest frequency you expect to use then for lower frequencies the tuner is used to add lumped reactances to make the aerial appear to the transmitter as a quarter-wavelength. If the aerial tuner is manually adjusted it will usually have two controls, one for matching of the transformer ratio (may be called "loading") and the other for "tuning." These controls will interact so the adjustment procedure will involve initially making a gross adjustment to each then final trimming for maximum signal. If the device is tuned electronically, the same basic procedure applies but tuning will have been continuously monitored by a phase detector comparing between input voltage and current so the equipment may appear to only make one initial tuning and matching adjustment.

In use it will help to know the approximate direction and straight-line path length to the station you wish to contact. The path length will indicate the vertical angle of the incoming signal in accordance with the table below. These angles are based on roughly average propagation conditions which are a virtual height for the F layer of 300 km and refraction effects not quite equivalent to the curvature of the Earth. The actual angles may vary by at least ±5° depending on the actual figures for those factors at the time of your use.

Straight-line length to other station (in Kilometers)	Vertical angle for F layer reflection (in degrees)
0	90
300	63.4
500	50
1000	30
1500	20
2000	15
3200	8

To make contact with the other station initially lay out the wire aerial from the socket or aerial base on the vehicle as high as possible in the direction away from the line to the other station. The wire can be thrown over a bush or small tree, if one is available, arranged so that the section nearest the vehicle makes a right angle with the vertical angle of the incoming signal. If any part of the wire must be allowed to come back close to ground level try to ensure that the section nearest the vehicle (i.e., that which carries the most RF current) is the high part and it is the open circuit end that is low. Refer to the diagrams of Figure 10.4 for explanation.

Once you establish contact you can experiment for maximum signal if you intend a long conversation or if the present arrangement seems to be giving a weak signal. Carry the open circuit end of the aerial around the semicircle from a right angle to the incoming signal in one direction to the equivalent position in the other direction as shown in Figure 10.5. Somewhere in that semicircle (and it could be in any direction) there should be a point of maximum signal. If you happen to find a point where the signal passes through a null the maximum is at a right angle to that position.

10.7 Directional Effects with HF Mobile Aerials

In the previous two sections it has been stated that with a couple of exceptions land vehicles are not big enough to support other than omnidirectional aerials. One of those exceptions applies to vertical whip aerials. There will be a small null spot in the pattern end on to the whip which is straight upward. Many operators report very restricted usable range for channels in the 2 to 3-MHz band for mobile-to-mobile contacts using road vehicles with HF SSB transceivers and vertical whips. When the vehicles are close together contact is by groundwave and the vertical whip is well-suited to couple into that mode of propagation; however, the power of approximately 100 W and the frequency of 2 MHz mean that the groundwave becomes unreadable over most soils at a distance between 20 and 40 km. At a range of 40 km the skywave, on the other hand, is making almost a vertical trip; the theoretical angle for a 300-km virtual height is 86.2°, which is probably being affected by the end-on null of the whip. If the vehicles can be stopped for the time of the conversation wire aerials will solve the problem. If the vehicles must be in motion some improvement may be possible by lashing the whip at an angle so that the null point

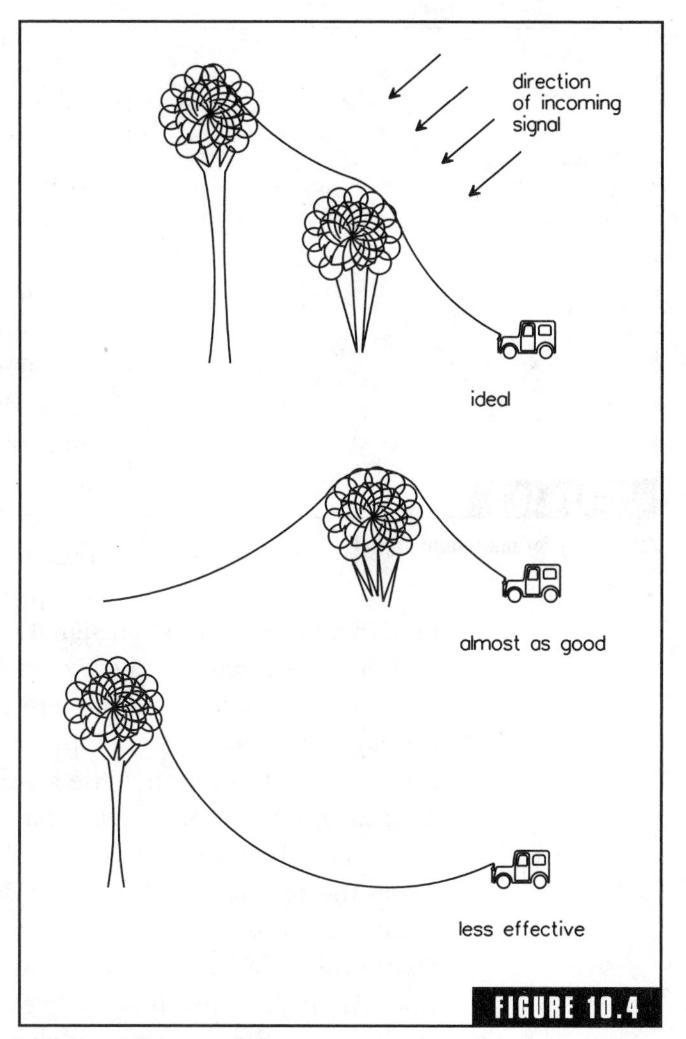

Wire aerial arrangements.

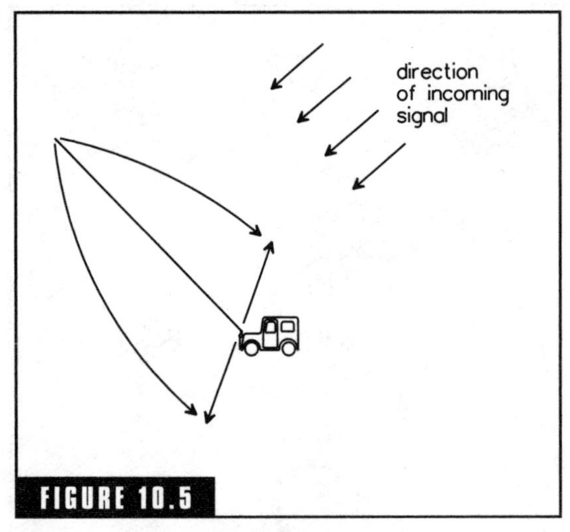

FIGURE 10.5

Adjusting for maximum signal.

is somewhere other than vertical and preferably so as to increase the electrical length of the whip/vehicle combination (i.e., if the whip is mounted on a bull-bar or front-bumper bar lash so that it is held forward to increase the total length of the combination.

Another exception applies to signals at the high-frequency end of the band (27 to 30 MHz) where the size of the vehicle is close enough to a quarter-wavelength for pairs of driven elements to show directional effects. One 27-MHz CB rig that is commonly used by long-distance heavy transporters is to install a quarter-wave whip on each side of the cabin and combine them in phase so that signal is increased along the line of the road.

For mobile stations the size of ships, or railway trains, or in remote areas where motor vehicles are in a string of several trailers (road trains) there may be some opportunity to build directivity into a mobile aerial but in practice the result will probably give no more than a 1- or 2-dB improvement in signal-to-noise ratio.

Aerial directivity can be useful in relation to HF mobile operations when the service is actually a base-and-mobiles network. At the base the power of the transmitter can be raised to improve the quality of the communication link but that only helps in the base-to-mobile direction. An aerial with directivity at the receiver can enhance the signal in the other direction to match the higher power of the transmitter. The receiver aerial must be rotatable and there must be some way for the base to keep at least a rough check on the location of each mobile; this could be as simple as an informal conversation between human operators every few days or it could be as sophisticated as a GPS receiver built into the mobile transceiver which transmits its location by a digital burst of information each time the selcall is used to initiate a call. At the base the digital coordinates are converted into azimuth and elevation figures which are used to control pointing of the base receiver aerial. It is also possible to use the more traditional form of radio direction finding, which requires the mobile to send a steady tone for a few seconds at the start of a contact during which time the base operator will swing the directional aerial for peak signal.

10.8 VHF/UHF Single-Channel Fixed and Mobile Nets

For a very large number of small to medium-sized businesses around the world their mobile communication needs can be met most economically by a network of transceivers within the specifications outlined in Section 9.2 carrying messages by voice. In its simplest form a single-channel transceiver is installed in each vehicle requiring mobile communications or carried in portable form by each person and a similar transceiver is installed at the company office with an aerial elevated on a tower or mast. Transceivers for this service are supplied either crystal-controlled or with a frequency synthesizer programmed to the required channel (or channels). The mobile operator has a push-to-talk button, and if the transceiver is programmed for several channels there will be a channel-change switch. Apart from that there are no controls that affect transmitter operation. The individual operator of the mobile station needs no technical qualifications to use the equipment.

All the electronics of the transceiver may be mounted in one solidly constructed box which can be either installed in the vehicle dashboard or instrument panel or on a cradle directly below it. Power is supplied from the vehicle battery ideally by a cable with a pair of stranded wires of substantial current capacity with direct connection to both terminals of the battery. The aerial for VHF services is usually a quarter-wave whip which may be mounted on the center of the cabin roof of a passenger car, sometimes on a gutter grip mounting, or in the case of a heavy truck there may be a mounting on a wing mirror bracket. The aerial feeder is coaxial cable and the outer conductor of the coaxial must make a firm electrical connection to the vehicle metalwork close to the aerial base forming a ground-plane reference for the radiating element. With UHF services gain whips may be used with an identical ground-plane reference.

There will be a flexible cord connection to a microphone which normally has the PTT button mounted on its side. There may be a loudspeaker built into the transceiver or the receiver audio output may be taken to an extension speaker assembly. The installation is normally designed to be operated by the driver of the vehicle so the extension speaker would be mounted close to the position of the driver's head when he/she is sitting in the driving seat. Figure 10.6 shows the general arrangement of all these connections. For the base transceiver, power supply will normally come from AC mains through a suitable power converter. That could be as simple as the battery shown in Figure 10.6 with a float charger keeping the battery fully charged. The base aerial would be mounted on a mast or tower and would normally be a structure with more gain than a quarter-wave whip.

A single-channel service may be allocated one frequency with all transmitting and receiving by both base and mobiles done on that frequency. That arrangement is common on the VHF band. On the UHF band two frequencies may be used, one for base transmitting/outstation receiving and the other frequency for outstation transmitting; this is known as "two-frequency simplex" operation. The significant difference between these is that with single frequency operation the mobiles can communicate directly with each other and, in fact, can relay messages to extend operating distance from the base. With two-frequency operation the mobiles do not normally hear each other's transmissions. The base can be set up to operate duplex to allow for mobile-to-mobile contact via the base as a repeater but in this case coverage area is absolutely determined by the signal from the base.

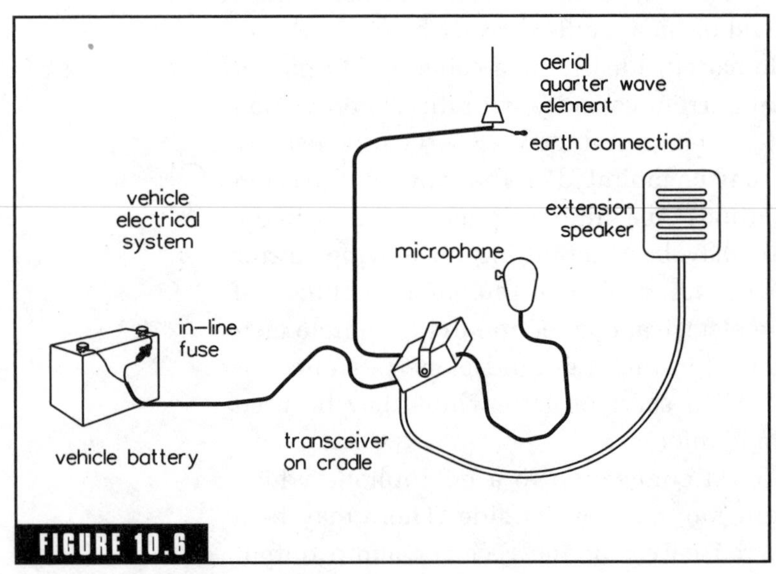

FIGURE 10.6

A mobile transceiver installation.

Calculation of signal strength for direct mobile-to-mobile contacts (for instance, the UHF CB service) is almost impossible. The location of both stations is changing all the time so if the calculation was being done on the basis of path profiles as is done for point-to-point links then the profile changes with each movement. In addition to that the cancellation and reinforcement "microclimate" effects described in Section 2.5 apply to the movement of each station independently so the total variability due to that effect can be in the range of 20 to 25 dB. The observed effect of that variation is to give the received signal a semiregular variation in strength which when the vehicle is traveling slowly causes a "woof-woof-woof" effect to the background noise or at higher vehicle speeds gives the signal a fluttering characteristic which is described by telegraphic operators as "mobile QSB." High-reliability operation mobile-to-mobile is really only possible over a short range; within the watershed area of a particular creek or up to a radius of about 10 km in open country. Although exact calculations are not possible for communications between mobile stations some general rules of thumb can be applied to indicate a likely degree of success:

Operating frequency has an effect on range, but that effect may depend on the size of the aerial as well as on other frequency-related factors.

For land-based stations experience over particular terrain is a valuable indicator.

Directional aerials are only a help if they can be kept pointed in the right direction, otherwise they will actually reduce the overall performance of the station (see Figure 10.2).

In cases where signals are marginal it should be possible to establish good-quality communication by the moving stations finding a spot where the microclimate is reinforcing and stopping there for the period that contact is required.

For signals which are referenced to a base, coverage area is more defined and could be mapped with field-strength measurements if that was economically justified. A coverage map could be plotted by calculation using the principles outlined in Sections 9.4–9.7, or in a less-formal way the mobile users can explore the major features of coverage limits simply by conducting a conversation with the base while they drive away from it and when the signal fades out that is the limit of coverage. In particular directions it could be anywhere from 10 to 150 km from the location of the base depending on the operating frequency of the network and features of the local terrain. In those services the variability of the mobile-to-mobile situation is reduced; there will be a substantial area near the base where signals are clear and steady but there will also be a considerable fringe area where mobile QSB effects are observed. New operators starting work on a taxi service or delivery van fleet that uses a radio of that type can learn within a few days the general coverage area of the service.

If you are one of the operators of a mobile-to-mobile or direct base-to-mobiles network and you suspect your equipment may be faulty Section 11.10 of *How Radio Signals Work* includes some tests that can be done in the field with no or very little test equipment. The aim of these tests is to initially localize the fault to a particular module or section of the network. If you are a technician who must do a full proof of performance of a single-voice-channel radio-communication system you will need test equipment to cover the following functions:

- directional RF watt meter to operate at the transmitter output frequency and power level
- FM monitor to cover the expected deviation at the transmitter output frequency
- frequency counter
- field-strength meter or tunable VHF or UHF receiver
- signal generator capable of being adjusted for minimum output of ~0.2 μV with deviation adjustable from 500 Hz to 15 kHz and modulating frequencies covering the range 50 Hz to 5 kHz.
- audio power meter
- noise and distortion meter
- DC voltmeter and ampmeter to cover from 0.7 V to at least 20% above the nominal voltage of the vehicle electrics and currents up to 10 A.

There are instruments called radio-communication test sets which can combine the functions of all those instruments in one box. They are expensive to keep if you only use them occasionally but if you are doing tests of that sort all the time they become cost-effective and save time. Test sets are very good for proving compliance with a set of specifications but can sometimes be less flexible in their operation than individual instruments. Once the faulty box is identified fault finding on the bench at the component level is best done either by using several different discrete instruments or with a "bed of nails" test jig programmed for that particular module.

The above list of test instruments is sufficient to prove the voice channel and the CTCSS facility (if one is fitted) of a 1 VF channel. If a digital signal, even as simple as a five-tone selcall, must be decoded extra digital test equipment is needed, and if the service includes multiplexed channels then you will be able to check the basic channel with the above list but some more sophisticated equipment will be needed to prove correct operation of the multiplexing.

It is increasingly common for manufacturers to use computer programs working on "bed-of-nails" test jigs for component-level fault finding, and this means that if a spare unit is available in the field the most economical way of fault removal is often to do no more than isolate the fault to a particular transceiver or module then swap it and send it to the manufacturer for repair. The economics of how much resources you commit to spare equipment is one of the questions for which there is no single right answer, but if you are in charge of maintenance of a network of base and mobile transceivers in a rural area the size of fleet that will make it worth having a complete mobile station on standby is probably a number somewhere between 4 and 10. A complete base on standby for a network of that size seems to be a bit excessive but it would be worthwhile to have an adaptor jig to make possible the substitution of the base for that spare mobile in the event of a failure. You will not need a spare station for every 10 mobiles; the size of fleet that would justify having two spare stations would be much larger, probably in the 100 to 200 range, but that would also depend a bit on the turn-around time for units being repaired. In a fleet of that size there may also be units of different ages which are not directly interchangeable with each other and in that case you may need a standby unit for each type of equipment in the field.

10.9 Increasing Operating Range

If you are the operator of a direct VHF or UHF base-to-mobiles net, you have satisfied yourself that all the equipment is working correctly in the electronic technical sense, and you wish to extend the coverage of the service there may be some things you can do. Before you make too many detailed plans, however, there is a simple test you should make. Climb to a high point near the base aerial (onto a roof or ladder up the tower, etc.), high enough so that you can look over local roofs and vegetation and see the distant horizon, and proceed with the following:

If in some directions the countryside is rising to a range of hills and the visible horizon is 20 km or so away, then you will get good coverage of a UHF transmission to a few hundred meters past the visible horizon and the usable range will not be affected much by anything you do at the base.

If you can see flat ground or a body of water in some directions, then the usable range can be calculated by a relatively simple formula and changes at the base will significantly affect the result. You can improve coverage by increasing power output, aerial gain, and tower height (in proportion to the square root; for instance, making the tower 4 times higher will double the range).

In directions where you can see low rolling hills, coverage will be patchy and it will probably still be patchy no matter what you do to the base aerial; although improvements in signal strength and tower height will raise the proportion of usable locations in a general area.

If you see continuous tall thick vegetation, such as a forest reserve in some directions, UHF signals will not penetrate very far into the forest; if the vegetation is in patches, such as a pine forest planted in blocks with fire breaks between the blocks, you may find there are usable locations in some of the clear areas even if there is a block of forest in the direct line of sight. In regions where most of the surrounding terrain is either rolling hills or carries tall thick vegetation, use of the lowest frequencies (30 MHz end of the VHF band) will give best coverage.

If you are using a low-VHF channel near the central business district of a large city and experience difficulty with communication in that CBD then moving to a higher operating frequency will significantly improve results.

If no more improvement is practical with the base located at the company office and more coverage is still needed then you should consider establishing the base at a hilltop site or using a repeater.

10.10 Repeaters and Remote Bases

A repeater is simply a receiver and transmitter operating side by side and usually through the same aerial connected so that the output of the receiver provides the input to the transmitter arranged as shown in the block diagram of Figure 10.7. The technical advantage of a re-

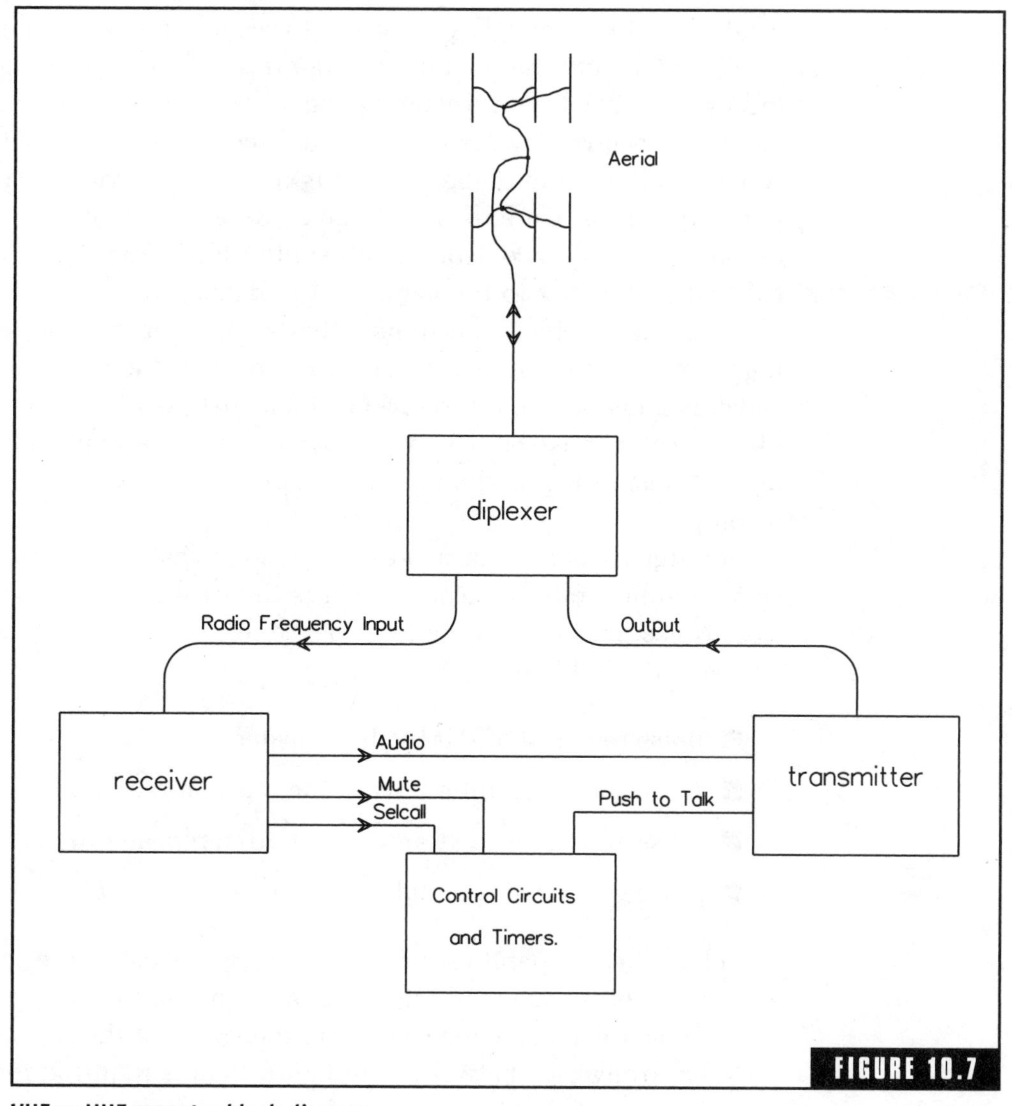

FIGURE 10.7

VHF or UHF repeater block diagram.

peater is that it can be sited for maximum coverage of the required service area without consideration of the constraints of needing to be physically close to the company's office. In modern times there does not even need to be land access to the site; a repeater can be constructed in a form that can be lifted by a helicopter so all that is needed is a small clearing on top of a mountain and power can be supplied by solar cells. For installations of that type, no large aerial tower is needed—the mountain provides the elevation and the aerial only needs to be high enough to clear local vegetation on the mountaintop.

To understand the operation of the repeater shown in Figure 10.7 follow this signal path. Incoming signal from the "trigger base" or a mobile is received by the aerial and routed by the diplexer to the receiver. Audio output of the receiver is connected to the transmitter input in place of a microphone signal. Receiver mute and signaling outputs are used to control the transmitter PTT. Transmitter output is fed by the diplexer to the aerial and radiated. To provide separation between the received and transmitted signals the transmitter carrier frequency is different by a certain amount from the frequency the receiver is tuned to. All transceivers which are to work through a particular repeater must be set to operate in the two-frequency simplex mode, transmitting on the repeater's input frequency and receiving on its output.

For signals going via a repeater a somewhat higher standard of signal-to-noise ratio is needed than is the case for a direct transmission. Noise produced in the loudspeaker of the receiver is the combination of the following:

- noise received off-air by the repeater
- noise generated internally in the repeater receiver
- noise added during transmission from repeater to mobile
- noise generated internally in the mobile receiver

If each of these separate sources of noise was exactly the same level as all the others each could degrade the signal-to-noise ratio by up to 3 dB. To achieve the same final ratio the signal at the repeater input must be stronger by between 4 and 6 dB than is required for a direct transmission. The repeater relies for its success on being able to receive a very clear signal at its input. It will have been initially sited at a spot with a low level of electrical noise and high enough to provide a line-of-sight path to all users as much as possible. Sometimes the ideal site is just not available in the required service area, and in those cases, the repeater mute must be desensitized so that local noise does not interfere with incoming signals. The result of that is to limit the service area; mobiles will find that near the edge of the range there will be places where they can hear the transmitted signal but cannot make a strong-enough signal to make the repeater hear them.

When the system is a direct base-and-mobiles network the base installation at the central office or depot must use a high-power transmitter with an omnidirectional aerial on a tower with as much aerial gain as is allowed. That is not needed for operation through a repeater; the equipment at the office or depot can have a directional aerial pointing in the direction of the repeater and sited only just high enough to clear local obstructions. The transmitter output can be quite low because it is only providing a link to the repeater. An installation of that type may be called a "trigger base" or "link." At the mobile the equipment for operation through a repeater does require to be set up for the two frequency simplex mode but apart from that is the same as is required for operation to a direct base.

To the operators, traffic between trigger base and mobiles can be carried on while almost ignoring the existence of the repeater. For mobile-to-mobile traffic the only difference noticed is that when mobiles are very close to each other (i.e., both in the yard of the same depot) the repeater signal does not overload the other mobile as happens with direct systems. Even if the two mobiles are side by side the signal must travel many kilometers to the repeater and back.

Fault finding on repeater systems is very similar to that for an equivalent base-and-mobiles or mobile-to-mobile system as explained in Section 11.10 of *How Radio Signals Work* and in the previous section of this chapter. If the system can be split into individual transmitters and receivers, each section can be treated exactly the same as for a direct system. It is an advantage to set the repeater up so that the output of its receiver can be diverted to a loudspeaker and the input of the transmitter section can be fed from a microphone or signal generator. On the radio-frequency side of the equipment the cable from diplexer to receiver and from transmitter to diplexer should be accessible for test equipment.

There is a potential for political problems with a repeater system that does not exist with a simpler net. The repeater is often owned by a management company and space on the channel is rented to users who have no commercial affiliation with the repeater management company. The users may observe a fault and have their own equipment tested and find it meets its manufacturer's specifications so they refer the problem to the repeater manager. If the repeater managers check their equipment and find no fault but the overall performance of the system is still defective, the stage is set for a first-class brawl. Political problems of this type are best avoided before they start; there is no sure way of preventing all such interfacing problems but one big factor that helps is that the technical people who do the actual hands-on testing should be allowed to talk directly to each other. These problems are much harder and more expensive to solve if the commercial administrators insist that technical information must be passed through nontechnical third parties.

Another system that can offer all the coverage advantages of a repeater but may avoid some of the penalties of retransmission is the remote base arranged as shown in Figure 10.8. If the remote site is only a few hundred meters from the speaking point this function could be arranged with a multipair cable to carry all the speech and control functions with no multiplexing at all. That would be appropriate if the site was such that an underground cable could be ploughed-in on a clear line down the side of a hill. The cost of that type of installation depends almost directly on the length of the cable run so is only good for short distances; however, it is practically all capital investment in the initial installation; the maintenance cost of a properly installed underground cable is very small.

The companies who manage telephone systems can provide dedicated landlines through their cable distribution. In that case, the cost may depend on the number of telephone exchanges the line must be routed through. There will not necessarily be a capital cost for setting up such a line but once contracted, the cost will be a fixed amount of money per pair of wires per year with no limitations on how much you use it. The landline can relay all the microphone, loudspeaker, control, and indicator functions as if the base transceiver was in the

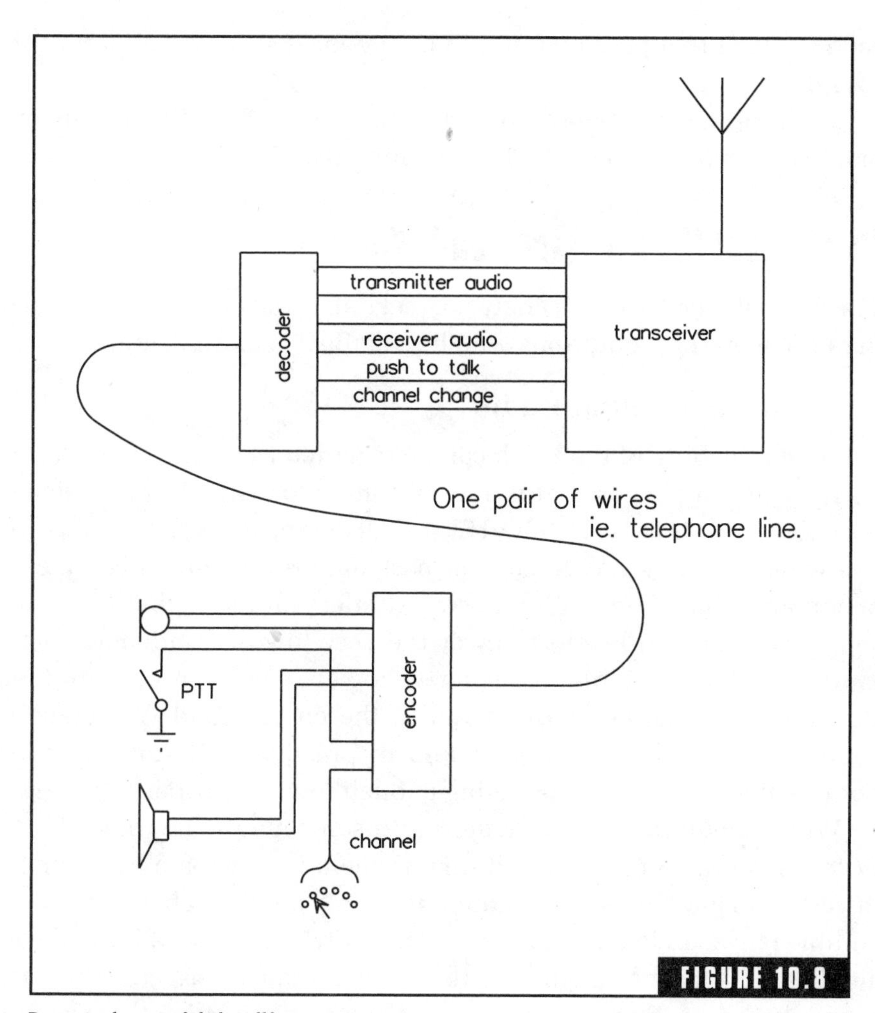

Remote base with landline.

same room as the operator, but all functions must be multiplexed onto one pair of wires so interfacing is required at each end of the line.

There are interface units made which can match a standard telephone line to the functions of a radio transceiver. If one of those is used at the remote base you can have communication with mobiles on the radio network from any telephone anywhere in the world. In this case, the capital cost is the cost of buying and installing the interface; there is a small time-related cost for telephone line rental, but if you use the system often you will pay for a telephone call at each use and that may turn out to be the most substantial part of the overall cost.

In all these cases the radio part of the system is a direct base-and-mobiles network as described in Section 10.8.

10.11 Earth, Air, Fire, and Water

The four "elements" of the ancient Greeks all indicate potential changes in radio operating conditions which will affect communications.

Earth—Underground Radio

Although the bulk of the Earth can be regarded as a conductor of electricity for most purposes, it is a lossy conductor as far as radio signals are concerned. Signals in the VHF and UHF ranges are completely lost below receiver-internal noise after passing through only a few meters of soil and rock. Attenuation has most effect on the higher-frequency signals; for a given level of power the very lowest frequencies penetrate most. The very low frequencies, however, have another problem in regard to coupling of the aerial to the energy field. Wavelengths commonly are many kilometers and in practical situations it is just not possible to make the aerial big enough to work properly.

When communication is required to someone underground in, for instance, a tunnel or mineshaft this problem is overcome by using the lowest-available frequencies, commonly only a few kilohertz above the audible range, and using very large coils of wire instead of normal aerials. The coils must be aligned so that both are square to a common axis; normally the one on the surface would be positioned directly above the one in the tunnel then both coils are horizontal. This system is not strictly speaking a radio device; coupling is by common induction between the coils. The radiated field of these systems is quite low so the signal doesn't go very far even in air but is sufficient to work over the relatively small distances involved in the vertical displacement.

There is one situation where this philosophy does not apply. Where the signal must go along a tunnel between stations that have close to a line-of-sight path between them all frequencies whose wavelength is more than twice the widest dimension of the tunnel will not "see" the tunnel as an opening and will be severely attenuated. Frequencies in the high-VHF or -UHF bands will obey free space or "propagation over an obstacle" rules and transmit along the tunnel with relatively little loss.

Air—Communication with Aircraft

The point to note is that aircraft use several radio systems in their normal operation for communication and navigation and when systems are used in close proximity such as in this case there are possibilities for interference due to overloading of receivers and transmitters operating in combination can cause the production of a range of extra frequencies due to intermodulation. In the initial safety certification of the aircraft, a great deal of testing is done to check all combinations of functions and prove that no significant interactions occur. Each time an extra transmitter is added the number of possible interactions is doubled.

Because of this fact, on commercial airline flights there is a complete and total ban on operation of all radio-transmitting devices by passengers. There are also limitations on use of devices such as electronic games which are not intentional transmitters but do produce radio-interfering signals. Do not be tempted to think that you can get well away from the flight deck and use a handheld cellphone because it is low powered. Radio modules are located in aircraft in all sorts of places for a variety of reasons and you may be sitting directly above or below the most sensitive module of the most vital receiver.

With cellphones in particular, they should not even be switched on; the transmitter can be triggered by a signal from a base even when you think you are not using it. All the airlines in Australia and most worldwide provide a message-handling service via their own communication links, and some of the major airlines have recently installed telephone connections to aircraft operating via satellite. It is recommended that anyone who needs mobile telephone facilities while a passenger in a commercial aircraft should use the facilities offered by the airline.

With regard to cellphones in private aircraft, the management companies of the cellphone bases do not encourage use of the system from very high altitudes. It would be possible for a mobile to cause responses from bases far enough apart to be reusing the same frequencies. A call under those conditions could cause interference with characteristics similar to a triple connection on a wired telephone service.

If you own the aircraft you can of course add facilities and observe the effect. A permanent installation should be done by someone with the appropriate aeronautical electrical qualifications. If the plane carries an instrument rating, initial testing of the added facility should be done with all instruments on but actually operating under visual flying rules so that the truth of instrument readings can be verified. There will also need to be a recertification of the instrument rating.

In operation, the pilot should, at all times, have the power of veto over use of any added facilities. This could be in the form of a main power switch or circuit breaker or it could be acceptance by all passengers that a command by the pilot to switch off the extra equipment will be obeyed instantly. The response must be instant and without argument; with aircraft traveling at several kilometers per minute when things start to get out of hand a dangerous situation can develop very quickly.

When you go to charter an aircraft you should mention any radio equipment you wish to use at a very early stage and allow the charter company to pass judgement on its use. If use of the radio equipment is the main purpose of the charter, as for instance, if you are to do a radio telemetry survey of some sort, and the company expresses reservations you may need to look for another charter company. Even if the company has no reservations about use of your equipment you must still give the pilot the right of veto as mentioned above.

Fire

A flame is a body of ionized gas and, as such, it is a lossy conductor of electricity which will have the same effect on radio signals passing through it as a body of earth/rock. In practice, the fire has to be very big before this effect becomes significant, but it has been observed to cause signal losses in some bushfire and forest fire-fighting operations.

The same mechanism causes a temporary loss of communication with spacecraft during the reentry phase when the whole craft is enveloped in the flame. This probably indicates that if a situation arose where an object was totally immersed in flame any radio link (as, for instance, a radio telemetry device operated at the top of a blast furnace) would be affected by it.

For fire fighters, the unfortunate combination of circumstances is that the biggest and fiercest fires where communication is most vital are just the ones most likely to cause loss of contact. The only thing you can do to avoid having your operation disrupted by this effect is to derate the range of your radio equipment so that contact is still possible even in the face of this extra loss of signal. Higher frequencies are most affected by this mechanism so if you have a choice between, for instance, VHF or UHF channels the lowest-available VHF channels will work best on the fire ground; the higher frequencies can be used for intercommunication between control stations.

On the Water

For commercial shipping, there has been a change in the basic principles of radio operation in the last few years of the 1990s. Until the advent of the Global Maritime Distress Safety System ships carried specialist radio officers highly trained in the techniques of passing messages by radio under the most adverse of conditions. Communication was via Morse code on channels associated with the international distress calling frequency of 500 kHz and voice communication using frequencies in the MF and HF bands. In the new system the basic link for reporting of distress messages uses data communication on a UHF channel to a network of satellites. All ship's officers are trained in use of the radio equipment so no specialist radio operators are needed.

In the case of a person who is on a commercial ship but not part of the regular crew the same considerations as for commercial aircraft apply in a less-urgent form. You should not attempt to operate a transmitter of any sort without the express permission of the Captain of the vessel. For receivers, battery-operated portables with self-contained aerials can generally be operated without restriction but may suffer interference due to front-end overloading in some particular locations. For fixing of external aerials of any sort the design and purpose of the installation should be notified to the officer of the watch before any construction work is commenced.

If the ship is carrying a flammable cargo a whole new set of circumstances apply. A transceiver may induce radio-frequency voltages into metal parts of the vessel which could cause arcing across corroded contacts and start a fire or if the cargo includes detonators they can be set off by induced RF power. If you are simply a passenger in a vessel with a flammable cargo it would be best not to have a transceiver in your possession at all; if you need to pass messages use the ship's commercial service. If you are involved in an operation that necessarily involves handling of a flammable cargo in association with use of portable radio transceivers you should at all times have someone on hand who understands the characteristics of the cargo and don't be too adventurous; once you have worked out a procedure that is safe stick to it. Radio-communication transceivers are required to use power levels that preclude inclusion in the "intrinsically safe" definition.

For small ships and the type of service classed as "inshore boating" the dangers associated with using added-on radios in clear weather are usually minimal. A steel-hulled motor vessel in which the plates have become loose and covered with oil or soot may present a danger of fire due to the induction process mentioned above; the danger is greatest when a temporary installation of a fairly powerful transmitter is arranged so that aerial currents flow through the plates of the hull. At the other end of the scale using a handheld set on a well-kept and clean wooden-hulled or fiberglass vessel presents almost no danger from that cause.

Portable radio transceivers may cause some interference with navigation instruments so should never be operated without the Captain's knowledge and as for aircraft the person in effective control of the vessel (helmsman, Officer of the Watch, etc.) must have the power of instant veto over all uses of nonpermanent equipment.

Some Technical Practicalities

Signal Propagation: At sea you may find the propagation of some signals is different than you may expect for a land-based service. For the lower frequencies (i.e., the CW band around 500 kHz and the 2 MHz radio telephone allocations) a motor launch is not really big enough to get the aerial a worthwhile distance above the ground plane so the signal tends to all go straight up. Signal strength in the direction of the horizon can be very low. This problem is less for a yacht where the aerial can be run up the back stay to a point which may be 20 m or so above the waterline.

Power Feeds: If you need to make a semipermanent installation and connect equipment to a vessel's electrical system be prepared for odd voltages. Thirty-two- and one hundred ten-volt DC or AC are not uncommon and the AC may be at 25, 50, 60, or 400 Hz. In most cases the voltage can be adapted to your equipment, but do not just plug it in and assume it will work even if the socket looks right.

Underwater

Seawater is a very good conductor of electricity and therefore acts as a very good shield for radiowaves. About the only time radio is used for contact with craft under water is the defense system for communication with submarines, represented in Australia by the installation at Northwest Cape. This requires the lowest-possible operating frequencies, enormous power, and enormous aerial structures. For submarines, the strength of signal depends mainly on their depth below the surface. A craft at periscope depth many thousands of kilometers away will receive stronger signals than one at maximum depth below a location within sight of the transmitter.

Pure water (distilled) is an insulator of electricity so it is tempting to think that fresh rainwater in Alpine lakes should allow use of radio underwater. In practice, however, the small amount of minerals picked up in the quick rush of a stream down the mountainside is enough to render the water conductive and therefore a shield to radio signals. Soundwaves propagate in water better than in air so the normal means of communication underwater uses ultrasonic sound as a carrier wave instead of a radio signal with modulation of the soundwave "carrier."

10.12 Installing for Minimum Noise

Motor vehicles in their natural state are very noisy in the electrical sense. Spark-plug ignition works on high-voltage pulses (up to 40 kV in modern cars) with microsecond rise-times. Engine management systems use digital computing techniques; instrument panel displays are in the process of adopting the same techniques. Alternator diodes generate electromagnetic radiation at every switching time. In addition to all the "intentional" radiation from those sources, mechanical movement of the vehicle over bumps in the road builds up triboelectric charges on every piece of plastic and rubber in all parts of the vehicle.

The need for suppression of electromagnetic radiation is well accepted by car designers. In modern times most are fitted with a broadcast-band radio receiver as standard equipment and many have audio systems. The customer requires that noise generated by the vehicle electrical system is inaudible in that receiver. The basic measures used to electrically mute cars are as follows:

- resistive ignition leads
- bypass capacitors as close as possible to all switching elements
- earthing straps to the hood to convert the engine bay into a shielded compartment
- use of a battery with very low internal resistance as a stabilizing element on the power distribution lines

These measures are helpful when you set out to add a communications transceiver to the vehicle but more may be needed as well. The point to bear in mind is that a modern vehicle has a great deal of suppression of electromagnetic radiation built into its original design and it is important that extra components added on do not defeat the effect of what is already there.

A factor that does not apply to the broadcast-band receiver is that electrical noise may ride into the transceiver on power or microphone leads and cause modulation of the transmitted signal. An installation that would be prone to trouble from that cause is shown in Figure 10.9. Measures that may help reduce noise modulation of the output are listed below:

- +ve and −ve power leads both go direct to the vehicle battery with no other connection to the vehicle electrical system
- transceiver electronics isolated from the transceiver case
- microphone audio lead arranged as a current loop or balanced pair with no earth connection at the microphone end
- aerial connection via inductive coupling with no DC connection to the transceiver electronics
- only one earth point on the transceiver case

Some of these points, such as isolation of the transceiver case and inductive coupling of the aerial, are facilities that are built into the

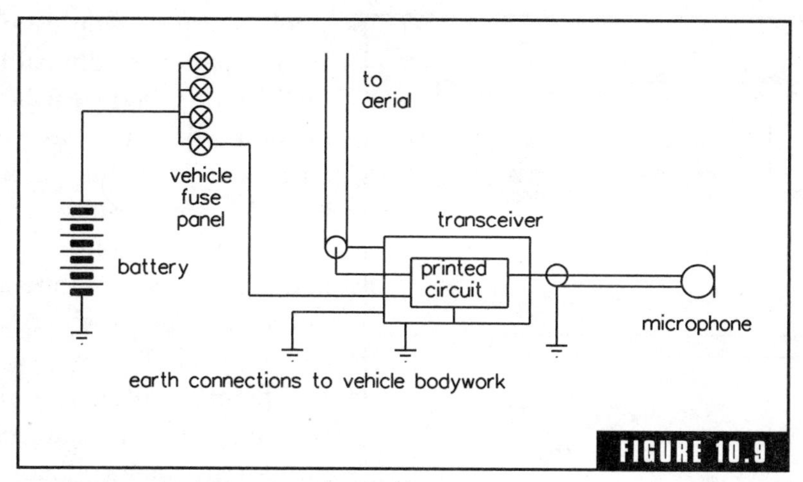

Installation with high noise modulation.

transceiver and the only control you as an operator will have over them is at the time of initial selection of equipment. For the microphone arrangement your choice of options will be limited by internal circuitry of the transceiver but the external circuit must match that to achieve a noise-free input. Figure 10.10 shows ways of connecting microphones for noise rejection; they can be compared with the relevant part of Figure 10.9 to show the features.

Balanced and current loop connections have the same aim, which is to make common-mode rejection as high as possible. The aim is that induction from external sources should induce equal and opposite signals into the two conductors of the microphone pair. Section 7.11 in *How Radio Signals Work* describes the cancellation process in relation to radio-frequency transmission lines; the process is exactly the same for audio signals. The significance of the two diagrams in Figure 10.10 is that dynamic and piezo sources generate an EMF so need no external power source, whereas the others require a DC feed to drive current through either a variable resistor (carbon microphone) or an amplifier. In the circuit of the top diagram the signal pair to the microphone could form a loop of conductor completely isolated from all other conductors. Note that transformer coupling may not be practical for a piezoelectric microphone due to the very high output impedance; special circuits to provide a "virtual transformer" function can be used in those cases. In the lower diagram, current entering the

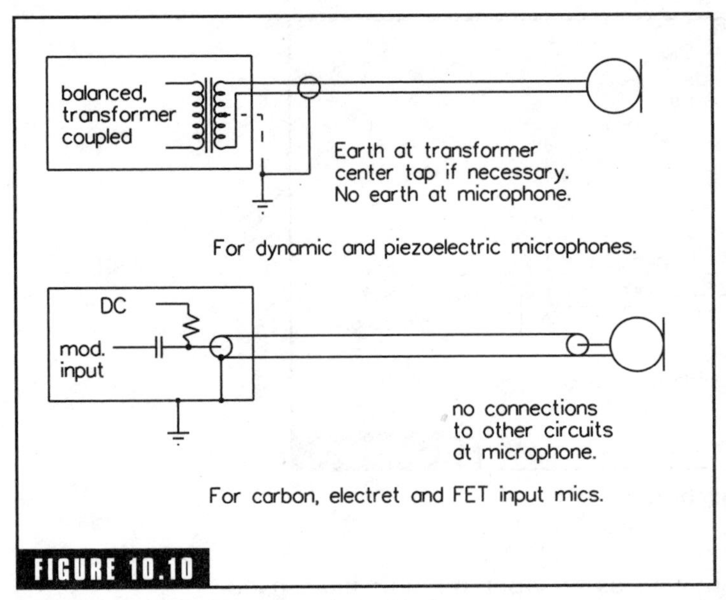

FIGURE 10.10

Microphone connections for noise reduction.

shielded cable should be exactly equal and opposite to that leaving via the earth connection with no leakage of current to any other circuit, and the earth reference for the input buffer of the modulator must be at the same point as the earth of the shielded cable.

The microphone connection may also be subject to radio-frequency feedback due to RF energy from the transmitter output being coupled onto the microphone input line at sufficient level to ride into the first amplifier stage and overload it. The overload drives the stage into nonlinear operation, and RF energy is detected and passed on to the rest of the transmitter as a genuine modulation input. This effect can cause oscillation if severe enough or affect modulator gain and frequency response characteristics in less severe cases. RF feedback to the modulator input is more of a problem in the case of amplitude-modulated and SSB transmissions than for FM and PM systems. The output power of an FM transmission is steady and when rectified by an overloaded amplifier it adds a DC component to the signal. That DC component can shift the operating point of the amplifier, which may result in distortion of its output but it does not produce an audio signal which can cause a parasitic oscillation.

RF induction into microphone leads is reduced by earthing of the transceiver case at the correct point. That earth connection should be a physically short piece of wire to an attachment point on the vehicle body; this is a case where minimum lead length is more important than low DC resistance. Shielding of the microphone lead and transformer coupling of a balanced audio pair also help reduce induction. Bandwidth limitation of the input amplifier will help but that must be built in to the transceiver at a point very close to the transistor or integrated circuit input. A system designed around an electret microphone or one with an amplifier built in can be an advantage to the extent that the transceiver can be designed with lower gain in the modulator input. On the other hand, carbon granules have a rectifying function of their own so that advantage may not apply to a carbon microphone.

Power supply and earthing connections have a potentially great effect on the degree of noise induced into the transceiver in both transmitting and receiving directions. The important factors for a VHF/UHF service are different from those for an HF or MF installation. Lead length as a proportion of signal wavelength is the factor of greatest concern. Skin effect can also be used much more effectively to control VHF signals than is possible for lower freqencies. For frequencies higher than 100 MHz it is feasible to have a shielding conductor with entirely different RF potential on each face; this is not usually practical for frequencies in the HF range. Figure 10.11 shows an arrangement for each end of the spectrum. For a range of frequencies in the low-VHF band where the size of the vehicle is comparable with half the wavelength both sets of conditions may be important for noise-free operation.

DC power from the vehicle is best supplied by a heavy-duty cable with a pair of wires going direct from the transceiver to the battery. Both positive and negative wires should be taken to their relevant battery terminal with no attempt to use the bodywork to provide a return path for the current. Transceivers may be designed so that a particular battery lead is connected to the transceiver case or they may have both battery leads isolated and kept floating. There may also be designs where the case is floating with respect to the DC supply but connected by capacitors so that audio and RF conditions are referenced to the case. For radio frequencies a choke in each lead will allow the earth connection at the battery to be regarded as a separate connection from that at the transceiver.

In VHF/UHF transceivers internal shielding around the tx output, t/r switch, and rx input section will allow the internal surface of the coaxial cable outer conductor and the internal surface of that shielded section to be regarded as the same surface as the outer surface of the vehicle body for radio frequencies. UHF aerial bases as shown in cross section in Figure 10.12 are designed to best promote those conditions.

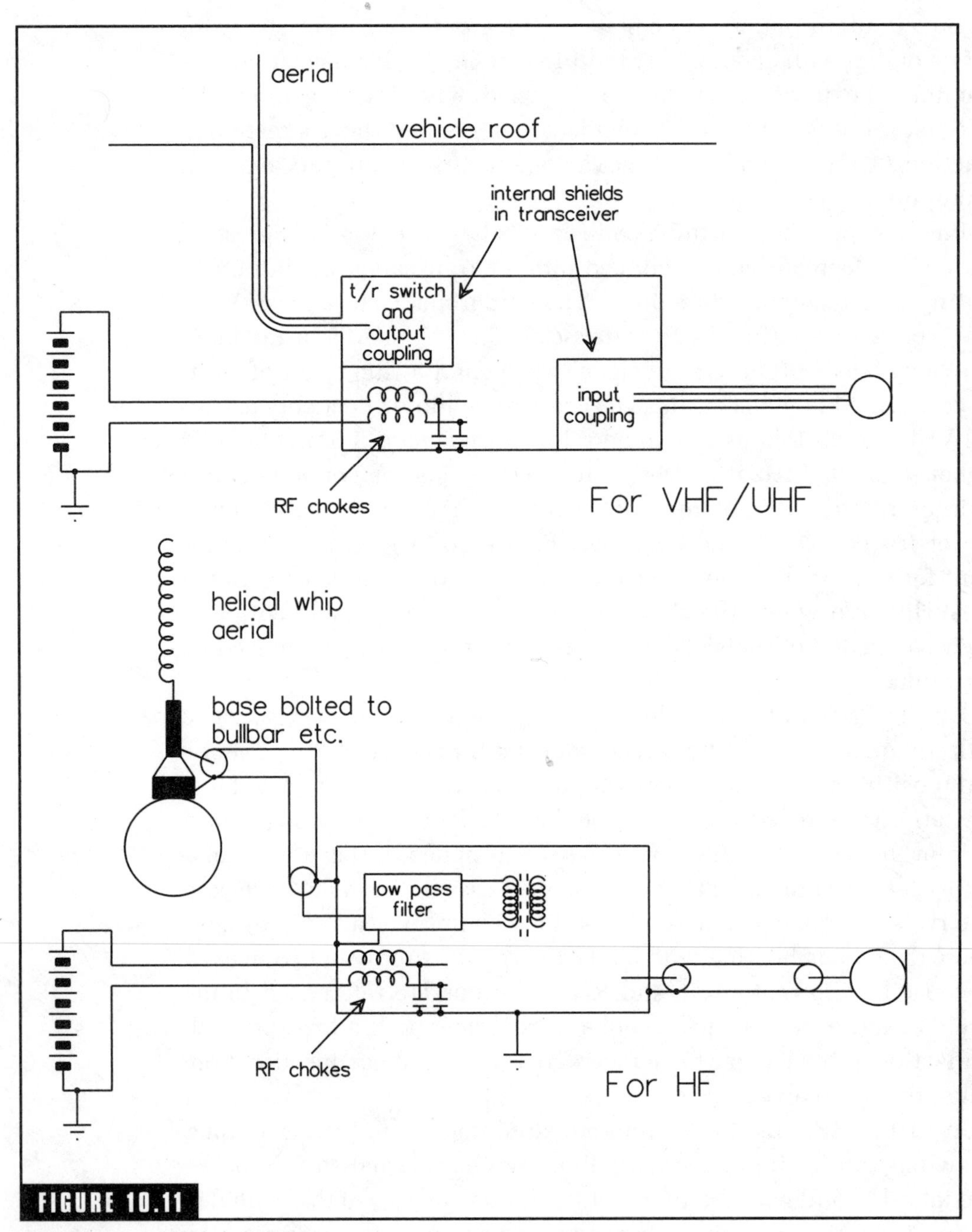

FIGURE 10.11

Earthing and power connections.

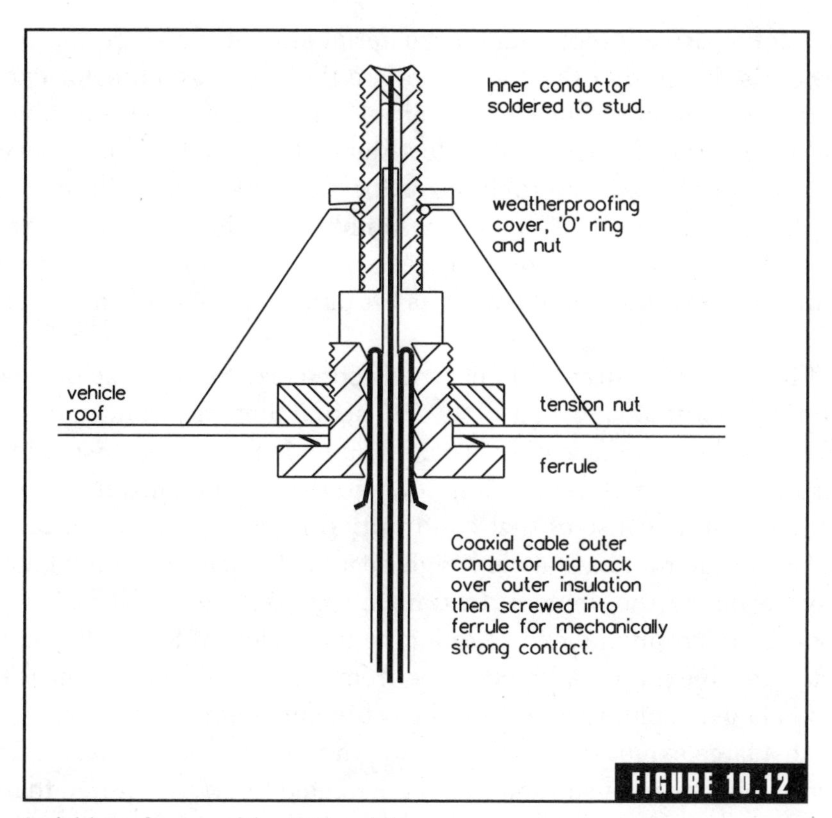

FIGURE 10.12

Aerial base for use with a UHF mobile.

In the same way an internal shield around the microphone input transformer or buffer amplifier will allow that in combination with the inner surface of the shield on the microphone cable to be regarded as a separate surface for radio frequencies. When those conditions apply the existence or otherwise of an earth connection to the transceiver case has no effect on the radio-frequency conditions; the existence and position of that earth can then be dictated by what gives best protection against audio-frequency electrical noise that is present on the rest of the vehicle electrical system.

For transistor-based HF transceivers the output stage is a push–pull pair with a ferrite strip-line transformer coupling to the aerial circuit. That stage is usually a wideband amplifier, and frequency selection for each channel is done by a low-pass or band-pass filter with components switched for individual channels. The earth reference for all RF components on the aerial side of the strip-line transformer is the case of the transceiver so the coaxial connector can be mounted straight on the case. Internal shielding in the transceiver is not as effective for HF as it is for higher frequencies; the theory of separate surfaces in shielded boxes for different RF conditions is not tenable so the whole electronic assembly must have one earth reference. The connection from the case of the transceiver to a nearby point on the vehicle body work is an important part of the installation.

Figure 10.12 is drawn about twice actual size for the sake of clarity. From the point of view of a VHF/UHF signal there is a continuous surface from the inner face of the coaxial outer conductor to the top of the ferrule, to the top of the tension nut, and then to the top surface of the vehicle roof. Because of that continuity of surface, signals on the top surface of the roof have a very high degree of isolation from whatever may be inside the vehicle. In transmit mode, the signal is isolated from the microphone lead so reducing the effects of RF feedback, and in the receive mode radiated noise from the rest of the electrical system is largely contained within the cabin and engine bay.

To a large extent the conditions required for a mobile transceiver to give a noise-free transmitted signal are related to those required to minimize noise in the receiver; both relate to minimizing coupling of signals in unwelcome directions. In addition to all that, there are some other things that can be done that affect reception only; these are changes that may be needed in the vehicle itself. What needs to be done to make the vehicle quiet depends on the source of the radiated noise so the first step is to trace it and identify its source. The following list may be helpful pointers:

A ticking noise that is present whenever the engine is running and changes the tick rate in time with engine speed probably comes from the ignition system.

If the tick rate does not change in step with engine speed it is probably coming from an electric fuel pump.

A steady heterodyne tone on one channel only may be coming from the clock oscillator of a digital system such as the engine management system or a digital instrument display.

A howling noise on all channels which is worst on the lowest frequencies may be coming from the alternator.

If a noise seems to depend on the operation of a particular control it can probably be traced by looking closely at the items moved by that control. In a related fashion, noises that appear to have some relationship to an instrument panel indication can probably be traced by looking back along the line to the sensor related to that indicator.

White noise (no particular tone) on all channels which is only evident when the vehicle is moving and increases in volume when the vehicle moves faster is probably due to triboelectric effects from many different sources all over the vehicle body and may also be due to air flow over insulating objects or the tires on the road.

Identifying the source of a noise is one thing but muting it may be a horse of an entirely different color. There are a multitude of ideas that have been tried for noise suppression and each has its place in relevant circumstances. Some of them have already been referred to in earlier sections of this book; Sections 5.9, 5.11, 5.12, 5.13, 5.14, and 6.9 may all have some useful information. Some other considerations that may be useful in the case of motor vehicles are listed below:

The field strength of noise signals varies quite markedly at different parts of the vehicle so search around to find a radio quiet spot for the aerial mounting. For passenger cars, a spot near the rear window may be a good place to start; in general, the further away you can get from all other uses of electrical energy the better.

If you have a choice of engine type, compression ignition (diesel) engines can be made electrically quieter than those with electrical ignition.

Shielded spark-plug leads can be used but they degrade the performance of the ignition system and in marginal conditions could be responsible for causing misfiring at high engine speeds.

RF radiation only occurs when electrical power is being switched so suppression as close as possible to the switching contact is an advantage.

Suppression slows the rise-time of the switching and in digital circuits there may be some components that rely on a fast rise-time for correct operation.

A marginal change of clock oscillator frequency may shift a troublesome harmonic out of the channel so that it can be safely ignored.

Parallel capacitance may not be a complete cure for everything; in some cases, series inductance may also be needed.

If a particular length of wire is forming a tuning component to concentrate energy into a particular channel or group of channels it may be possible to change the electrical length using series inductance or damp the resonance using a variation of the "anode suppressor" circuit of Figure 5.11.

Triboelectric effects normally only occur when resistivity of the insulating material is very high (over 10^{10} Ω/m) so using an antistatic spray on the surface of plastics may cure some of those problems.

Amateur radio operators who have an interest in mobile operation represent a source of a very large fund of practical suggestions; the *Handbook of the American Radio Relay League* contains many useful suggestions for particular circumstances.

The above notes were written specifically about motor vehicles but could be adapted with little change (mainly taking extra notice of corrosion hazards) for installations in boats and small ships. For larger ships and light aircraft where communications equipment has been installed as part of the original design it is likely that all that is useful has been originally designed in. In the case of aircraft in particular it could be very dangerous to make modifications to the electrical wiring without the most extensive consultation with the appropriately qualified experts and the original manufacturer. The installation should, however, be in accordance with the earlier parts of this section as well as meet the electrical standards that apply to that class of craft.

CHAPTER

11

Cellular Mobile Telephones

11.1 Mobile Phone System Types

Cellphone and trunking radio systems are variants of the general field of mobile radio communications. This chapter follows directly from Section 9.3 and should be read in conjunction with that section and also Sections 9.4–9.7.

When a single-channel system such as described in Section 9.3 is required to service a larger number of subscribers, some extra facilities must be provided in the technical equipment. At the exchange-end relay set a multitone selcall signaling function must be included to identify which subscriber is to be connected and in the subscriber's end a selcall receiver for each separate subscriber's line must be connected in the ringer circuit. There must also be a lockout facility to provide privacy for calls intended for other subscribers. For calls from a subscriber to the outside world, there must be a coded tone selcall sender at each mobile and the base (exchange-end relay set) must either have a tone detector capable of reliably discriminating between calling tones from different mobiles or the service requires a manual telephonist to arrange outgoing calls. For well-engineered systems of this type, a single-channel base can service up to about 6 to 10 mobile subscribers.

When there are more than 10 subscribers needing a mobile service, single-channel systems of this type can be duplicated, but duplication of single-channel services rapidly becomes uneconomic when the number of subscribers rises to hundreds or thousands. There is also a limit on the number of channels available which would limit total availability worldwide to something less than a million subscribers. A radically different philosophy has been developed to cater for the needs of multiple millions of mobile subscribers. The essential features are listed below:

- broadband channels arranged for demand assigned multiple access (called "trunking" in some parts of the world)
- reuse of frequency channels
- multiple bases with provision for automatic handover of traffic from base to base

Cellular mobile telephone systems can be described as either analog or digital and within each of those two classes there are a number of different standard protocols of operation. The analog and digital descriptions refer to the method of modulation of the RF carrier; the factors listed above are more basic than that and are used by all cellular systems no matter what method of modulation is used. You will need to understand what is meant by "trunking" as applied to a radio service and know how the multiple base and reuse of frequencies factors operate in order to follow the thread of later sections of this chapter so the next two sections will explain these points.

I recommend that you not proceed to later sections of this chapter until you understand the basic concepts of the next two sections.

11.2 Radio Trunking

The principle of demand-assigned multiple access or trunking is adapted from the internal circuitry of an automatic telephone exchange, and it is worthwhile to look at what trunking means in the exchange to understand how the same principle works for radio signals. Figure 11.1 is a block diagram of part of a typical telephone exchange with some of the trunk circuits identified.

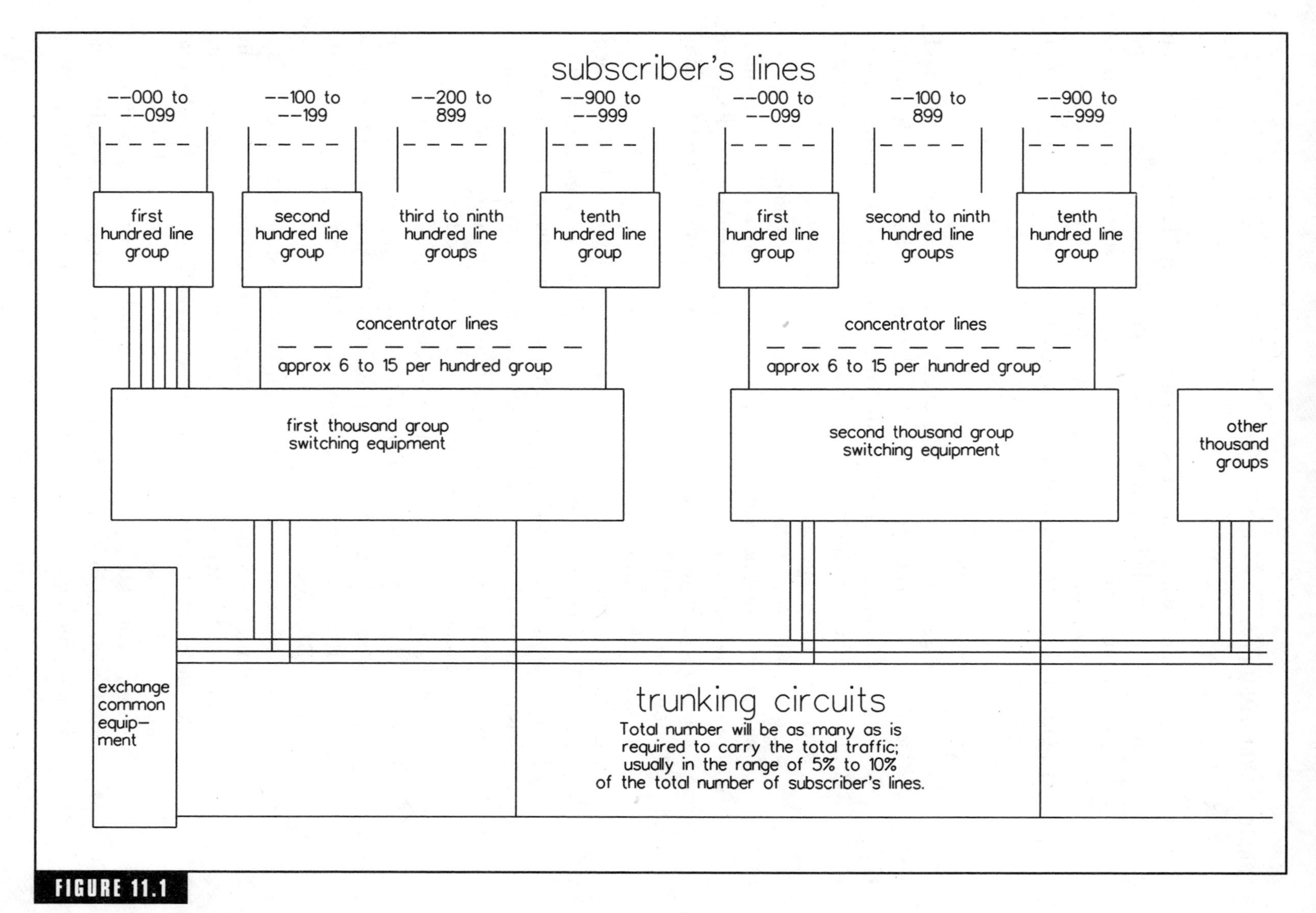

FIGURE 11.1

Trunk circuits in a telephone exchange.

The only wiring in the exchange shown in Figure 11.1 that can be identified as belonging to a particular subscriber is the incoming pair of wires of the subscriber's line and the first few contacts in the hundred-line group of switching equipment. When a handset is picked up to initiate a call the exchange controller connects that subscriber's line to whichever of the concentrator lines is free. In the thousand-group switching equipment the concentrator line is connected to one of the trunking circuits.

The decision about which trunking circuit is connected to which subscriber is made by the exchange controller based purely on the technical needs of the equipment. There is in no sense any ownership of particular trunking circuits by any particular subscriber or group of subscribers; all subscribers have an equal share of access to all trunking circuits. Any of the trunk circuits can be switched to any of the subscriber's lines as required provided they are not already being used by another subscriber, and under normal circumstances successive calls by a particular subscriber would probably use different trunking circuits to carry the call.

When that principle is adapted to a radio system the physical trunking circuits are replaced by channels in a band of the radio spectrum. The intermediate stage of concentrator lines between the hundred-line group and thousand-line group switching does not exist but the basic concept that a group of channels is available to a large group of users but not owned by any specific one is the same as for the telephone exchange. Radio systems of this type are based on repeaters which have capability for multiple-channel operation and a central controller which does the actual allocation of a particular channel for a particular call. All channels are available to each user but are allocated only on demand, hence the name "demand-assigned multiple access" (DAMA).

To initially make the connection there must be a control or data channel for communication between the mobile outstations and the controller. All mobiles switch to the control channel when they are idle, the mobile commences contact with a short burst of digital signal (like a selcall) on the control channel, the controller then uses the control channel to call up the receiving mobile, and when both are on line it instructs them by a digital signal to switch to one of the voice channels for the actual conversation. All this digital negotiation takes only a few seconds and sounds to the nonspecialist listener not much different from a selcall being followed by a revertive tone.

There are "trunking radio systems" that have no connection to the telephone network which use the DAMA principle to allocate repeater channels to a large group of mobile transceivers and that, from the user's point of view, work almost exactly the same as any other mobile-and-repeater transceiver network. The differences that may be noticed are that for each call there is a few seconds of tone signaling between the mobile and base to establish the channel and the waiting for a clear channel that is so common for single-channel systems is almost nonexistent with these. When the demand-assigned multiple access principle is used for a mobile telephone service the controller connects each mobile to a telephone circuit leading to the public switched network rather than to a repeater channel to call up another mobile.

A point worth underlining in this section is that the common layman's understanding of "trunk line" telephones as being those with long-distance operation has no relevance to the concept of "trunking" as applied to a radio system. A single repeater of a trunking radio system has exactly the same coverage area as would a single-channel repeater of the same power and operating frequency located at the same spot. For a cellular mobile telephone system, the coverage of a single repeater (using the radio trunking principle) may be quite small; in some parts of mature systems, individual cells may only be a few hundred meters across.

The advantage of this method of use of channels is that for a given expectation of an individual user being able to find a clear channel, the total traffic carried when all users have shared access to a group of channels is much higher than is possible when each user only has access to one channel. For instance, if a group of users shares one channel and the system manager specifies 50% expectation of a clear channel as a grade of service condition, that means the channel can only carry half of its maximum traffic capability. If the expectation of a clear channel must be 90%, then the channel can only carry traffic to 1/10 of its capacity and if 99% availability is required, then usage can only be 1% of capacity.

If two channels are available, the system can offer 50% expectation of a clear channel if both are carrying 75% of their maximum capacity and the 90% grade of service can be achieved even if one or other of the channels is in use almost all the time. This factor becomes more favorable as more channels are made available to the extent that when a group of about six to eight channels is used, the system can offer 90% expectation of a clear channel even if all except one channel are carrying almost 100% of their maximum capacity.

Time factors also are made more favorable by multiple-channel operation. If requests for service can be stored as is normal for processor-controlled systems, then if conversations last for an average of 1 min each and six channels are available, a user can expect a clear channel within 10 s even if all channels are carrying traffic to 100% of their capacity.

11.3 Cells and Frequency Reuse

All of the technology described in Section 11.2 can be used at a single base and there are radio repeater systems designed around a single repeater which use DAMA simply to gain the high traffic volume and channel availability functions that it offers.

A cellular mobile telephone system, however, is planned around multiple bases with low power and short range for each base. The systems use frequency modulation and carrier frequencies higher than 900 MHz, where signals that are not over a line-of-sight path are severely attenuated. Under these operating conditions it has been found possible to rely on the capture effect of the FM system to allow frequent reuse of channel frequencies. Providing that all the cells in a region are about the same size, bases that are separated by two other cells can be operated on the same carrier frequency so that an infinitely large group of cells can be provided by using a minimum of seven sets of carrier frequencies. The diagram of Figure 11.2 shows the theory of how that can be done. In Figure 11.2 the capital letter (A to G) in the center of each cell indicates the frequency set used for that cell.

In the larger group of Figure 11.2 it is possible to show that every cell is surrounded by six cells which have the other six frequency sets and that in all cases there are two cells between any particular cell and the closest reuse of its frequency. If the Earth were smoothly spherical and population were evenly distributed the scheme of Figure 11.2 could be extended to cover the whole Earth with evenly sized cells (which in the original theories were expected to be about 5 km across) using only seven frequency sets.

In practice, however, cell sizes do vary. In the central business district of a big city, traffic volume may be enough to make very small cells (perhaps only a couple of hundred meters across) economically justified. At the other end of the scale, a cellphone base on top of a high mountain or even a moderately sized hill with a clear sight over a body of water may cover a radius of up to 40 to 50 km and in fact may cover all the operating area of a group of smaller cells in a densely populated city. The total band allocated for cellphone use needs to allow for a few more than the theoretical minimum of seven sets of frequencies.

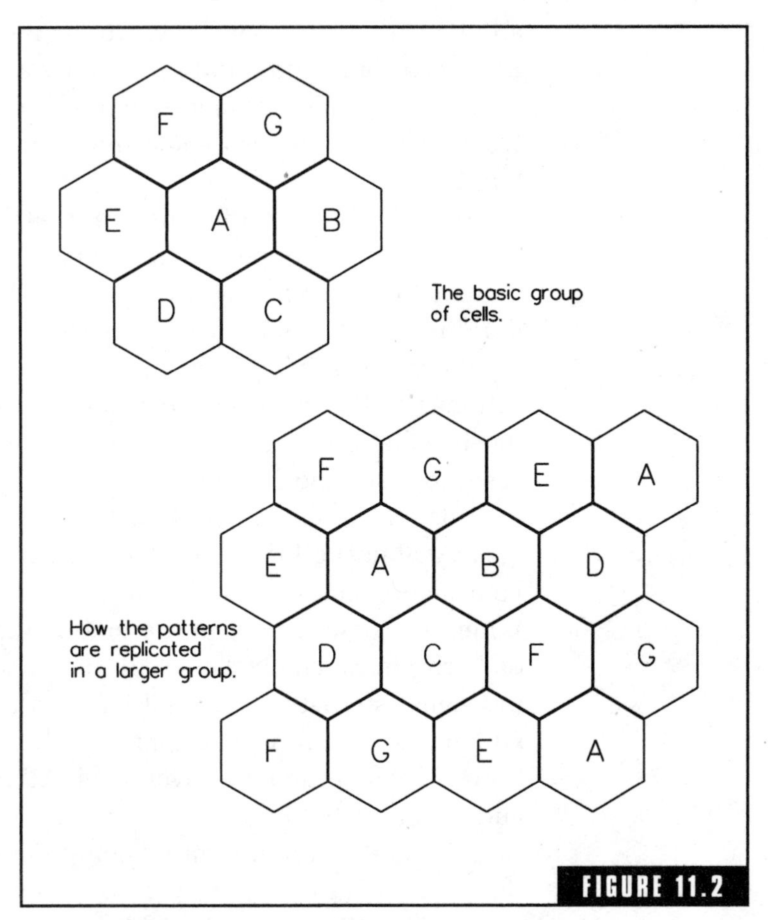

FIGURE 11.2

Radio cells plan showing how frequencies are reused.

It needs to be stressed that this repetitious reuse of frequency sets is only workable with frequency-modulated transmissions and relies on the capture effect of that modulation method. Capture effect results from the rejection of amplitude variations inherent in the FM detection process in the receiver; in all cases where there is a few decibels difference in signal strength between the two signals the detector output can only follow one signal so there is a tendency for the other to be silenced. In cellphone systems it has been found that roughly 12 dB of difference in field strength is sufficient to give interference free reception of each transmission in its own cell. In order to guarantee that the signal from the desired base in each cell is at least 12 dB above all others there is a power control/gain control function in the mobile transceiver and some tailoring of aerial illumination pattern at each base. Capture effect is mentioned in Section 4.5 in *How Radio Signals Work* and there is also some relevant information in Section 9.7 of that book.

As time goes on traffic volume in each base grows. (At least the operating company hopes it will!) When traffic is close to the maximum volume that the existing cells can handle, extra cells can be added by building new bases in locations halfway between those that are carrying the most traffic. If extra frequency sets are available in all those locations no other changes are needed but if not there may be a need for the installation project to include some frequency changes at other bases to accommodate the extra cells.

From the point of view of the mobile user most of this can be ignored almost all the time. The allocation of which base will contact your mobile and on which channel is decided by a central computer. When you first switch on the mobile the system goes through a "log on" procedure which checks all the functions of the mobile, compares the signal strength in several bases, allocates a channel on the base which has the strongest signal, and finally commands your mobile to transmit at a particular power level. All that happens in a few seconds and is completed by the time the mobile displays the welcome message. From then on the central computer keeps watch on your connection and makes changes, even to the extent of changing the base you are working through in the middle of a call.

11.4 When Cellphone Reception is Weak

The question of what to do when the signal becomes weak and noisy has at least three different possible answers depending on where you are and the local environment.

First, if you are in a big-city building the masonry and in particular reinforcing rods in the concrete as well as such things as metal window and door frames and networks of electrical wiring will severely absorb signals in the 900-MHz range. If you have a signal strength problem in a place like that you will make a significant improvement if you can take the phone outside the building to a balcony or such in the direction toward where the signal is coming from. If you cannot get outside moving to the center of a floor-to-ceiling picture window will do almost as well.

The fly in the ointment is that in a big city you will almost certainly be within range of several bases and you will not normally know which direction the signal is coming from. If you have access to windows on each side of a corner of the building you may be able to do a comparison test by trying a window in another direction if the first trial is not satisfactory.

A different situation is the fringe area. You may be out in a boat close to the horizon or traveling in a car toward the service area and wanting to make an urgent call as soon as the signal is strong enough. In places like that, height is the key; there may be a good strong signal only a few meters above your head and a relatively sharp reduction in signal strength as the mobile aerial is lowered.

In the vehicle, if the road is at all undulating, do a test at the top of each rise and as you drive along keep testing at the top of each rise until you can pick up a signal. On the boat, get as high as you can. (This only applies either if the service uses analog modulation or if digital you are located within the workable radius from the base; refer to Section 11.5 for more information.) Note that it is only the aerial that needs to be raised; if it is detachable and you have a coaxial cable that can be used to connect between the aerial and the phone you could tie the aerial to the top end of a fishing rod and use that to raise the aerial into the strong signal while you stand on the deck to talk.

Third, you may be attempting to carry on a conversation while you are driving (actually, while someone else is driving is safer!) through a suburban area that is a bit hilly. You may find the signal fading out while you talk. You are probably sort of within range of several bases, each of which is covering the area with beams of strong signal alternating with shaded areas. You only need a workable signal from one base and in almost all locations there is a strong signal from at least one base but as you drive along the system must keep swapping bases to keep in contact.

When you first switch your phone on and it goes through the log-on procedure the central computer will poll several bases to find the one with the strongest signal, but once the call is established it is left connected through that base for as long as the signal is strong enough to satisfy that base even if your trip takes you well outside its intended cell area. There is also a time delay, the signal strength must be low for several seconds before a switch is initiated. If the call has a high priority, the way to get the best signal is to drive slowly, watching the signal-strength indicator as you drive; when you find a good spot stop there and make the call.

If you must talk on the go and you find signal strength is gradually fading you can force a switch by artificially making the signal ***weaker.*** For instance, if you are using a handheld phone in the passenger's seat, you can switch to another base by moving the unit down to between your feet for a few seconds.

11.5 Living on the Fringe

In geographically large, sparsely populated countries like Australia, terrestrial mobile phone systems will never cover all of the population. Systems in use in Australia in 1998 advertised coverage of between 90 and 95% of the population, but that is achieved by covering only about 4% of the area—which consists of the major cities and some regional centers and the highways connecting between them.

As with all radio systems there will be a zone at the edge of the coverage area of each base where signals are weak and patchy. In most of these zones the cellular aspect of multiple sources of signal with the possibility of handover from one base to another is absent; the system takes on most of the characteristics of a UHF repeater-based radio service. The system designers will have attempted to place most of the area of these fringe zones in agricultural, pastoral, or other sparsely populated land.

If you are a farmer or pastoralist whose land is on the edge of a cellphone service area and you wish to use a telephone in the tractor, truck, harvester, or other machinery that you have on the property you will be repeatedly covering the same ground day after day so you will soon find that the phone is usable at some locations but not at others. Try to find out the location of the cellphone base that covers your land—that will be a help in predicting where the good signal spots are. The rules of UHF propagation described in Chapter 9 apply to this service exactly as they do to other radio-communication services.

The foregoing applies more to analog or CDMA ("code division multiple access") than to digital cellphone systems. For cellular systems using digital modulation with time-division multiplexing (such as GSM) there is a limitation on distance even if the signal is strong enough to be workable. Digital systems use signals fitted into time slots and if the total round-trip time delay from base to mobile and back again is too great, the signal can end up in the wrong time slot. For the GSM protocol, the maximum workable distance from the base is 35 km. If at the stage of signing up for a mobile phone service you have a choice between analog/CDMA or digital systems, get a map of your local area, plot the position of the digital base on it, and draw a 35-km radius. If your land is all safely within that radius you will probably get generally better service overall from the digital system. If a portion of your land is further than 35 km from the digital base, the analog or CDMA system will give better results.

For signals in the frequency range used by cellular telephone systems, reflection is the major mechanism for signals to get beyond line of sight. Section 5.6 of *How Radio Signals Work* shows what happens. Diffraction, groundwaves, and refraction only have a minor effect in the form of making a slightly fuzzy boundary on the shadows.

11.6 Cellphones by Satellite

The restrictions on coverage area in less densely populated regions can be overcome by using a satellite-based repeater. The terrestrial cellphone systems are not directly compatible with satellite links but the differences are not so basic that a unit cannot be built that covers both services. To allow the mobile units to have simple aerials as is required for contact with terrestrial services the satellite links must be to low Earth orbit repeaters which means that for continuous coverage a large cluster of satellites is needed. The facility of the terrestrial service for automatic handover of the mobile from base to base solves one of the most basic problems of using LEO satellites. Multichannel operation, which implies a broadband signal, reduces the relative importance of the Doppler effect, which is another side effect of LEO satellite use. See Sections 8.3 and 8.7 for information on the mechanics of these processes.

In 1998 a service using LEO satellites became available under the brand name "Iridium." It was a huge project that required an annual budget equivalent to the GDP of many small nations for several years.

The service used 66 satellites arranged in six clusters to offer continuous coverage of all points on the Earth's surface and offered telephone contact with mobiles that could be made compatible with terrestrial cellphone services. Satellites traveled in polar orbits at altitudes of 780 km (above sea level). The radio link from the satellite to the mobile phone used carrier frequencies in the 1.6-GHz range with enough power and sensitivity for the mobile to use a nondirectional or very low gain aerial. The Iridium Corporation was declared bankrupt in March 2000 but it had shown the way for LEO-based mobile telephone systems and in the twenty-first century there are several large multinational companies offering services based on that principle.

The LEO satellite-based system does not have the approximately $\frac{1}{4}$-s time delay that is characteristic of geostationary services; there is still a time delay but too short to be noticed by human users. The delay is long enough, however, to render digital services based directly on the GSM principle unworkable. Each satellite has an aerial system that divides its footprint into many separate cells and the communication link for each cell uses a combination of time division and frequency division multiplexing to give connection to a large number of individual mobile units. A variety of special-purpose modulation methods are used for the links to the mobiles and between the satellites and Earth stations. Systems are engineered for a huge volume of traffic; worldwide they may be handling several million two-way conversations and paging calls at any one time and there are not busy and quiet hours as is characteristic of terrestrial systems—there are zones of high traffic volume which sweep around the Earth in time with its rotation.

Another difference between LEO and geostationary services is that in the geostationary case the mobile only needs connection with one pinpoint spot in the sky, usually at a moderate elevation above the northeastern (in the Southern Hemisphere) horizon. LEO services, however, have satellites moving across the sky and require clear access to a much greater proportion of it. The satellite previously in use may be quite low on the horizon before the next one is available. In the worst case when the Iridium mobile user was halfway between the subsatellite tracks of the orbits each satellite was never more than 17° above the horizon and could have gone as low as 4.8° elevation before another satellite became available; in that case, clear line of sight to the northeast, southeast, southwest, and northwest horizons was an advantage. In the case where the user was very close to the subsatellite track the satellite would follow a north–south line and the one previously in use was very close to the horizon before the next one rose.

A clear line to the horizon in all directions such as would apply to an open ocean or treeless plain is ideal for these services. Nearby obstructions such as buildings, trees, and hills will all sooner or later cause a temporary interruption to the service. The most recent players in this market have recognized this hazard and have engineered satellite numbers, orbits, and altitudes so that a figure for minimum elevation above the horizon is included in the system specification. Figures of 20 to 40° are commonly used.

Terrestrial mobile services all use short-range links (maximum 40 to 50 km), and each base represents a considerable financial investment so they will never be used to give coverage to remote agricultural, pastoral, and wilderness areas. The smallest of the mobile units for contact with a geostationary satellite telephone is a unit the size of a briefcase with a directional aerial which must be aligned to establish contact. The specific feature that LEO mobile telephone systems can uniquely offer is worldwide access to the public switched telephone system with a handheld device.

11.7 Technicalities

If you are operating a general electronics servicing business and a cellular phone is dropped in with a "please fix it" message attached, be aware of the following points:

The system is duplex so if you attempt to measure receiver functions you will always have transmitter output power to deal with.

There is a power control function in the working system so it is possible for a properly working unit to give a power meter reading anywhere from a few milliwatts up to about 3 W.

The exact operating frequency is determined by software at the base and may be changed at any time.

Because of all this, proof-of-performance tests using basic instruments can be quite difficult. In service depots testing is done with radio-communication test sets equipped with adaptors for the specific protocol. There is, for instance, an adaptor for a particular brand of test set to work with the AMPS protocol, and AMPS mobile phones of all brands can be tested using that adaptor. If a phone which uses the GSM protocol is to be tested, a different adaptor must be used. In the field, however, the correct combination of test set and adaptor will not be available very often and only "first-in" maintenance is possible. There are some notes on first-in maintenance in Section 11.10 of *How Radio Signals Work*.

If the phone is a handheld unit and a same-make-and-model copy is available, the simplest way of proving a fault is by substitution. You will need to do the test by setting the known good unit in a particular position and noting indications then removing it and testing the suspect unit in exactly the same position. (A side-by-side test is not good enough.) When you can identify a particular function that is different try swapping the batteries and see if the fault stays with the phone or goes with the battery; then try the same with the aerials. If the fault follows either the battery or the aerial it can be cured by scrapping that item and replacing it. If the fault stays with the electronics package then you will probably have to send it to a service depot for repair.

When the equipment is built into a motor vehicle the direct substitution test may not be conclusive and there are liable to be several modules with cable looms connecting between them. For the aerial and its associated coaxial cable, a directional power meter will give definitive measurements if you can get enough forward power to give a significant meter deflection. For cable loom connections, most of the signals on individual wires will be DC (may be switched), audio, or low-speed data. All these can be displayed with an oscilloscope. If any faults can be traced to a handset, selcall, data reader, or similar module there is a good chance that they can be fixed in the field. If the fault can be definitely traced into the RF module that will need to be removed and sent to a suitably equipped service depot.

Mobile transceivers operating in a trunking radio net are not much different from those used for single-channel operation as personal mobile radio systems as described in Sections 10.8–10.10. Except for the specification of operating frequency all the proof-of-performance tests that can be done to a single-channel transceiver can be done equally well to these. However, as the internal functions may be extensively affected by the same microprocessor as controls the channel settings it is recommended that you not attempt to change any components without first having read through the service manual for that specific model of transceiver even if the component to be changed appears to be well-separated from the frequency control function.

For systems operated through LEO satellites the proportion of clear sky available (see previous section) is a factor to be taken into account before any electronic tests are commenced. Also if the mobile has a dual role with access to a terrestrial cellular network the terrestrial equipment should be tested in isolation from the satellite link equipment. If a newly evident fault is proved into the satellite portion it would be worthwhile to wait about half an hour and try it again. In that time a different satellite will have moved into position, and if the fault is still evident on the second test the assumption can safely be made that it is in the mobile equipment. Any fault proved into the electronics of the mobile unit will require that it be sent to a fully equipped service depot for repair.

The economics of service work is a factor that has to be kept in mind. In Australia in the late 1990s terrestrial telephones are being offered by the service providers at little or no cost and it is tempting to regard them as throwaway items. That can be true providing that the unit is not subject to an airtime contract which must continue to be paid even if the unit is scrapped. If present economic conditions continue there will probably be many units that have no special functions and are not subject to a continuing contract and for those the most economic form of repair may be to use them as trade-in items for a new service.

CHAPTER 12

Data, Codes, and Selcalls

12.1 Technical Basics of Data Transmission

The essential differences between data signals and voices or video are:

The data signal uses square waves (i.e., the modulating signal switches from one state to the other in the minimum time as determined by the bandwidth of the system).

The data signal may include steady-state signals which must be transmitted so that the output can be held at a particular DC level.

Radio systems are normally not capable of transmitting a DC level so some processing of the modulating signal and its resulting receiver output is required to transmit data over radio. The data signal can be made compatible by presenting it in a form that ensures a return to a reference voltage level at frequent intervals; examples are Morse code, in which dots and dashes are separated by spaces of zero voltage, or data presented in the "RZ" format, in which a period of zero voltage is interposed between successive bits of the digital signal. RZ ("return to zero") format reduces the rate of transmission of data to about half the maximum rate allowed by the bandwidth of the channel. In frequency shift keyed systems DC levels can be reconstituted if the original modulation is arranged so that the frequency of the transmitted signal is directly controlled by the DC level of the input data. Amplitude modulation cannot be used to transmit a DC reference except in the form of a subcarrier tone modulated onto the carrier at a particular modulation percentage.

At each instant the data is presented as either one of two logic states which when frequency-shift-modulated onto the RF carrier will mean that at any instant the transmitter output will have one of two possible frequencies. Output on other frequencies in the passband of the channel is not possible. That fact can be used to provide a form of noise immunity and, in fact, some reconstitution of clean data from a noisy signal is possible. The detector in the receiver is equipped with a circuit of the Schmitt trigger type, which merely senses signal above or below particular levels and sees all intermediate states as noise. The process is illustrated in Figure 12.1; in this diagram the signal received off-air is shown as the voltage output of a discriminator or ratio detector. In circuits of this type the exact setting of trigger levels and the hysteresis between them is one of the trimming adjustments that has a very large effect on the success of the overall installation.

If the noise level on the channel is low enough the rate of data transfer for a given RF bandwidth can be increased by allowing for multiple-level outputs. If the receiver demodulator can resolve four

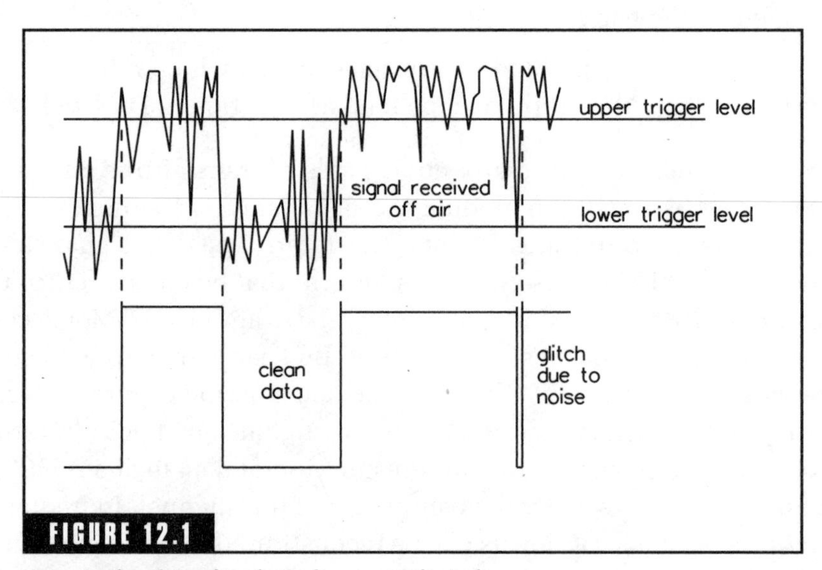

FIGURE 12.1

How a noisy data signal can be reconstituted.

separate voltage levels two bits of data can be signaled at each change of state. If eight voltage levels can be resolved three bits can be signaled and if 16 levels can be resolved then four bits of data can be signaled at each instant and so on. The need to resolve multiple voltage states causes a drastic reduction of the noise immunity of the system. For VHF and UHF services engineered for data only, and using a variety of FSK modulation, noise level can be made lower by increasing the level of the received signal (putting limits on the total allowable path loss) so there is a tradeoff between maximum-range services operating at low data rates and higher-speed services with more restricted range.

Radio systems used to carry data signals can be operated in either of two ways. The data can be modulated onto a subcarrier tone or the system can be arranged so that the instantaneous logic state directly controls the RF carrier frequency. The subcarrier option is used when the radio link must also be able to carry voice, video, or some other signal with analog components or when the radio link uses single-sideband modulation. A modem of the type used to connect personal computers to the Internet via telephone lines transmits the data using subcarrier tones. If properly configured, a modem of that type can be used to transfer computer data over a normal voice link operated by radio transceivers. Directly modulated data links offer higher data-transfer rates but must be dedicated to data operation only; they cannot handle analog signals except in the form of the output of an A to D converter. In the case where a subcarrier is used, the RF channel must have sidebands to generate the subcarrier tone as well as those used to transmit data information. Figure 12.2 shows the sideband structure of each of the two systems for the simple case where each interval only signals one bit of information.

12.2 Bandwidth

For wired circuits the only consideration given to bandwidth is "as much as possible for the cost!" Integrated circuit designers are continually working to develop faster and faster circuits because there are no intrinsic technical limitations on how much bandwidth can be used. It is not so for radio, however. The sensitivity and operating distance capability of a radio system is limited by how much noise is present

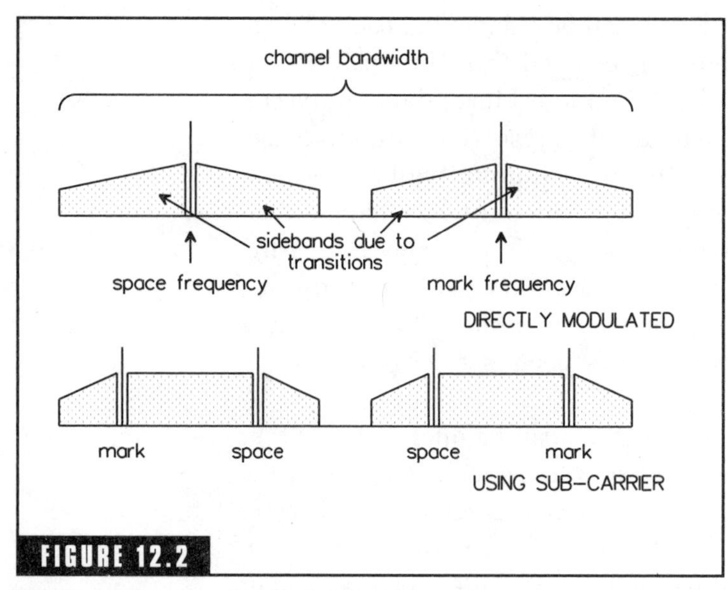

FIGURE 12.2

Sideband structure of radio data links.

with the wanted signal and for most noise sources the distribution of energy in the spectrum is wideband so that a wider bandwidth almost always means more noise and therefore lower sensitivity. The actual relationship between noise power and bandwidth may vary depending on the type of noise source.

For certain types of thermal noise such as that typically generated by the first amplifying stage of a receiver the relationship is that noise power is proportional to bandwidth so if sensitivity is limited by internally generated noise then a particular change in bandwidth can be calculated to give a certain change in sensitivity. In terms of signal voltage sensitivity is inversely proportional to the square root of the bandwidth.

When the noise source is impulsive (distant thunderstorms) a different relationship may apply between frequency and noise power. Power tends to be concentrated at low frequencies and those which are favored by any resonances in the generating mechanism. For radio receivers when the channel width is a small proportion of the carrier frequency the power per unit of bandwidth will tend to follow the thermal noise relationship fairly closely. Wideband systems such as VHF television may show some frequency-dependant factors in the noise/bandwidth equation.

When the noise source is energized by repetitive power, such as AC mains or a computer monitor deflection circuit, the noise is radiated as harmonics of the frequency of the source with broadband spectral characteristics as for the impulsive sources described in the previous paragraph. In those cases the noise power to the receiver will be related to the number of harmonics that fall within the receiver passband.

For example, with actual numbers an AM receiver requires a 6-kHz bandwidth to detect a communication-quality voice signal and a particular receiver may be able to give an output with a 10-dB signal-to-noise ratio with 1 μV in 75 Ω applied to the aerial terminal. The power of that signal is 0.0133 pW or 138.75 dB below 1 W so actual noise power in the first amplifier is about −150 dBW. If the same receiver is reconfigured to receive a Morse code signal using a bandwidth of 100 Hz instead of 6 kHz (60 times less) the level of noise power drops by 17 dB to −167 and the minimum signal power required for a 10-dB signal-to-noise ratio drops to −157 dBW. The voltage required for the minimum readable signal is 0.13 μV, which is the square root of 60 times lower. There are several practicalities which may modify that calculation; for instance, the receiver may not have enough extra gain to amplify the smaller signal to a usable level or the receiving operator may be a Morse code hotshot who does not need anything like a 10-dB signal-to-noise ratio for a readable signal.

In the other direction if the same electronic components are used as the input of a television receiver with a 7-MHz bandwidth (1167 times greater) the signal voltage required for a 10-dB signal-to-noise ratio is 34 μV; however, the minimum signal required for a comfortably watchable picture requires nearer to a 30-dB signal-to-noise ratio so the practical minimum signal for that television system is in theory about 340 μV. For a 40-dB signal-to-noise ratio close to 1 mV is required.

Radio data systems with speed buffering are not intrinsically confined to any particular speed of data transfer or bandwidth. There are wideband systems which can transfer several megabytes per second but these rely on links that are very restricted in range, strictly only line-of-sight paths, and even then there is a limit to the distance that can be covered in one hop even if free-space path loss is the only relevant factor. At the other end of the scale, bandwidths of only a few Hertz can be used to give a very great increase in working distance for one hop. Channel allocations for telemetry in the low-frequency end of the VHF range or at the extreme top of the HF band can be used with fair reliability up to distances of several hundred kilometers with 25-W transmitters if the receiver bandwidth is reduced to the 30- to 100-Hz range. There are also systems operating with that bandwidth over distances up to about 10 km with transmitter power of only a few milliwatts with the remote station designed to be operated by solar power.

12.3 Morse Code Technical Specs

Radio signaling by data transmission actually has a longer history than does the use of radio for transmission of voice signals. The first telegraphs used a source of electrical power, a switch, a length of wire, and a bell or sounder of some sort; the level of technology was equivalent to that of a doorbell circuit. The only information that such a system is capable of transmitting is the presence (called a "mark") or absence ("space") of a signal. The coding into short and long pulses of signal devised by Samuel Morse in 1838 is used to give that level of technology the ability to transmit alphabetic characters and punctuation. The first publicly acknowledged use of radio was for safety cover for international shipping using spark transmitters and Morse code. Apart from the advent of CW transmitters to replace spark and superhet receivers with very narrowband IF amplifiers to give the overall system greater sensitivity the basic field of data transmission by radio changed little for most of the 20th century until the growth of computer networking.

Morse code and the range of telegraphic codes related to it are designed to be read conveniently by a human operator. They are formulated from "dots" and "dashes"; a dot is a burst of signal that is the shortest burst that can be definitely identified (compares with a pixel on a video screen) and a dash is a burst of signal 3 times the length of a dot. A letter may be coded with anywhere from one to four elements and numbers are coded with five elements. The space between elements within a letter or number is the length of one dot and the space between characters is the length of a dash (three dots). Figure 12.3 shows how all that fits together in practice.

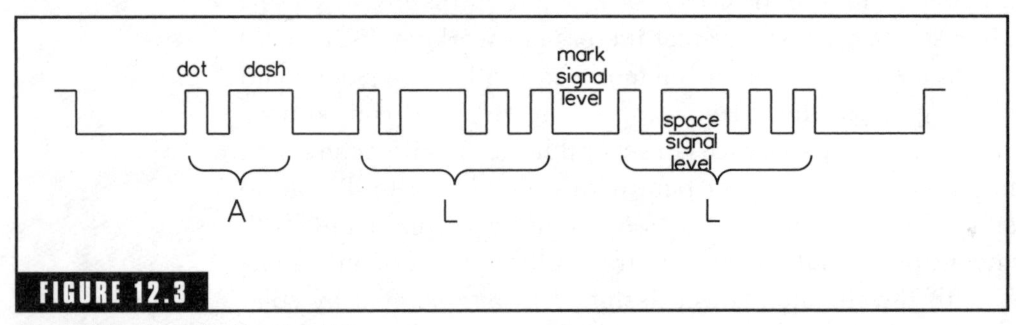

FIGURE 12.3

Morse code transmission of "all."

Morse code signals can be generated by machinery without too much difficulty. Mechanical devices using coding wheels similar to an incremental shaft position encoder have been used for many years to provide the identifying signals for navigational beacons. In more recent years circuits which step sequentially through the addresses of an EPROM have been more commonly used for that task. Detection of received signals by automatic machinery is, however, very awkward to arrange. The fact that characters do not have a constant length is one major problem. The way in which the coding of some characters is identical with the start of others is a difficulty (for instance, compare the "A" and "L" in Figure 12.3. A noise pulse received near the middle of the intercharacter space results in the detection of two characters run together as one and Morse code has no provision for error detection.

In the modern world most of the data transmitted by radio is intended to be received and decoded by a computer so other codes which overcome these difficulties have been developed and Morse has become less important. CW transmission and Morse code will never entirely disappear, however; when any form of modulation must be detected the signal strength required is above the minimum needed to just detect the presence of the carrier wave itself. Our ear/brain combination also has an intelligent filtering facility which allows us to concentrate attention on one signal in the presence of interference of about equal or greater strength from other intelligent signals. The combination of these two factors means that the minimum signal strength required for reception of a very weak CW signal by a human operator is about 20 dB lower than the signal of equivalent bandwidth required for error-free detection of data by a computer.

12.4 ASCII

This section describes a code which is more suitable for computers to use and explains some of its implications in relation to transmission of data by radio. The letters stand for "American Standard Code for Information Interchange." ASCII was devised specifically for transferring data between computers, microprocessors, and other data-processing devices. It is not specifically designed for radio transmission; in combination with a connection standard such as RS232 or V25 it can be used in relation to wired connections between computers or for connection to a modem or storage device (i.e., tape recorder).

ASCII uses eight binary digits (bits) to carry information. Of these, seven bits are used to encode 128 separate states which include the alphabet in both lowercase and uppercase forms, all the decimal digits, all the punctuation marks of the English language, and some nonprinting characters which have been assigned to internal control functions. The eighth bit is a parity bit (in transmission systems) whose state is controlled by the checksum of the other seven bits. A checksum is obtained by simple binary addition of all the bits ignoring the significance of the position of the bit in the word. If parity is defined as even and there are an even number of bits (two, four, or six) set to logic 1 then the parity bit is set to logic 0. If there are one, three, five, or seven bits set to logic 1 then the parity bit is also set to logic 1 to make the combined result of the eight bits an even number. That group of seven information bits and a parity bit is named an "ASCII word" even though what it transmits would be shown on a screen display as a single character. If a word is received in which the combined checksum of the eight bits is an odd number, the receiving computer detects that as an error and asks for retransmission of that word.

If parity can be dispensed with the eighth bit can be used to specify a second group of 128 characters. In some computer peripheral devices such as printers, the second field can be used for graphic characters such as horizontal or vertical lines of particular widths, the shapes of faces, cloverleaf patterns, and so on. In languages which are based on different characters the second field can be used to define those characters. For example, in Japan the first 128 characters are used for the original letters, numbers, punctuation, and so on of the English language and the second group may be allocated to a selection of the most commonly used characters of the Kana script. In almost all countries where ASCII is used, the first 128 characters carry the same significance; ASCII 41 is always "A" and 61 is always "a," for instance, but if the eighth bit is not used as a parity bit then the significance of the second field of characters is less standardized.

A full ASCII word includes another three bits which are a "start" bit and two "stop" bits. The start bit is always logic 1 and the stop bits are always logic 0 and if there is idle time between words the channel is held at logic 0. The logic 0-to-logic 1 transition is the signal for the start of the word. There is no return to a zero state between bits as is the case for Morse code; after that start signal the significance of each bit is determined by time from the start which is defined by the baud rate setting of the transmission system (Figure 12.4).

An important difference between Morse and ASCII is that in the Morse system information content is determined by whether the element is a dot or a dash, both of which represent a "mark" or "signal on" condition; the "space" condition serves no purpose other than to delineate the elements, whereas in ASCII both the mark and space conditions represent logic states which carry information, with the mark voltage level normally corresponding to logic 1 and the space voltage level corresponding to logic 0. The equivalence of logic states in Morse code would be related approximately to logic 1 = dash and logic 0 = dot but even that is not an exact relationship. In a comparison between Morse code and ASCII the features which make ASCII more suitable for use between computers are as follows:

- characters have a constant length
- the "non-return-to-zero" format gives greater information density
- parity coding gives an error-checking function

When a code such as ASCII is transmitted via frequency shift keying the transmitter is radiating power all the time the data channel is open. There is a frequency which corresponds to the computer's logic 0 and another frequency which corresponds to logic 1. The transmitter spends most of its time radiating one of those two frequencies, sidebands are only present during switching times, and the transmitter almost never transmits a signal exactly on the nominal center (unmodulated carrier) frequency of the channel.

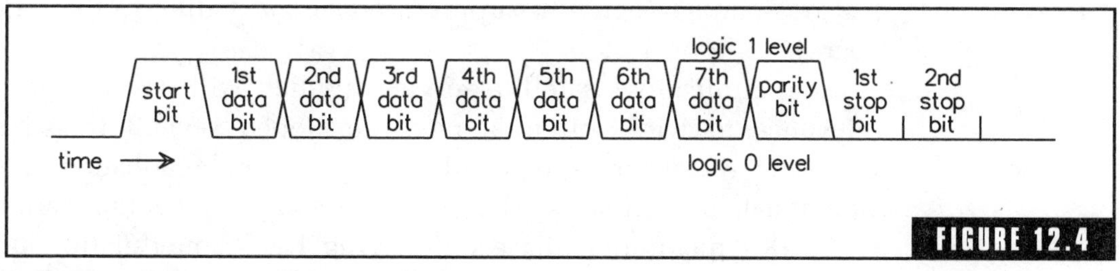

An ASCII word.

At the end of the twentieth century there are many transmission protocols in use for data links over radio systems, ASCII coding of 1VF channels with frequency shift keying is one of the simplest. There is a broad division between synchronous systems and asynchronous transmission. ASCII is an example of an asynchronous protocol because the timing reference for each ASCII word is the start bit within that word and there can be a random amount of time after the stop bits are completed before the next word commences.

There have also been advances in noise immunity and error detection over the relatively simple parity bit system. A couple of more advanced error-detection protocols are described in Section 12.5.

12.5 Broadband Data Links

When the volume of data to be transmitted is greater than can be accommodated on a single VF channel broadband links dedicated to high-speed data must be provided. The majority of such links are wired circuits associated with either the public switched telephone system or a LAN ("local-area network") or WAN ("wide-area network"). Transmission protocols are often selected to maximize performance of the wired circuit with radio links being treated as nonstandard additions to the wired system which require some adaption of the protocol to cater for the special conditions of the radio link. In many cases of that type operating conditions of the radio link have been less than ideal so the results have been relatively poor compared to what could have been possible with a system designed specifically to favor the operating peculiarities of the radio link. Deficiencies usually are seen in relation to weak signal performance and are overcome either by restricting the operating range of the radio link or by building extra power into the transmitter to provide a strong-enough signal to suppress noise and fading to the required degree.

Communications satellite services (which will never have much relevance to wired circuits) have developed several transmission protocols which are specifically designed for broadband transmission of data by radio link. The most commonly used of these are based on QPSK ("quadrative phase shift keying") as the modulation method and use of a very large number of carriers (several hundred) which are each modulated with quite slow rate data. The data streams are subdivided from the main flow by an encoding computer and the high rate stream is reconstituted at the other end of the link by a related decoding computer. Allocation of data to particular carriers is done on a semirandom basis to minimize the effect of noise and interference on the final output. Digital buffering of the data gives a worthwhile protection against short-term fading. The data streams on the many carriers are synchronized with a clock signal which must be transmitted on one of the carriers and there must also be provision for hand-shaking information between the computers. The randomizing code and some other housekeeping information must be included in the data streams.

These advanced coding schemes specifically developed for use with radio systems have an immunity from amplitude changes; as much as possible all the information is carried in the instantaneous frequency of each carrier. That feature gives the receiver opportunity for a useful degree of noise rejection but also has another advantage—the output amplifier of the transmitter can be driven to the limit of linear performance and slightly beyond into the nonlinear range with greater power output for a given size of amplifier and greater electrical efficiency. Both those factors are critically important in relation to signals transmitted from satellites (see Section 8.4).

12.6 When the Signal is Noisy

One of the characteristics of a radio link which is different from what is expected for a wired circuit is that when the signal weakens noise is introduced. On links carrying digital data the exact effect the noise will have depends on the type of noise, the signal-to-noise ratio, and features of the data signal. For example:

If the data stream is an executable program and the noise is a few high-level impulses the operation of the program can be completely disrupted by a noise level that would not even be noticed if the channel were used to transmit an audio signal.

The same rare high-level impulses in a fax transmission would cause a few small dots on the printed page.

Thermal noise of sufficient level will cause a random or semirandom background on a fax page but with a more comprehensive data-regeneration and error-correction algorithm clean data at a slow rate can be transmitted through even severely noisy conditions.

A heterodyne tone in the received channel can force the demodulator to give a permanent logic state (can be either 1 or 0) which totally prevents all transmission of data.

For error-free reception of data in severely noisy conditions the data is sent by two paths and the result compared. There are a wide range of protocols by which that may be done, but two of the most common which illustrate the principles involved are ARQ and FEC.

The ARQ protocol involves sending the data in small portions. When received, each small portion is loaded into a buffer and copied and the copy is then sent back to the source and compared. If no errors are detected, the next portion is sent; if there is an error, the previous portion is resent and will be sent as many times as required to achieve error-free transmission. With small portions (may be only a couple of dozen bits) ARQ is capable of error-free data transmission over very noisy channels; the only effect of the noise is to slow the rate of data transfer. ARQ relies on very rapid transmit/receive switching and no propagation delays; it cannot, for instance, be used for transmission via geostationary satellites; it is commonly used over HF communication links but requires that the transceiver is designed for t/r switching within a few milliseconds.

The principle of FEC can be explained as an extension of parity coding. The letters stand for "Forward Error Correction" and the name covers a range of techniques in which enough extra information is sent to allow identification of errors to a particular bit in the stream. For instance, with a group of seven of the ASCII words described in Section 12.3 an eighth word can be constructed as follows:

- bit 1 of the eighth word is a checksum of all bit 1s of the previous seven words
- bit 2 of the eighth word is a checksum of all bit 2s of the previous seven words
- bit 3 of the eighth word is a checksum of all bit 3s of the previous seven words

- bit 7 of the eighth word is a checksum of all bit 7s of the previous seven words.
- the eighth bit can be a checksum of the parity bits, which if there are no errors should also give a parity check on the other seven data bits of that word.

In this scheme of information incorrect transmission of a bit will show as a parity error on the relevant word and also as a parity error on the particular bit of the eighth word. That information is enough to identify and correct a particular bit in a particular word. FEC works best in relatively noise-free channels; in the above example if noise was severe enough to cause two errors in a particular group of eight words the detection system would be confused; however, the amount of error-detection information sent with the data stream can be tailored for the noise level expected on the channel. FEC is used where propagation delays are expected as, for instance, geostationary satellite links.

A whole range of error-detection schemes and a range of variations on both of the above principles are available. The exact scheme most suitable for a particular application will depend on the type of data (specifically the importance or otherwise of a single uncorrected error) and the expected error rate. From the particular interest of the radio technician the exact details of the error-detection scheme are often not needed to be known; the system works from the data-generating device or computer at the input to the link to the computer or device at the link's output. The technician must know of any implied effects such as rapid t/r switching if they apply, but apart from that it is only necessary to bear in mind that some bits of the data stream will be transferring information and some will be carried for the purpose of error detection and correction.

From the operator's point of view in the case where the signal is suffering gradually increasing noise or interference (as, for instance, a mobile station moving into a weaker-signal area) the observed effect should be that the data transmission rate slows and eventually stops before incorrect data is transmitted. The transmission of errors does not indicate faulty equipment; it indicates that the error-correction protocol is not matched to the conditions of operation in the field and faults of that type can only be corrected by redesign of the error-correction facility.

12.7 Selcalls, Paging, and Telemetry

A radio link can be used to remotely operate machinery; all that is required in principle is that the data output be used to trigger a relay which switches a circuit which can then do anything that can be done with electrical or mechanical energy. The simplest circuits of that type use a tone generator at the transmit end and a frequency-selective tone detector at the receiver arranged in the circuit of the single-channel r/t described in Section 9.3 CTCSS (see Section 9.2) is based on that principle. Performance of these simple devices is adequate for noncritical uses but subject to noise and interference; the receiving tone detector must include a time delay to ensure that the received indication is a genuine signal. Systems can be adjusted to give a reliable response so that no calls are missed but at that setting are susceptible to false responses due to steady tones (heterodynes) on the receiver output.

To improve the discrimination between true and false calls the system can use a short sequence of coded tones with the characteristics of a data signal transmitted by FSK as the operating signal. Signals used on VHF and UHF services have a lot of the characteristics of an ASCII word as described in Section 12.3; selcall systems can be constructed using the standard tones of a FSK 1VF data-transmission channel and in the same way as the ASCII word has a start bit, a body of data, and a stop bit, the selcall transmission has a lead-in tone, a series of coding tones, and a tail or status tone (sometimes). Digital signaling may also be based on schemes in which the frequency of the tone burst indicates a decimal digit (like 2VF telecommunications signaling) so that a group of five tones has the capacity to signal any one of 10,000 stations.

There are several "standard" protocols for selcall systems, each with its own standard tones and timing. Within a particular class of operation all digital protocols work about equally well but the receiver must be set to receive and correctly decode the protocol being transmitted. There are significant differences in the technical requirements for selcall for an HF SSB service compared with that for use with VHF/UHF transceivers. HF selcall requires several seconds of lead-in signal (which is not actually a single tone) and multiple repeats of the data; the total signal takes about 8 s to transmit and would be unwieldy to use on VHF/UHF. On the other hand, the VHF/UHF selcall signal would not work at all well with a sideband transmission and impulsive noise on the HF band would cause false codes to be received.

In modern digital selcall systems the receipt and decoding of the call is signaled back to the transmitter by a signal called a "revertive tone," which is actually a version of the digital selcall signal sent back from the called transceiver to the caller.

"Paging" is a general term used to describe several different functions. Simple tone paging may be used to alert a large group of people such as an emergency response team all at once. Similar tone paging may be used to signal a predetermined message such as "Go to a telephone and dial the switchboard" to an individual. Tone paging has the same operating conditions as selcall and in fact when the paging facility is used in association with a transceivers network a group call on the selcall system will serve the same function as a paging call. Paging may also be used to transmit an alphanumeric text message, often in the form of ASCII data on a 1VF channel.

Paging is a one-way-only transmission of messages; the receivers are receivers only, so the calling party has no means of knowing that the message has been received. If the paging receiver has been carried out of range or switched off or has a flat battery, the person with the receiver is the only one who will be aware that calls may have been missed.

Wide-area broadcast paging uses frequency allocations at the low end of the VHF range, around 40 MHz. An individual paging transmitter can offer significantly wider coverage than is available to a VHF/UHF voice radio-communication service operating at the same location because of the lower frequency and also because the paging service is not tied to a VF channel bandwidth so can offer a better signal-to-noise ratio by reducing the receiver to the minimum necessary bandwidth. Wide-area alphanumeric messaging services can also use multiple transmitters on the same allocated frequency with messages sent simultaneously to all transmitters. In all except a narrow zone where field strength is about equal from adjoining transmitters the capture effect of an FM system will greatly enhance one over the other and cleaning up of the received signal using a Schmitt-trigger function as shown in Figure 12.1 will very largely eliminate interference from the weaker field strength. Because there is only a one-way flow of messages, paging services must use forward error correction.

Telemetry and remote sensing systems can have a very wide range of technical standards based on how much data must be transmitted, how often, and at what rate. At one end of the scale a stream-gauging station or remote monitor for an isolated piece of machinery may only need to transmit the equivalent of one ASCII word every few hours and use a system based on a very low power transmitter teamed with a receiver with bandwidth of only a couple of hundred Hertz. At the other end of the scale a real-time television camera mounted in a satellite may require a link with a 5- to 7-MHz bandwidth to get the signal back to Earth. If more than a few bits of data are to be sent or the output of several different sensors must be multiplexed onto a single carrier the tendency in modern times is to present the message in a format that is easily used by computers with one of the standard baud rates (300, 1200, 2400, or 9600 Bd are common) and using ASCII or similar characters.

For many environmental monitoring duties, such as stream gauging or automatic weather stations, the value of the measurement is unlikely to change significantly for many hours at a time. In these cases it would be grossly uneconomic in terms of both electrical power and use of spectrum space to run the telemetry transmitter continuously. If there is only one remote station in contact with a base, the receiver at the base can be run continuously and the transmitter operate with a time switch; on a regular schedule the transmitter switches RF carrier on, transmits an identity code much like a selcall, transmits all the data collected to that time, then switches off. When there is more than one remote station the base must control timing; all the outstations have receivers running continuously and at the scheduled time the base sends an identity code to each in turn then switches to receive. The outstation responds with an identity code in reply then sends all the data it has collected. When the base has received error-free data from a station it disconnects that one until the next scheduled time and sends the code for the next station. This process is called "polling."

If you are a technical person charged with the maintenance of a paging or telemetry service of any type, it is suggested that before you accept the contract you or the owner of the equipment take possession of a service manual for that particular system. Essential information required for any effective servicing work includes the following:

- assigned carrier frequency for the radio system
- type of modulation, bandwidth (deviation), and maximum data rate
- level and frequency of all pilot tones and subcarriers
- significance of each frequency in the modulation bandwidth
- expected signals at input and output test points for each subsection
- sense of logic (i.e., does logic 1 correspond to +ve or −ve voltage and higher or lower frequency), baud rate of data, and error-correction scheme

If the source of the data is a remote station which forms an integral unit with the radio equipment you will also need information on the type and location of the sensors supplying the original information and details of any preliminary data processing that may be done before the input to the radio link.

12.8 E-mail, Voice Mail, and Packet Switching

If the data must be transmitted in real time and received immediately, the radio link is required to offer a commercial grade of reliability (i.e., up to 99.9% for major backbone circuits) and so must be substantially as described in Chapter 9 and Sections 12.1, 12.2, 12.5, or 12.7 of this chapter. If, however, a time delay is allowable, the data can be recorded and transmitted at some later time when the channel is available and when it arrives, can be stored for retrieval at the receiving operator's convenience; then a whole new set of operating conditions become available and channels become usable down to a much lower level of reliability. For radio circuits, a lower requirement for reliability translates into longer maximum distances for links of a particular power level and receiver bandwidth and if speed buffering can be used with reduced receiver bandwidth quite dramatic overall increases in path length can sometimes be designed for.

The equipment for these classes of service uses the functions of a radio transceiver and a management computer in one interconnected unit. T/r switching, transmitted frequency, modulating signals, receiver frequency, and received bandwidth are all under software control as shown in the functional block diagram of Figure 12.5. When the channel is normally clear but reliability is low due to vagaries of propagation the system can be configured so that the transmitters at each end send a beacon signal. The receiver at each end listens for the beacon and when it is detected the computer sends first hand-shaking information and then data from the file and continues to send file contents as long as acknowledgment of receipt is being received and there is data in the file to be sent.

Systems of that type can be used on the HF, VHF, and UHF bands with operating bandwidth, data-transfer rate, and time delay all variable over a very wide range. At one end of the scale, the principle can be used in full duplex links with data rates of megabits per second and typical time delay of only a few microseconds and at the other end of the scale the equipment used is HF-single-sideband with nominally 1200 Bd rate and many hours or up to a day or so time delay. Systems operating in the low-frequency end of the VHF range can use propaga-

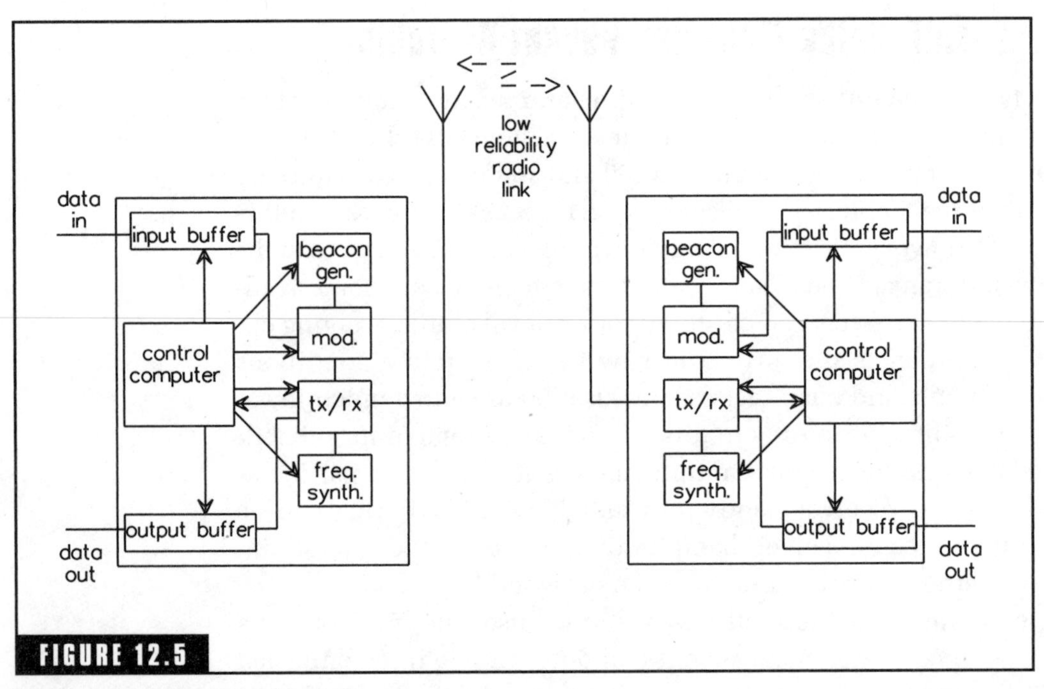

FIGURE 12.5

E-mail system functional block diagram.

tion via meteor trails and other intermittent phenomena over distances of several hundred kilometers with time delays of typically seconds to a few minutes.

Packet-switching systems use a refinement of this technology adapted for the case where the limiting factor is congestion of the channel, not propagation. No beacon signal is needed; the receiver monitors the channel measuring the degree of congestion. Whenever the channel is observed to be clear and a file is waiting to be sent the system initiates communication with the intended receiver and dumps a packet (or frame) of information to the transmitter. There is no monitoring while the packet is being transmitted, but at the end of each packet a check is made for a clear channel before a new packet is started. Packet switching is used on wired circuits as well as radio channels to increase the usage rate in those situations where several data devices must share a common cabling network of restricted bandwidth and real-time data is not needed. Packet radio is used extensively on the amateur bands.

In amateur use a standard one-voice-channel amateur transceiver is connected through a modem to the serial port of a standard home computer and all the functions required to assemble the packets and control their transmission are provided by software in the computer. The protocol uses 8-bit frames assembled into fields which may vary in total length. The start and end of a field is signaled by a unique "flag" frame which consists of a 0, six 1s, and a 0. The rest of the field is constructed so that no more than five 1s in succession are ever transmitted except in the flag frame.. Each field also contains an address which transmits the callsigns of the sending and receiving stations and any repeaters being worked through as well as a control frame which specifies the type of information being transmitted and the total length of the field. Each field is therefore a separate packet which for as long as it stays complete can be transferred through many terminals and repeaters before being eventually received by the addressee.

The AX.25 protocol used by amateurs is derived from and closely similar to the CCITT recommendation X.25 designed for commercial use. The differences are due to the requirement that for amateur use there are a very large number of potential addressees (all the licensed amateurs in the world) and all are peers. In commercial nets the size of the group is usually much smaller and the net is arranged so that one station is the master in control of the net and all others are slaves. In modern systems access to the radio channel and checking that it is clear is the function of "terminal node controller" (TNC) software in the computer. For some earlier experimental systems those functions had to be done manually. These systems can transfer any information that can be reduced to digital data including voice messages, images, and programs. In operation the sending operator loads data into the input buffer, the system transfers it at a rate depending on channel availability, and when the file is complete and error-free the receiving operator is notified and collects the file.

12.9 Servicing Data Radio Links

When you are asked to find a fault on a radio data system the sensible first step it to try to define whether the fault is due to the radio link itself or the attached data-processing equipment. In some cases of complete failure a transmitter or receiver giving no output can be easily identified using basic test instruments. In more difficult cases the link may appear to be intact but data transfer is being impeded by errors resulting from distortion of the modulation; these cases will be harder to define. For data systems a useful tool for breaking the system into sections is the OSI seven-layer model of the process of data transfer. OSI stands for "open systems interconnect." This model is most useful in the design stages but if the design is properly done the designer will have provided flags (or definitions of expected signals at test points) to signal the successful transfer of data from one layer to the next. The diagram of Figure 12.6 shows the OSI model and the path of a signal through it.

The OSI model was not originally defined for the purpose of fault finding but it gives a useful concept to break the problem into sections for testing. For effective data transfer, the signal at all seven layers on the receive side must match the corresponding signal on the transmit side. The layers become progressively less intelligent in moving from top to bottom on both sides. For instance, the "application" layer is concerned with whether the program is a data file, an executable program, a word-processor document, or such, whereas the "data link" layer is concerned with which voltages correspond to the logic states, whether data is synchronous or asynchronous, baud rates, parity scheme, and so on; the "physical" layer is interested in whether the link is a switched telephone line, a dedicated pair of wires, or a radio link, and such things as carrier frequency, modulation type, and bandwidth and ensuring that the receiving end is actually switched to receive when the transmitter commences sending data.

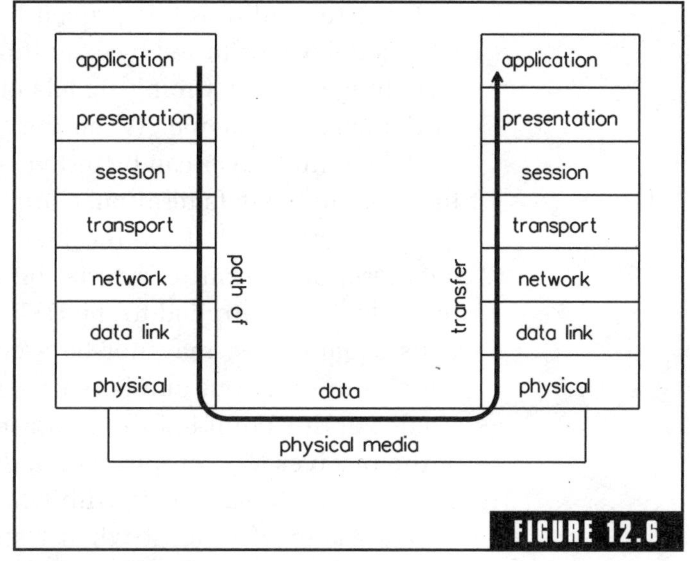

The OSI model of data transfer.

One significant point to be gained from the OSI model in relation to fault finding of data radio systems is that testing of the actual radio link should use test signals that are as unintelligent as possible. If you are working on a system that uses only one logical bit per baud (that is, only two voltage states) and you can get direct access to the modulator input of the transmitter and the detector output of the receiver, you should be able to completely define the performance of the radio link with the following three test signals:

- a DC voltage at the logic 1 level
- a DC voltage at the logic 0 level
- a square-wave alternating between these two voltages at the designated baud rate of the system

Modern data systems commonly use modulation schemes with greater information density using multiple voltage levels and/or quadrature modulation (in which the phase and amplitude of the carrier or subcarrier are both varied to carry the data). For these, a test procedure which is more complex but still not with intelligent data is needed. The test must cycle through all the possible voltage levels and must also test transitions of various sizes. Frequency response and group-delay errors in the radio link can cause overshoots or undershoots, which only are significant for transitions of particular sizes.

Once the radio link is proven by itself the "physical" layer of the OSI model can be tested by examining such factors as transmit/receive switching and any timing signals or pilot tones that may be included in the complete signal. When the "physical" layer is proven correct the "data link" layer can be tested. There should be some point where the input to the data terminal equipment is a serial bit stream on a single conductor, and at the receiver there should be a point where the data terminal output is a serial bit stream which is an exact copy of that which was applied to the DTE input at the transmitter. A continuous square-wave may not be a complete test at these points; data which send random numbers in a group of either 128 or 256 may be needed at this point to fully exercise the equipment under test teamed at the receiver DTE output with a decoder which can receive, record, and compare the numbers with what was sent.

The factor of interest when testing links from input to output of data terminal equipment is the "bit error rate" (BER) and BER testers are built for that purpose. The link should be tested in two conditions: with error correction active and also with the facility turned off. The figure obtained with error correction on is the proof-of-performance figure quoted when negotiations are required on systems whose parts interface with other authorities and the industry standard for most types of service is a BER of 1 in 10^9. The rigorous definition of BER at that level is a very slow process; even on a 2-MB/s link a BER of 1 in 10^9 corresponds to one error every 500 seconds and a test lasting at least 3 times that period is required to define the figure with any reasonable confidence. A fully rigorous test of a 1200 B/s link would indicate one wrong bit every 6 to 8 months in a test taking about 21 months to complete. The figure for BER with correction off will often be more useful to a service technician, particularly if the actual characters sent incorrectly form any sort of repetitive pattern.

In some large networks the "transport" and "network" layers of the OSI model may be represented by separated physical items of equipment such as routers or packet-switching devices. The top three layers are always related to software functions. In simple systems such as a personal computer connected via a modem to a radio link all the layers down to "data link" will apply internally to the computer (either hardware or software) and the radio link will directly relate to the physical layer.

If the system is big enough to employ specialists and if you can prove the fault is at a higher level than the "data link" layer on either the transmit or the receive side you are justified in handing the problem to the Information Technology specialist. At the other end of the scale you may be able to prove that the fault is in the transmit unit of a small remote sensing station whose purpose is to make environmental measurements of some sort at a particular (usually very awkward to get at) location; the OSI model may still apply and be the foundation for a useful fault-localization procedure. Figure 12.7 shows a practical application of that concept.

If your tests show the fault is in the radio equipment, localizing it is not greatly different in principle from the equivalent process on a link carrying voice or video information. The radio link can be broken into sections by testing the transmitter and receiver separately. The transmitter can be broken into "frequency generator," "input buffer," "modulator," and "output buffer/power amplifier," sections. For total failures, transmitter checks such as "output power," "reflected power in the aerial feeder," "output frequency," and "deviation" tested against published specifications for that particular piece of equipment should definitely point to fault location if it is in the transmitter. For degraded operation, more detailed checks of some

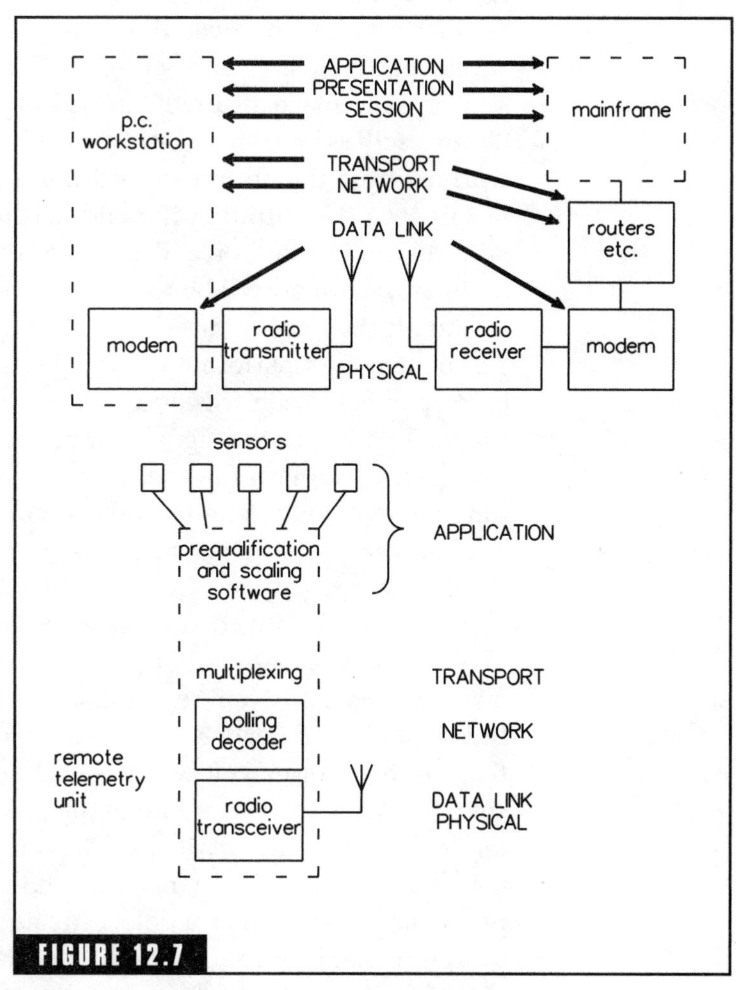

FIGURE 12.7

Relating the OSI model to items of equipment.

aspects of modulation performance such as frequency response and group delay may be needed in addition to the above for you to be sure of fault location.

At the receiver, "RF stages," "local oscillator," "IF amplifier," "demodulator," and "output buffer/line driver" are the sections. For most of these sections test procedures the same as for most other radio links can be used; the greatest difference is the nature of the modulating signal. The actual numbers for such things as IF bandwidth and signal-to-noise ratio may be different and there may be different degrees of emphasis on some aspects of the testing. By comparison with an audio signal, data places much greater emphasis on relative timing of signal components and reduces the importance of amplitude-related measurements such as linearity of gain. For most receivers, a sensible test strategy is to place a suitable measurement device on the output (may be an oscilloscope or logic analyzer) and test each section in turn working from the output toward the input. A function generator will usually test the output buffer and may also work on the demodulator; an FM signal generator will work for the IF amp and the RF stages; a frequency counter will be required to check the output of the local oscillator. If the receiver uses Schmitt triggering on the output a critical parameter of its performance will be the actual radio frequency that corresponds to each triggering level; a check of that will require analog modulation of the signal generator in the form of either a sine or sawtooth wave with the oscilloscope display calibrated in terms that indicate frequency. The test setup will be very similar to the sweep-and-marker testing used to check and adjust IF amplifier passband response.

With a radio-linked data system in field conditions you are unlikely to have access to both ends of the link at the same time. This factor can be a problem in cases where the performance of the overall link is slightly degraded and no clear cause can be definitely identified. In those cases you will naturally check first and set to standard specifications whichever end of the link is most accessible, and then if the fault is still present check the other end. If the overall link is still below specified performance after both ends have been checked the problem is described as a defect in "netting." In those cases, carefully set the transmitter exactly to its standard specifications (particularly with regard to output frequency) then go to the receiver and make minor adjustments to tuning of the early stages (aerial, RF section, and local oscillator) and triggering levels of demod output to bring the signal received off-air to the maximum sensitivity point of the passband of the IF amplifier. Receiver tuning should then be left at those settings even if they do not exactly agree with published specifications.

When the data system uses multiple carriers (for example, digital TV or satellite mobile telephones) the RF, local oscillator, and IF amplifier sections of the receiver will be the same as any other receiver with similar operating frequency and bandwidth. The FM demodulator may be replaced with a decoder which often may be a single VLSI-integrated circuit. The source of definitive information about the integrated circuit may be the IC manufacturer rather than the equipment supplier and often the only fault-clearance procedure available is to remove and replace the defective IC. This operation should not be attempted in the field—it requires desoldering equipment of a particular type (may need a vacuum or be hot-air-operated) and temperature-controlled tools for soldering in the new component.

Appendix

A1 The Theory of Fresnel Zones

When radiated energy arrives at a receiving aerial by more than one path the components of the received signal may add strength to each other or they may subtract and cancel or partly cancel each other. The effect depends entirely on the relative phase of the signals by the different paths, and that phase depends on the exact length of the path each signal has traveled over. For every transmitter/receiver combination there is one path which is shorter in time than all others. In free space that path is the direct line connecting between the two aerials, and if there are no objects to reflect the energy the receiver aerial will pick up energy from one signal which follows that direct path.

When a reflective surface exists in the radiated field, as illustrated in Figure A1, it will result in reception of a second signal component which can either add to or subtract from the received strength of the other component. Whether it adds or subtracts will depend on the extra path length due to the reflection. In terrestrial situations many reflective surfaces are possible and signals arriving by a number of different paths are so common as to be regarded as a normal event. Reflective surfaces are possible at any direction or distance in relation to the aerials and some of them will add strength to the main signal and some will subtract.

There is a 180° phase change when the signal is reflected so if the

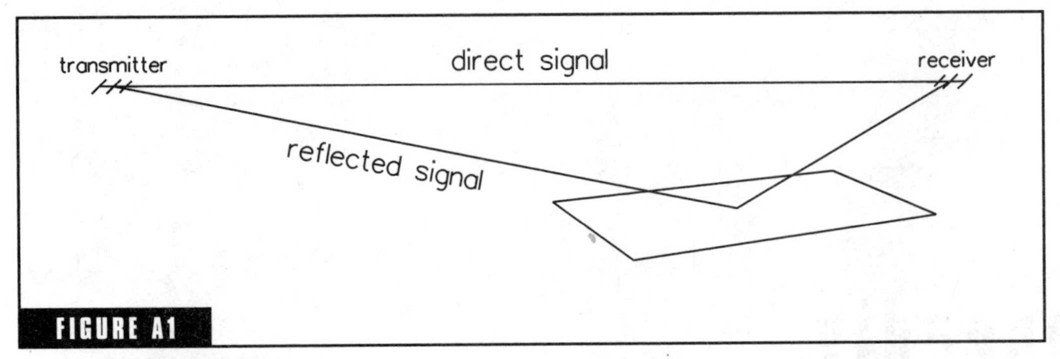

FIGURE A1

Direct and reflected signals.

extra path length is very small the reflected signal will subtract from the direct one. If the extra path length is equal to half the wavelength of the signal the reflected signal will add to the strength of the main signal. For paths in which the extra due to the reflection is between zero and one-fourth the wavelength the reflected signal will subtract a component whose strength will depend on the exact length (in wavelengths) of the extra path traveled by the reflected signal. Reflection points which involve the signal in extra path length between one-fourth and three-fourths wavelengths will cause addition of strength to the direct signal with the strength of the component added maximum when the extra path due to the reflection is equal to one-half the wavelength and zero at each end of that range.

The zone which includes all points that involve the signal in less than $\frac{1}{4}$ wavelengths of extra path is a three-dimensional ellipsoid whose foci are the centers of radiation of the transmit and receive aerials. A "three-dimensional ellipsoid" is a shape like a long, thin football (rugby or Australian Rules) or a dirigible. Outside that zone there is one in which a reflection would involve the signal in between $\frac{1}{4}$ and $\frac{3}{4}$ wavelengths extra travel whose outer boundary is also a (slightly wider) three-dimensional ellipsoid. Any signals from reflections in that zone will add strength to the direct signal. Outside that zone there is another zone of subtraction for reflections which produce extra path lengths of between $\frac{3}{4}$ and $1\frac{1}{4}$ wavelengths and a zone of reinforcement for extra path lengths of $1\frac{1}{4}$ to $1\frac{3}{4}$ wavelengths and so on.

For each particular combination of transmitter aerial, receiver aerial, and operating frequency there are an infinite number of these zones, all in the form of three-dimensional ellipsoids. Zones which involve an even number of half-wavelengths of extra path will result in reduction of the received signal strength and zones which involve an odd number of half-wavelengths will be associated with aiding signal strength. The illustration of Figure A2 is a two-dimensional representation of the first few of these zones but is a considerable simplification of the complete picture. It must be remembered that the full picture is three-dimensional and there are other zones outside of those shown.

The existence of Fresnel zones depends on the exact difference between the length of the direct path and the length of the reflected path. The explanatory diagram of Figure A2 shows the part of the signal path involved. The two reflection points R1 and R2 are both on the same Fresnel zone line because the same amount of extra path length is involved. The extra path length can be calculated by trigonometry; it is the difference between the tangent and cosine of the angle between the direct and reflected paths multiplied by the path length.

Fresnel (pronounced "fraynel" or "freenell") zones are named after Augustin Jean Fresnel, an 18th-century French physicist who did much of the early experimental work which defined light as a wave phenomenon. They are derived from a particular application of the general principles of wave motion interference. Interference can be much more easily studied in relation to visible light than with radio signals so most of what is known is based on optics; however, the principles of optics can be directly translated to the larger wavelengths associated with radio waves.

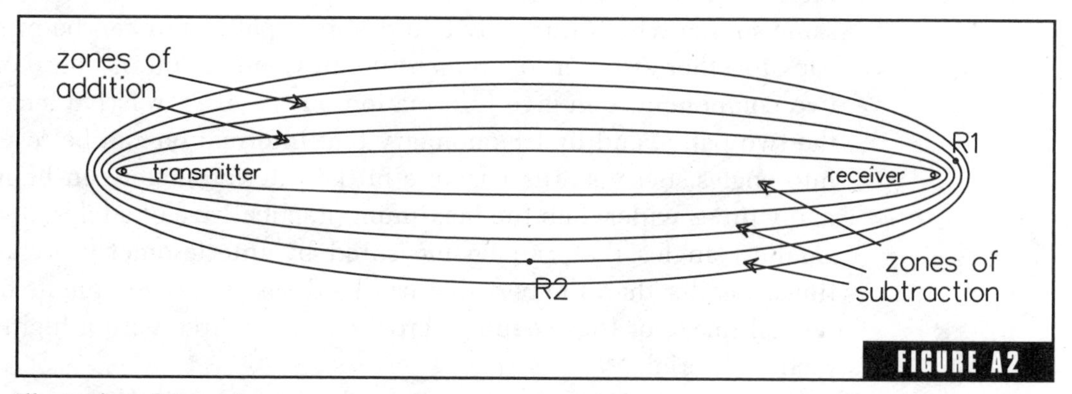

Illustration of Fresnel zones.

Calculations involving Fresnel zones are most commonly done during the design of fixed microwave links. If there are no reflective surfaces in a particular zone there will be no signal due to that cause so the designer's aim is to ensure that there are no reflective surfaces at all in the closest (subtractive) zone and that the closest surfaces in the first additive zone are slightly more distant than the surface that gives half a wavelength of extra path when atmospheric conditions produce an effective Earth radius of $\frac{4}{3}r$. Changes in atmospheric conditions can be calculated as if they caused the apparent radius of the Earth to change, and even with that clearance there will be some times on long links when the apparent surface of the Earth will move closer to the line of the direct signal and may intrude into the subtractive zone. Relevant information on the effect of changes to weather conditions are in Section 9.7 of this book and Section 5.5 of *How Radio Signals Work.*

The principles of Fresnel zone theory can be usefully adapted for fixed links at the longer wavelengths; there are references to it and related subjects in Sections 3.8 and 4.4. Considerable calculation is involved in Fresnel zone plotting so the theory is not usually closely applied to mobile communications. The effect of driving through alternate zones of aiding and subtracting signal strength is, however, observed in practice, mainly affecting signals in marginal areas, and is described by the name "mobile QSB."

A2 How an Interferometer Works

Interferometry is a process of comparing signals or images from the same source which travel by two different paths and can be brought back together at the measurement point. Comparison of phase of the two components can give information on the exact relative length of the two paths, and by trigonometry that information can be resolved into angles. Because the baseline of the interferometer can be made many times wider than the maximum practical size of a telescope the smallest angles that can be measured by interferometry are many times smaller than the best that can be done by measurements on an optical image or by sweeping across a radio source with a highly directional aerial.

An optical interferometer uses mirrors to bring the two signals together, arranged as shown in Figure A3. The outer mirrors can be adjusted for different distances; the angle to be measured can be calculated from the distances apart at which the image is brightest or darkest. Note that the comparison telescope is not an ordinary instrument; interferometry only works when the light is monochromatic and its wavelength is accurately known.

When the principle is adapted to radio wavelengths it is realized as two aerials with feeder cables to a comparison point or as aerials with phase detectors and a highly accurate timing signal distributed to both aerials. When the signals are directly compared the electrical arrangement is exactly the same as a very widely spaced colinear array, and in fact the principle of operation is the same as the process that gives a colinear aerial its forward gain. In the second case the comparison is done digitally working on the data output of the phase detectors. The general arrangement of each of these types of interferometer is shown in Figure A4.

The interferometer sets up a directivity pattern as shown in Figure 8.11. A point worth noting is that the directions of the beams are not controlled by the directivity of either of the aerials but by the exact electrical length of the two signal paths and the spacing between the aerials. There can be a mismatch between the interferometer beam directions and the center of the beam of each aerial, and in fact the aerials can be pointing in slightly different directions. In either of those cases the sensitivity of the interferometer may be reduced if the difference is

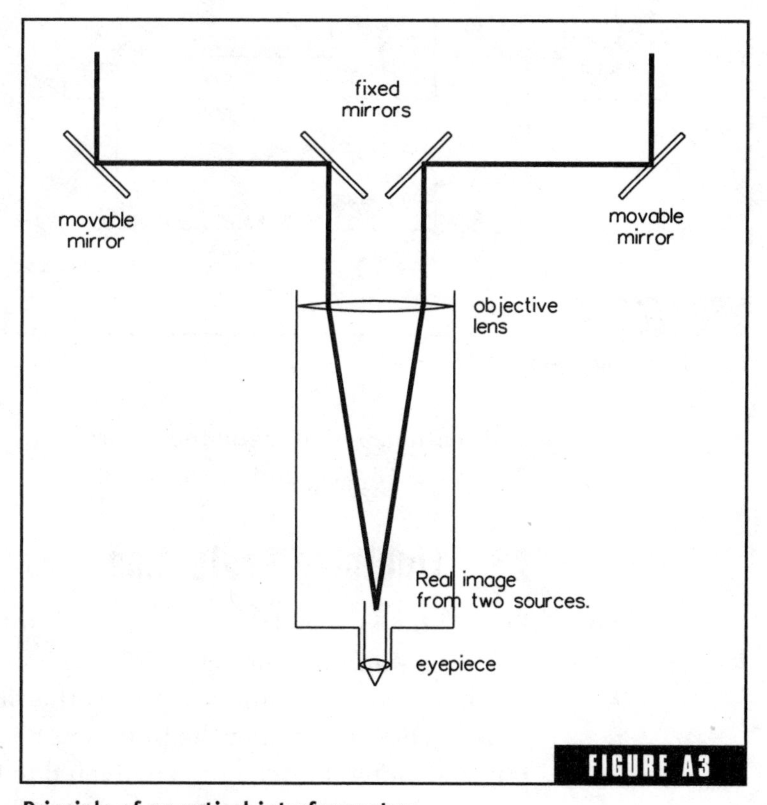

FIGURE A3

Principle of an optical interferometer.

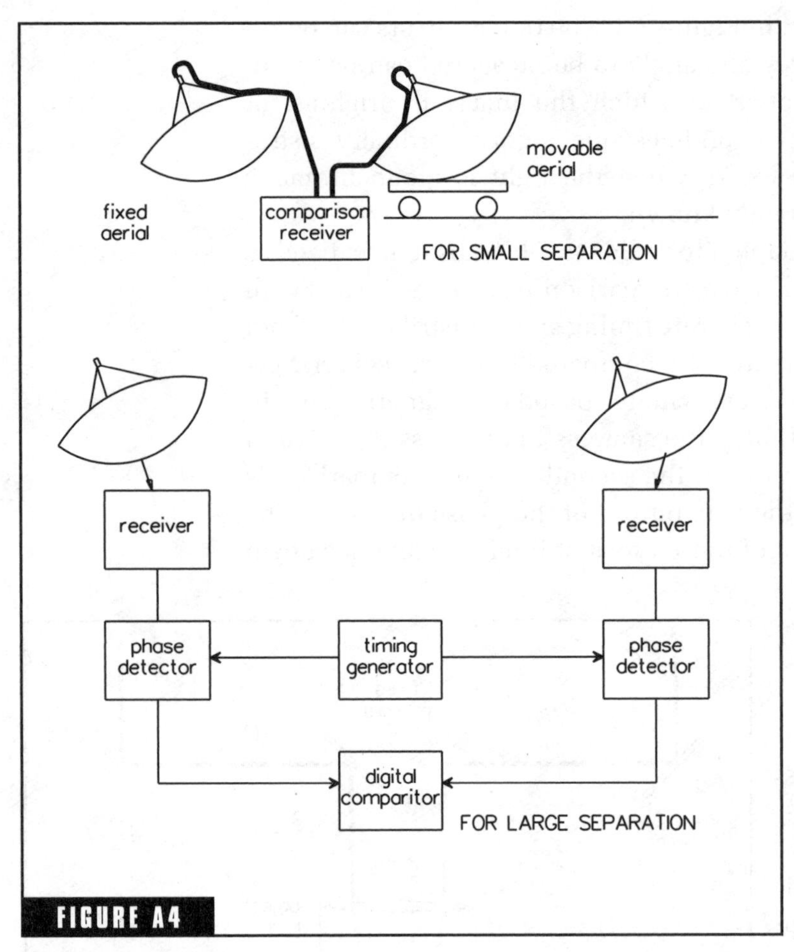

Radio interferometers.

great enough to place the interferometer beam on the side of the lobe of the aerial pattern but the direction of the beam is not changed. On the other hand changing the position of either aerial, changing the electrical length of either feeder, or changing the relative timing in the case of the phase detector version will change the direction of the interferometer beams irrespective of which way the actual aerials are pointing.

The principle of the interferometer is closely related to the mechanism by which most directional aerials achieve their directivity and also to the mechanism that causes a standing-wave pattern in cases of multipath fading. There is related information in Sections 3.8, 4.4, 4.5, 4.7, 7.2, 7.5, 8.13, A1, and A5 of this book.

A3 Lightning Protection

For radio receivers the risk of a direct hit by lightning on the aerial is much smaller than the risk of damage due to an induced pulse from a nearby strike. For induced pulses the danger comes from two main causes. One is that after the pulse has been absorbed by the aerial and traveled along the cable it is delivered to the receiver with most of the characteristics of a very large genuine signal. The other major source of damage is that the momentary current of thousands of amps passing through the resistance of the bulk of the Earth can cause voltage gradients between points of earth connection which are normally intended to be at the same voltage.

The electromagnetic pulse generated by a lightning stroke is a single spike of very high amplitude and very short duration and usually vertically polarized. Radio aerials are resonant structures (even short untuned pieces of wire have resonances somewhere in the spectrum), and when the single spike is absorbed its energy causes a shock excitation of the resonance which reduces its level and smears it over a longer time. Then when that signal is passed through a transmission line the components which propagate best are those whose impedance match that of the line. The signal that arrives at the receiver input is mostly composed of energy whose frequency is in the range the receiver is most sensitive to and whose impedance is about right to be absorbed by the receiver front end and so does maximum damage.

The major thrust of lightning protection is to either absorb the energy and dissipate it in a resistive element or divert its path away from the receiver. These measures are most effective as early as possible in the system so that they work on the pulse before it is modified by resonances and other such phenomena. In many systems several stages of protection are used before the signal is delivered to the first amplifying stage of the receiver, with the first stage being a "lightning rod" placed at the highest point of the supporting tower or mast as illustrated in Figure A5. A lightning rod is simply a conductor placed vertical in a position where its top is higher than any other point of the structure with the top brought to a fine point. Its major purpose is to produce a silent discharge whenever storm clouds are near to reduce the number and severity of actual lightning discharges in the immediate area; its effectiveness is determined by the geometry of the tip—a needle point would give the best ability to discharge clouds but would be melted if it ever actually received a direct hit. The design of a practical lightning rod is a compromise between these two factors.

The second line of defense must still be a robust, brute-force device whose major purpose is to protect the next stage protective device from itself being damaged. Spark gaps or gas-discharge tubes placed as near as possible to the first accessible point on the aerial are commonly used at that stage of protection. For vertical aerials where the

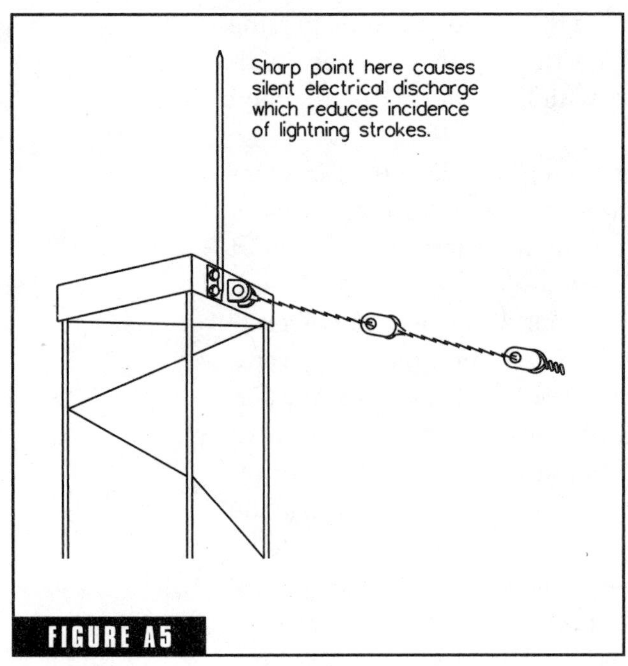

FIGURE A5

Lightning rod installation.

connection to the feeder is close to ground level a spark gap with connection to a solid earth point gives good energy diversion.

For horizontal aerials of the dipole type with a coaxial cable feeder to the center of the aerial wire prefabricated baluns are available with spark gaps built in to divert induction spikes to the outside of the coaxial outer conductor. Baluns of that type work best if the outer conductor of the cable is connected to a good earth as close as possible to the aerial so that surge currents can drain away freely. These systems are illustrated in Figure A6.

Due to the combination of several factors the surge voltage applied to the feeder input with spark-gap protection is liable to be from several hundred up to a couple of thousand volts. In theory spark gaps can be made to fire at any voltage just by adjustment of the gap; the breakdown voltage of air insulation is about 20,000 V/in of gap, which corresponds to 787 V/mm. In practice to make a gap to fire at, for instance, 20 V would require a clearance of about one-fortieth of a millimeter, and that would be easily bridged by dust, spider webs, and insects and so would be unworkable. For prefabricated baluns as shown in Figure A6 dust, spiders, and so on are not such a great problem, but the commercial reality of them is that the major market is among amateur radio stations where the aerial is used for both transmitting and receiving with transmitters of possibly up to 1-kW PEP output. The working peak output voltage of such a transmitter can be in the range of 750 to 800 V and the spark gap must be set to at least 2 or 3 times that voltage to give a working safety margin.

Stray inductance in series with the spark gap will also cause the peak voltage of the spike across the feeder to be raised in the form of an overshoot above the actual breakdown voltage of the gap. The actual voltage will be related to the multiplication of that inductance by the rise time of the applied pulse as shown in Figure A7. Stray inductance is always present but is minimized by using a connection in the "four wire" form with the lead length from the connection point to the actual spark gap kept as short as possible on both sides of the gap. (Common impedance in the earth side counts too.)

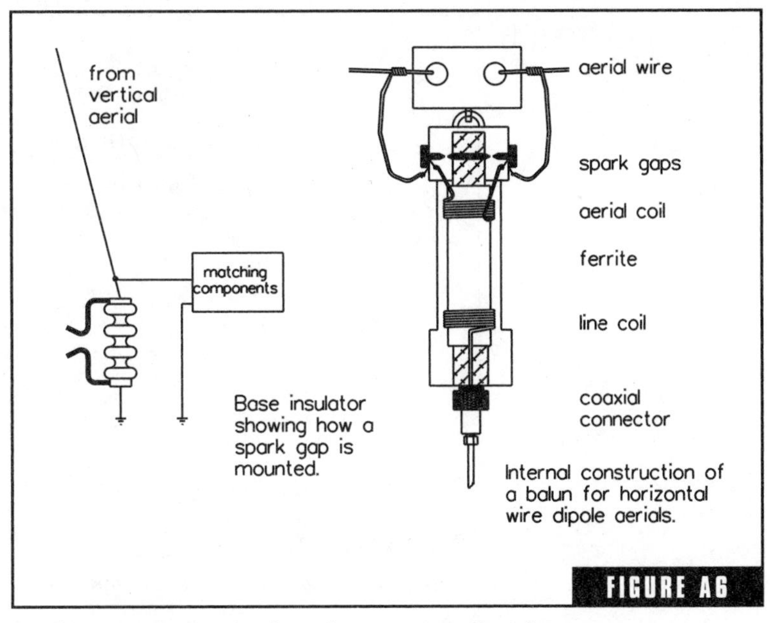

FIGURE A6

Spark-gap arrangements for primary energy diversion.

The spark is initially struck by voltage stress across the cold gas in the gap causing ionization. Once ionization is started current flowing through the gas heats it and conduction is maintained as an arc through the minute cloud of hot gas for as long as there is sufficient current to keep the gas hot enough to be ionized. Recombination of the ions does take a finite time as the gas is cooling so if the energy flow has been sufficient to keep a big enough cloud of gas ionized the conductive state may persist for the low-voltage instant between half cycles of the resonant (at radio frequency) applied voltage. These states are all illustrated by the thicker line in Figure A7. The primary purpose of a spark gap at the aerial is to reduce the energy level of the induced pulse and divert the excess away from the feeder input.

Gas discharge tubes are made with lower breakdown voltage than is practical with spark gaps; there are some designed for use on overhead telephone lines whose firing point is set to about 80 V. They do add significant parallel capacitance (several picofarads) and lead length is longer than is possible with a four-wire-connected spark gap; if these factors can be allowed for in the design of the aerial and matching network then a gas-discharge tube may give improved protection on some installations.

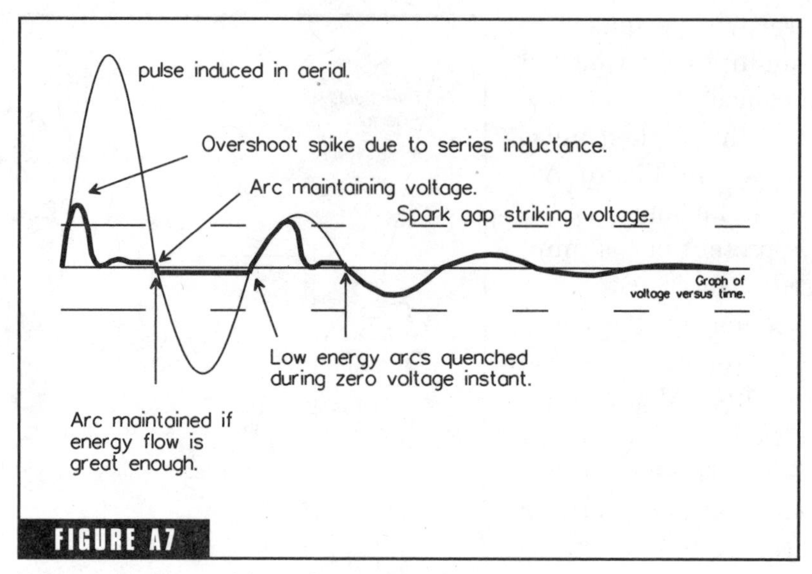

FIGURE A7

Voltage waveforms with spark-gap operating.

At the output of the feeder there is still a need for spike reduction using a less robust but quicker-acting clipper circuit. Parallel connected signal diodes can be used but even with these stray series inductance is present. In this case, however, there may be an opportunity to place inductance in series with the feeder in such a way that the spike voltage is divided across the feeder inductance in series with the stray component; with of course the receiver input connected across the clipper and stray inductance only as in the circuit of Figure A8. This group of components is so small that it can often be built into the receiver case as the first few components in the circuit. In most cases the junction barrier voltage of the diodes is sufficient to provide a working range for the wanted signals; if minimum cross-modulation at very high signal levels (i.e., 0.1 to 0.7 V) is required a reverse bias of up to several volts DC can be applied to each diode to widen the linear range of the clipper.

Component values used in the circuit of Figure A8 will vary with the operating frequency. The series inductance will need to be such that it gives no more than about 10 to 50 Ω impedance at the operating frequency, but that could vary from a few tens of microhenrys at the low-frequency end of the MF band down to a matter of nanohenrys in the UHF range, and in some cases the output impedance of a coaxial cable may be all that is needed. The two capacitors should be as big as can be used without causing problems due to stray series resonance at the high-frequency end of the wanted signal range and with no risk of a resonance in association with the series inductance. The diodes need to be signal frequency or VHF/UHF switching diodes and the resistors normally will carry little or no current so would have values in the hundreds of kilohms range. If the junction potential voltage of the diodes is sufficient to provide a clipping level on its own then the capacitors, resistors, and voltage supply can be dispensed with and the diodes connected directly from the signal line to earth.

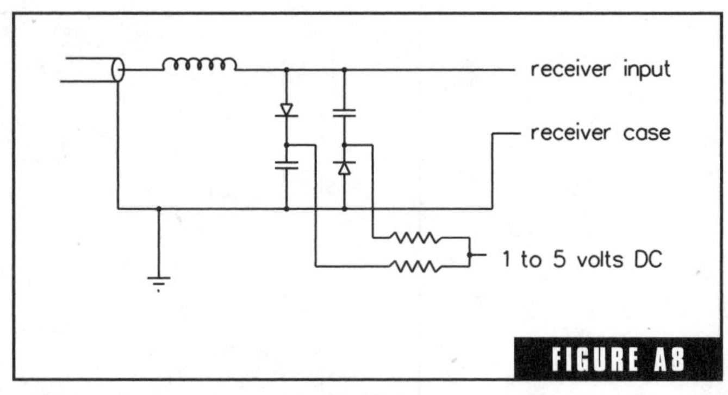

A diode clipper arranged for lightning protection.

The risk of damage due to voltage gradients in the bulk of the Earth is in theory prevented by the tried-and-true techniques of single-point earthing; however, that is complicated by the need for a short direct earth connection at both the aerial and receiver. Coupling of the signal by mutual inductance will allow for single-point earthing conditions at both points but it must be done in such a way that there is no significant conductive path (even by stray capacitance) across the coupling. The circuit of Figure A9 shows how an aerial and feeder can be arranged for maximum lightning protection. Note that in some circumstances the requirements of lightning protection may be in conflict with the arrangements for maximum interception of very weak signals.

A4 Clean Power Supply

There are a multitude of possible sources of noise built into power supply and distribution systems. Power from a battery which is not connected to anything else should be fairly clean, but as soon as a charging system is added there are possible sources of electrical noise—even if the charging current comes from solar cells! AC mains as distributed in

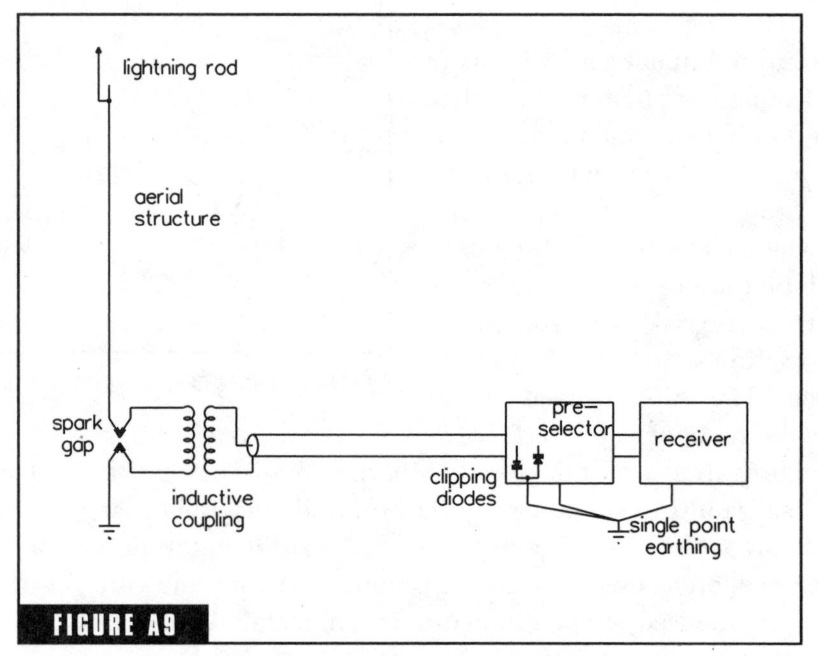

FIGURE A9

An aerial and receiver system with good lightning protection.

towns and cities will always have so many noise-generating devices connected to them that it is not practical to control noise at its source. (That action may be more possible in rural areas or small villages where there may only be a few items of each type of equipment.) For mains-powered devices there are four major paths for electrical noise to get into the signal chain and be amplified to produce an audible output:

- The noise could be due to hum (whistle for an SMPS) which is able to reach other circuits due to a defect in filtering or regulation.
- The noise may be a radio-frequency signal being induced or radiated into the early stages of the receiver.
- It may be at intermediate frequency and be coupled through a conduction path into the circuit somewhere between the mixer and the detector.
- It could be at audio frequency being coupled through a common-mode path into the low power stages of the audio amplifier.

For modern solid-state receivers designed for operation as fixed stations powered by AC mains the receiver includes a power converter which may have several sections for driving the different sections of the circuit. The section which drives an audio output amplifier supplies the bulk of the power used in the whole receiver and is liable to operate at a higher voltage than earlier stages but its needs for filtering and/or regulation are minimal. Amplifying stages closer to the front end may only take a couple of milliamps but filtering must be impeccable because any minute AC signal introduced to earlier stages is highly amplified; in solid-state receivers the supply to these stages is usually via a regulator circuit configured to maximize the filtering and low source impedance aspects of the regulating process. In a working receiver if there is evidence of a defect in the suppression of hum and noise from the power-conversion process the regulation of supplies to these early stages should be suspected first. Exact definition of fault conditions may not be simple because you may require to measure AC components of only a few microvolts in the presence of several volts DC and also the internal resistance of the regulator output may only be a few milliohms but that figure also must be measured in the presence of the DC component.

If you observe electrical noise without an associated level of mains hum when only the power supply from public mains is connected to a receiver that indicates a signal coming from outside the receiver and somehow being coupled around the internal filter either by another conduction path or by radiation into the front end of the receiver. The only effective cure for a conducted signal is to filter the mains leads close to the receiver. For RF and IF signals on the mains a filter such as that shown in Figure A10 will give protection against almost all ***conducted*** interference. It may not be necessary to have a filter as complex as that; each inductor/capacitor section is capable of giving about 30 dB of rejection for signal frequencies higher than ~400 kHz so if you only need 20 to 30 dB of rejection one section may be enough. [Note, however, that switching spikes can have amplitudes up to about 10 times (= 20 dB) the rated voltage of the AC supply.]

The component values shown in Figure A10 are about right for the MF broadcasting band but may not be appropriate for higher frequencies. The problem of stray components puts limits on how wide a band of frequencies can be filtered. Each of the inductances has a cer-

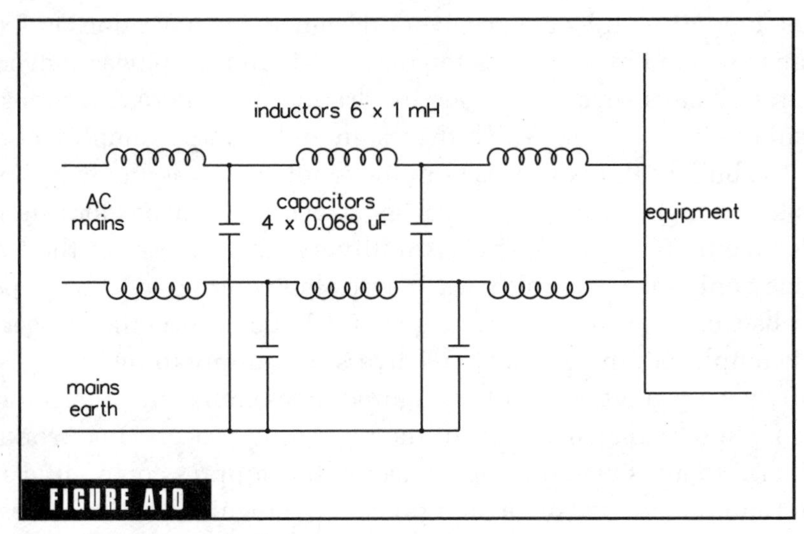

Mains filter schematic diagram.

tain amount of stray capacitance built into it and each of the capacitors has some stray inductance. Every component has a self-resonant frequency at which inductances behave as parallel tuned circuits (very high impedance with resistive phase) and capacitors behave as series-tuned circuits (almost zero resistance). For frequencies higher than self-resonance each will behave as a component of the opposite type (i.e., inductances have the effect of capacitors on the circuit and vice versa). In some cases the self-resonant frequency can be quite low; for a 1 millihenry inductance constructed as a single pie wound on an inert former the self-resonant frequency may be only a few hundred kilohertz. If filtering is only required over a limited band of frequencies (a 2:1 or perhaps 3:1 ratio) the circuit of Figure A12 is most effective if the self-resonant frequency of the inductors can be placed near the center of the band and the capacitors are chosen so that their self-resonant frequency is a little higher than the top end of the required range.

Radiated signals will not be defeated by filtering the mains wiring. The path of the coupling by radiation is from the mains wiring to the receiver input and any voltage which appears between the input socket inner and outer pins (for coaxial cable) is treated by the receiver as a genuine input. This voltage could be due to a signal picked up via the center pin and compared to an earthed case but it could equally well be due to pickup on the case itself being compared to a center pin which is held at a constant potential. Radiation as a method of transfer of the interfering signal can sometimes be hard to definitely identify but in most cases if the inner and outer conductors of the aerial socket are connected together by a screwdriver blade or other very short piece of conductor the radiated signal will vanish. As well as that in most cases a hand touched on the case of the receiver will change the signal in some way, not always to make it weaker, but even if the interfering signal is strengthened that is still an indication that radiation is the mechanism of coupling. Pickup of radiated interference is usually cured by proper shielding of the receiver and connection of a short direct earth. Shielding and earthing are part of the subject considered in detail in Chapter 3.

Radiated signals are almost always at the radio frequency. If a signal is being received by radiation directly into the IF amplifier that indicates a fairly severe deficiency in the technical performance of the receiver itself and if not due to a correctable fault may be taken as an indication that the model of receiver is not appropriate to the intended use. If you are a service technician called to an existing installation at which you find that situation applies you may be able to improve the overall system performance by adding a prefilter stage to the aerial input in the form of a notch or narrow-bandpass reject filter tuned to the intermediate frequency. The principle of operation is dealt with in Section 2.4 and relevant circuits for the MF and HF bands are shown in Figures 2.11 and 2.12. For systems which use an IF in the VHF or higher frequency bands parasitic absorption traps as shown in Figure A11 may be more effective.

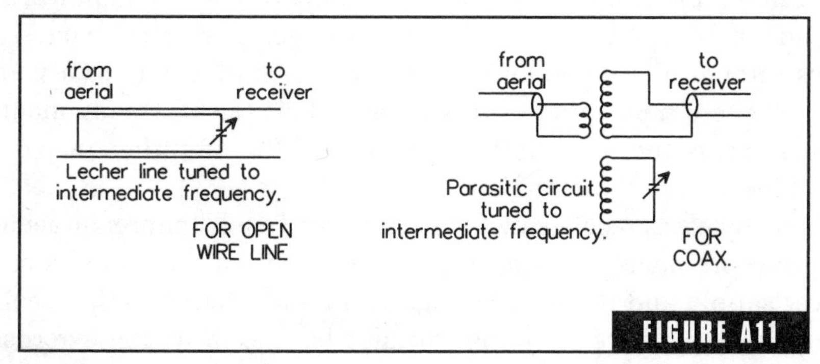

Parasitic absorption traps.

For noise or interference which is being generated in the power supply and radiated to the receiver input, suppression at the source is worthwhile. Section 5.14 gives information on spark suppression in the context of switching contacts such as applies to commutators and relays; semiconductor switching components such as high-speed diodes and thyristors can generate waveforms similar to that shown in Figure 5.15 but in this case there is no visible spark. Suppression of radiation from these circuits can be achieved by treating the semiconductor element as the switching contact shown in Figure 5.14, but a more refined method of selection of the capacitor value must be used because the rapid transition of the voltage waveform may be an essential part of the operation of the circuit. In most cases an oscilloscope connected in differential mode directly across the switching component and triggered to show the transition will display an overshoot and ringing if the component is causing troublesome radiation; the capacitor value should be chosen to give best possible removal of the overshoot without slowing the major part of the transition.

When the interfering signal is in the audio-frequency range the method of coupling will usually be via the effect of stray components and unplanned connections. Filtering of the mains would in theory be possible but there are practical problems in doing so. One potential problem is the stray resonances mentioned above; another is that the frequency of the interfering signal may be so close to the frequency of the mains that effective discrimination between them may require a filter with a lot of components. The coupling mechanisms are known by the generic name "power supply common impedance problems." Figure A12 is a partial schematic diagram to illustrate the mechanism of common impedance coupling. It shows part of a simple power supply and a low-power audio amplifier stage. The interfering signal flows through the diode and filter capacitor and generates a voltage across "R common." That voltage is carried via the bias components to appear across the base/emitter junction of the amplifier in parallel with the genuine signal.

In many cases receivers which are sensitive to common impedance coupling problems are also affected by rectified AC hum from the power supply and those cases suggest a fault internal to the receiver. Common impedance coupling can also be due to wiring external to the receiver. In most of those cases there is no undue sensitivity to rec-

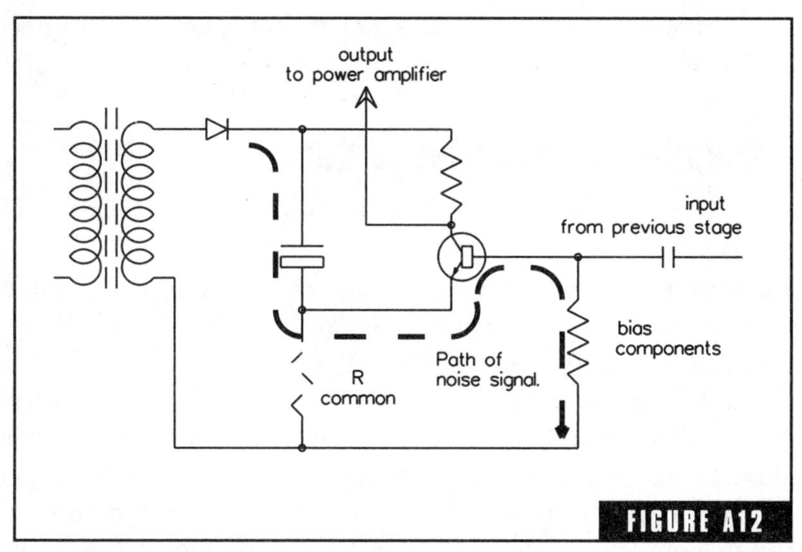

Example of a power-supply common-impedance fault.

tified AC from the receiver's own power supply and sometimes the overall effect of noise can be reduced by arranging that an equal amount of the same noise is introduced in the opposite phase to cancel out the troublesome component. Exact cancellation requires that level and phase are both adjusted to a correct value; phase adjustment becomes more difficult to control at higher frequencies so this technique is most useful when the frequency of the noise is close to that of the power mains.

For systems which are receiving signals other than audio (i.e., video or digital data) the measures described above are all equally relevant but the indicated effect of particular types of noise will be different. The source of the noise gives more information about control measures than comparison of the observed effect; in general whenever noise on a digital receiver can be identified as coming from a particular source then the same measures of suppression as are used for the audio receiver can be used equally well for other forms of signal.

In addition to all the foregoing there may be some cases where a video or digital receiver is particularly susceptible to disruption because the frequency of the noise has a particular relationship to the frequency of a clock signal or line rate. In those cases a slight shift of frequency of either the noise or the clock rate may be all that is needed to cure the problem.

A5 Coupling between Tuned Circuits

When energy must be coupled from one resonant circuit to another the degree of coupling is a critical factor in the overall performance of the equipment. The effective degree of coupling is in turn dependant on the Q factor of the resonant circuits and loading conditions at the point where the energy is to be used. The condition at which energy transfer at the exact resonant frequency is maximum is called "critical coupling." When coupling is looser than critical, energy transfer is reduced but selectivity is increased. When coupling is tighter than critical the reactance of each tuned circuit has the effect of detuning the other so that energy transfer at the exact resonant frequency is reduced but at slightly different frequencies new combined points of tuning are created with maximum energy transfer at those frequencies. Consider in detail energy transfer through the circuit shown in Figure A13.

Explanation of the circuit of A13 is simplest if we assume for a start the somewhat artificial conditions of two exactly equal tuned circuits; the load is correctly matched to L2 without any reflection of stray reactance by auto transformer action and an equal and opposite autotransformer matches the L1–C1 tuned circuit to the source also with no stray reactance to upset tuning. The initial set of calculations can be done using tuning capacitors to cover the MF broadcasting band with actual capacitance (tuning gang plus strays) variable over the range from 20 to 180 pF to tune to frequencies from 533.3 to 1600 kHz. The inductance required to tune that range (consisting of L1 or L2 plus strays) is 495 μH for each coil.

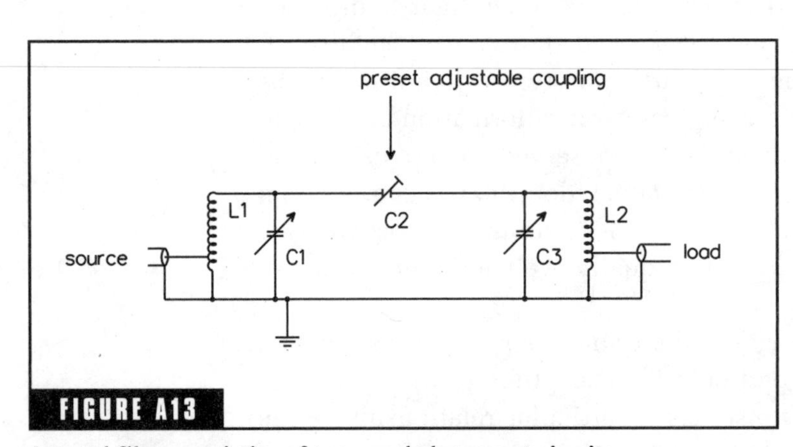

FIGURE A13

A tuned filter consisting of two coupled resonant circuits.

A figure for Q factor of 50 can be guessed at; that

would be about right for air-cored coils such as would be used in a prefilter of the type shown in Figure 2.9; for high-fidelity reception a Q factor much higher than that would involve loss of treble response. The condition for critical coupling is that the ratio between the impedance of the coupling element and the reactance of one of the tuning components is the same as the Q factor. In this circuit C2 must have 50 times smaller capacitance than either C1 or C3. One difficulty is that C1 and C3 are variable so for a fixed capacitance only one frequency can be correct—the actual capacitance value required varies from 0.4 pF at 1600 kHz up to 3.6 pF at 530 kHz. If a coupling element were chosen to be right at one frequency in the middle of the range it would result in coupling less than critical at the low-frequency end and coupling over critical with a double-humped frequency response at the high-frequency end. These conditions are illustrated in the frequency-response graphs of Figure A14.

The change in bandwidth that goes with adjustment of frequency is working the wrong way around in the circuit of Figure A13. At the

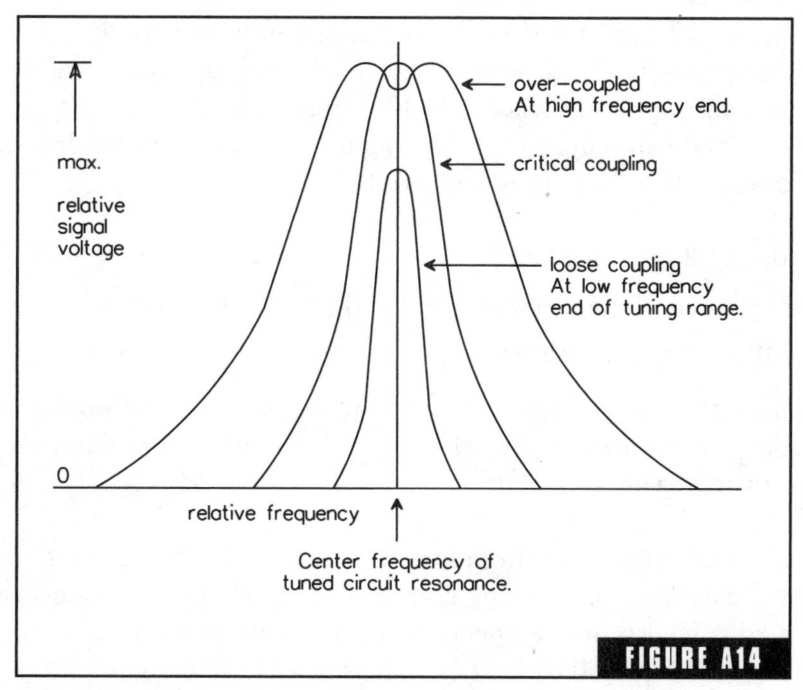

Comparison of frequency response for the three states of coupling.

low-frequency end coupling is least so bandwidth is narrowest, but the bandwidth required for a particular treble response is greatest in terms of percentage. At the high-frequency end bandwidth required in terms of percentage of the spectrum is least and coupling is over critical, giving a double-humped response. If tuning could be achieved by keeping C1 and C3 constant and varying the inductance, that defect would be much reduced but permeability tuning units designed to cover the 3:1 frequency ratio of the MF broadcasting band are rare and expensive because of mechanical difficulties in their design.

When the band of frequencies required is narrower a circuit similar to Figure A13 becomes a very simple and practical way of building extra selectivity into a receiver. In the case of the amateur bands and broadcasting allocations in the HF range where the band required is only a fraction of a megahertz at a frequency in the 10- to 30-MHz range, a filter based on two tuned circuits slightly overcoupled is a very effective way to cover a particular band without the need to retune the prefilter. In that case tuning components are preadjusted for the correct response to a particular band then switched for changes between bands.

The circuit of Figure A13 can be made workable over the 3:1 range of frequencies if C2 is made adjustable. That is simple enough to arrange; appropriately sized variable capacitors are available but the process of adjustment can be a bit tricky. For all retuning the following procedure would have to be followed:

1. Adjust C2 to minimum.
2. Set receiver to required channel and identify wanted signal.
3. Peak L1/C1- and L2/C3-tuned circuits.
4. While watching a signal-strength meter (multimeter on the AGC line if no built-in S meter) advance C2 until signal strength just stops increasing.

All adjustment of tuning must be completed with C2 set at minimum capacitance and tuning components must not be touched after C2 is advanced unless a spectrum analyzer or panoramic adaptor is available to display the effect of changes.

In practical circuits the neat and tidy situation of exact equality does not apply so if all three adjustments are trimmed for a peak the result as illustrated in Figure A15 will be a lopsided response in which coupling is over critical but the two humps of the overcoupled response are not equal in height so the observed response is a single sharp peak with the other peak showing up as a slight bulge on one side. Skirt selectivity of a filter tuned in that way is very little better than that of a single tuned circuit.

To interpret the graph of Figure A15 compare it with the "overcoupled" curve of Figure A14. Overcoupling can be used to give a filter characteristic with a flat (with slight ripple) passband; however, initial alignment must be done with a sweep generator or spectrum analyzer so that the shape of the graph can be displayed and adjusted for the required flat top. Filters provided by a pair of overcoupled tuned circuits cannot practically be tuned to a range of frequencies; they must be aligned for a particular passband then left set. Overcoupled filters are regularly used in the IF amplifier of many types of receiver—in those cases they are initially aligned then kept unchanged for the working life of the equipment. A pair of overcoupled tuned circuits can also be used for RF stage tuning in receivers designed specifically

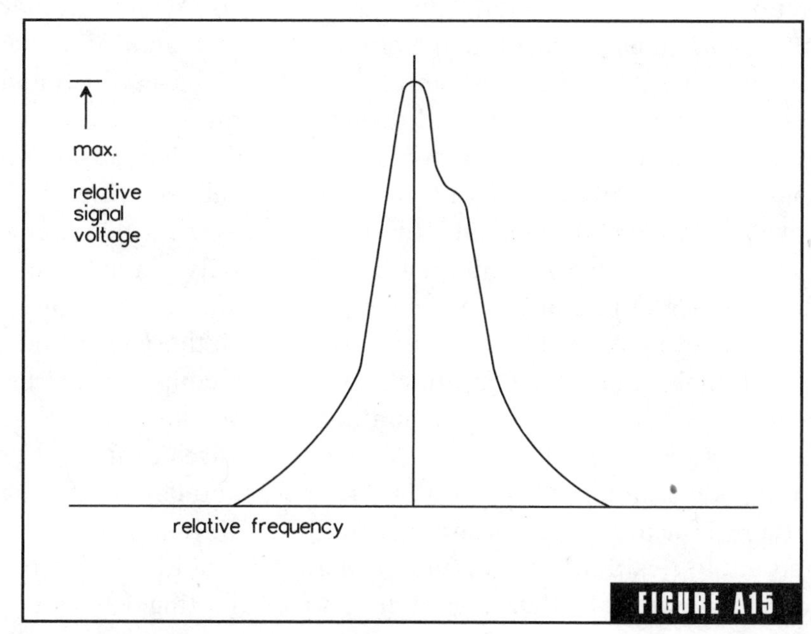

FIGURE A15

End result of incorrect tuning procedure.

to cover amateur bands so that the whole band can be covered without retuning of the RF stage.

In theory the coupling element can be any sort of impedance. In practice, however, an inductance in place of C2 in Figure A13 would need to have a value Q times (50 times) that of L1 or L2. In relation to the calculations above that would mean a choke of close to 25 millihenrys, which on the MF broadcasting band would almost certainly have sufficient stray capacitance to be operating above its self-resonant frequency. The impedance actually presented to the rest of the circuit would be a capacitive reactance whose value varies rapidly with changes in tuning. A resistance could be used and the effect of change in bandwidth would be reduced, but the resistor forms a voltage divider in association with the dynamic resistance of L2–C3 and dissipates power from the signal. Use of a resistor in place of C2 would reduce the output signal level by 6 dB. The resistor does avoid the possibility of VHF/UHF parasitic responses so may occasionally be used in circuits where the main design consideration is prevention of overloading.

The type of coupling shown in Figure A13 and described in the previous couple of paragraphs is called "top coupling." Other circuit configurations are possible and there are ways in which inductive coupling can be used with tunable circuits. When mutual inductance is used the two coils are placed in physical locations as illustrated in Figure A16 such that the required degree of coupling is achieved between their magnetic fields. The required separation depends on the Q of the tuned circuits; for high-Q circuits critical coupling may be achieved with the coils separated by several times their own diameter and at opposite ends of a shielded box; for heavily loaded coils the coupling may require the coils to be almost touching and in some cases have some of the turns of one coil interleaved with those of the other.

For mutual inductance coupling the control of coupling coefficient is related to both the distance separating the coils and their relative angle. When coils on the same axis have their turns exactly aligned coupling is maximum; if one coil is turned 90° and kept on the same axis there is a null point where coupling is minimum. If coils are placed at 90° positions but off-axis so that they are in effect "around the corner" from each other there will be some coupling between them but its magnitude will be a bit unpredictable.

Bottom coupling in which a small portion of the inductance at the earthy end of each tuning coil is brought together and made common to both circuits is a method of inductive coupling in which the degree of coupling is better defined but more difficult to change if adjustment is required. In theory bottom coupling of the capacitive element of the tuned circuits should be equally possible, but in practice the frame of a ganged tuning capacitor is connected to the earth side of each variable capacitor so that option is not normally available. The diagram of Figure A17 shows how bottom coupling is arranged; L2 is the coupling component. Note that the combination of inductive bottom coupling and tapped coil for input and output connection is possible but not generally used due to the straight through path for VHF/UHF parasitic signals.

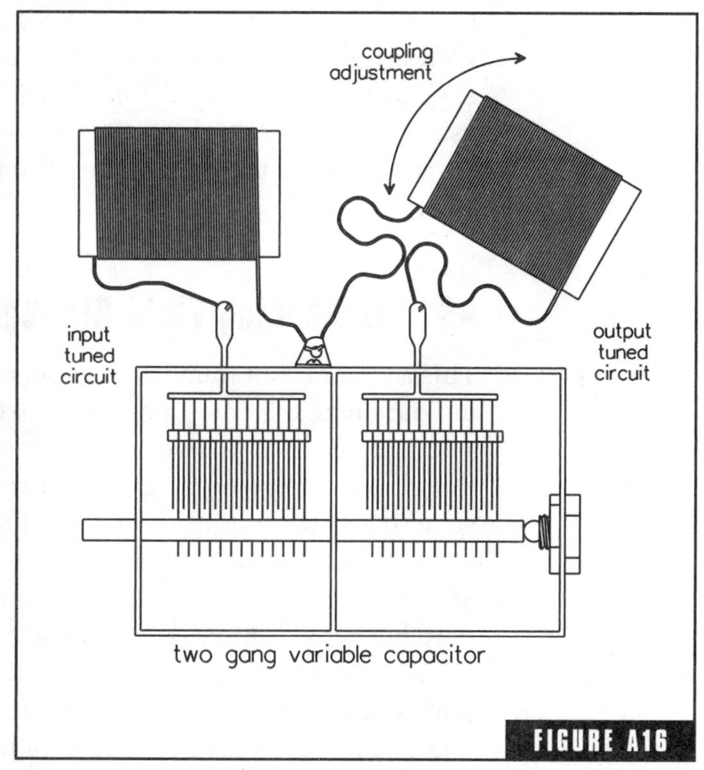

Mechanical arrangement for mutual inductance coupling.

Link coupling has the predictability of bottom coupling combined with easier methods of adjustment and works well in association with

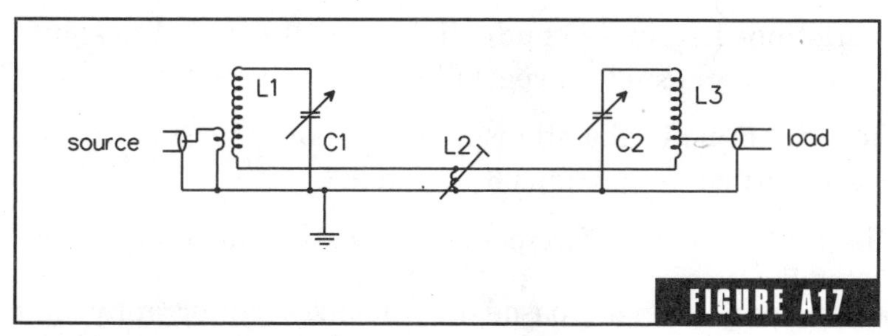

Inductive bottom coupling schematic diagram.

high-Q-tuned circuits but the maximum degree of coupling available with a link is limited so it is less useful in heavily loaded circuits. Link coupling is very useful in VHF/UHF equipment where it can be used to transfer signal energy from one shielded compartment to another without compromising earthing conditions on each side of the shield.

A6 Measuring Field Strengths

This section is not intended as a set of instructions for taking practical measurements; a comprehensive text for that would be a book by itself. It is intended as an outline of the subject to make you aware of the limitations of presently available measurements and able to understand the implications of published reports and maps.

There are instruments manufactured which combine the function of a receiver with calibrated gain with a signal-strength meter and a portable aerial which purport to give measurements of field strength to an accuracy of ± 1 dB or less. That these instruments can be calibrated in a standard field to an accuracy of closer than ± 1 dB is justified in a laboratory sense, but to transfer that resolution to meaningful measurements in the field is one of the most difficult jobs a radio technician is ever required to do, and if it is done by an inexperienced person the errors that make a particular reading meaningless can go unnoticed.

The relevance of particular readings of field strength is degraded by the following:

- the reflection and standing wave effects described in Section 3.8 of this book and Section 5.6 of *How Radio Signals Work*
- local variations in ground conductivity and absorption of signal due to local factors such as vegetation
- weather variations in the atmospheres of both the Sun and the Earth as described in Sections 6.13 and 9.7
- near-field effects for measurements close to the transmitter aerial

An expected field strength based on idealized topography and standard weather conditions can be calculated with fair accuracy; the purpose of actual measurements in the field is to check the relevance of those calculations. One of the complicating factors is that the effect of major factors such as mountain ranges or city-central business districts is to be measured but there are a host of smaller-scale variations due to essentially the same causes which are to be ignored. In most cases reflections, refractions, or sources of absorption which cause standing-wave patterns comparable in size to the wavelength of the signal are counted as local and therefore to be ignored; factors to be measured and accounted for are significantly larger in scale than that. Be warned, however, that there are some exceptions to that rule.

With regard to weather conditions attempts are made to avoid the extremes but if you try to avoid measurement in particular conditions the only thing you can do is wait until the weather changes and that is not always possible. There are times when an intelligent guess must be made of a factor to be added or subtracted to the present reading to give the figure expected in average conditions.

To limit the scope of variability the report of a field-strength meter reading should also include the following:

- date and time of measurement
- type of aerial used, its heading if directional, and the height to its center above local ground level
- exact location of the measurement in relation to nearby buildings or other structures
- local soil type and degree of vegetation cover
- local weather conditions and any known variations over the signal path

Even when all the known factors are adequately controlled there will still in most cases be uncontrolled variability which can combine to make total differences of up to 6 to 8 dB in either direction. If a measurement at a single spot gives a reading which can be reconciled with a calculated figure to within 8 dB that should be taken as supporting the accuracy of the calculation. If a series of measurements are made over an area of a few square kilometers some measurements will be above the calculation and some below it; the average discrepancy should be closer than the figure for a single measurement. A series of measurements over an area of that size that consistently showed an average greater than 6 dB different from a calculated figure should ***not*** be taken as supporting the calculation.

Technicians engaged in field-strength measurements class particular sites as either good or bad depending on the depth of the standing-wave pattern present. A technique that is helpful in assessing that factor is to make three measurements in a triangular pattern as shown in Figure A18 with the sides of the triangle being about half- to one wavelength each (on the MF and HF bands). If the metering equipment can be moved while displaying the measurement carrying it for a couple of wavelengths along the line of the incoming signal will give similar information. For VHF/UHF equipment in which a tower with continuously variable height (pump up or wind up) is used watching the meter reading while the tower is being lowered will also show the effect. If the variation between the three readings is less than a couple of decibels that would indicate a "good" site and the average of the readings could with confidence be taken as indicative of a larger area. If the readings showed a total variation of more than 10 dB an attempt should be made to find a better site. Total variations between the measurements of from 2 to 10 dB indicate a site whose reading could be recorded but preferably checked against another nearby site if possible.

The measurement site must also be checked for relevance to real

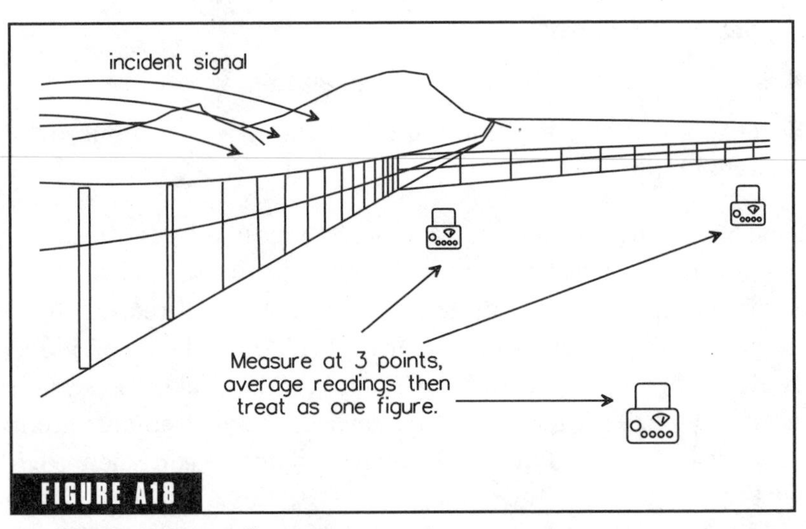

FIGURE A18

Checking for local standing waves.

users of the service. For radio-communication links there is really only one area of interest, the exact location and height of the receiving aerial, so measurements must be made in relation to that location. For broadcasting services the location of groups of listeners or viewers is the controlling factor and measurements should be made with them in mind. For instance a check of television reception in a small town or village nestled in rolling hills could be done on a local hilltop. The procedure of making three measurements in a triangle would probably show the site was almost perfect from the local standing waves point of view but the result would be of little relevance to the real viewers. Measurements on an oval or a patch of vacant ground near the center of the town may indicate a worse site for standing waves but is much more relevant to the real users so that is the measurement which should be reported.

Field-strength measurements on the LF and MF bands usually use a calibrated portable receiver with a tuned loop aerial attached to the top of the receiver. If directional effects must be checked that is done using the null part of the loop's polar pattern. A similar instrument can be used on the HF band but field-strength measurements on the HF band are usually only done close to the transmitter aerial to check its radiation pattern or for the purposes of research into propagation. Signals which have propagated via the ionosphere are so greatly affected by atmospheric absorption that a single measurement done at a particular time and place would have almost no relevance to any other location or time.

On the VHF- and higher-frequency bands the aerial may be a center-fed dipole with adjustable length elements or it could be a directional array. The aerial is equipped with its own calibrated length of coaxial cable and a calibration chart. The factors that affect the actual signal (in millivolts or microvolts) that arrives at the receiver input are as follows:

- tuning of the aerial
- height of the aerial
- attenuation loss of the cable
- length and velocity factor of the cable
- input impedance of the receiver at the operating frequency

Because of all these factors the calibration is really only complete if a calibrated receiver, aerial, and cable are used as a group. There are standard heights for measurements; one that is common worldwide is 9.1 m or 30 feet above local ground level. In some countries 3 m is also used as a standard for some purposes.

When a field-strength survey is being conducted with the aim of producing a map similar to the one shown in Figure 4.3 the final line on the map is a statistical comparison of a number of local readings over a range of distances from the transmitter and is expressed as a proportion of sites and a proportion of times that actual signal received will be stronger than a particular value. For example, on the map of Figure 4.3 the inner line may be described as the "1 mV/m F90,90 line" and this means that in 90% of locations the signal will be at least 1 mV/m for 90% of the time. The outer line could be designated "250 μV/m F50,50" and this means that in 50% of locations the signal is at least 250 μV/m for 50% of the time.

The foregoing applies particularly for measurements well distant from the transmitter aerial where the radiation field is the only one to be considered. Electromagnetic fields consist of two vector components which in the radiation field exist in a well-defined relationship. One vector is an electric field (designated "E field") which can be expressed in either volts per meter or {$(\text{volts})^2$ per square meter}. The other vector is a magnetic field (designated "H field") which is usually expressed as {$(\text{amps})^2$ per square meter}. The power flow in the radiation field may be expressed as units of power density per unit of area as for instance watts per square meter and the expression of vector fields in terms of square measurements per area is a mathematical operation used to establish their relationship to power flow measurements. Well away from the transmitter aerial the power flow and its two vector fields are tied together in well-defined relationships and measurements made in one form can be freely translated into any of the others. Figure A19 shows the relationship between the vector components and the power density.

Close to the aerial there are an extra set of conditions which must be accounted for due to a factor which can be described in terms of variations of impedance at particular points. In the radiation field a particular power density sets up E and H fields in which the E field is numerically 377 times greater than the H field. There is nothing magi-

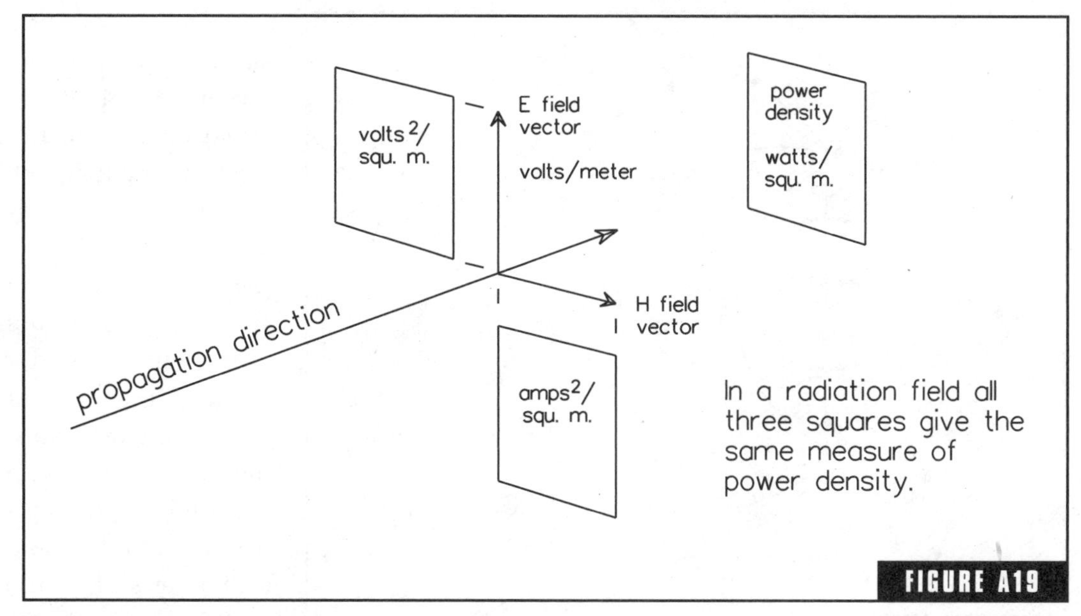

Relationship between E and H vectors and power density.

cal about that number—it is just what the figures happen to work out to with volts and amps as they are currently defined. That quantity has the mathematical properties of an impedance similar to the characteristic impedance of a transmission line so it is sometimes named "the impedance of free space." The impedance of free space is 377 Ω. Close to an aerial that impedance relationship is modified by the presence of the conductor; toward the center of the conductor where current surges back and forth the H field predominates so the "impedance" is reduced. At the conductor tips there is almost no current but voltage is concentrated so the E field predominates and "impedance" is raised. Note that these high measurements of a particular field do not indicate high power flow; this separation of fields all occurs within the induction field where energy is flowing back and forth between the field and the conductor. These separated vector fields have the characteristics of reactive impedances. Figure A20 indicates how the field vectors are distributed around a resonant dipole aerial element. For more complex aerials each conductor sets up fields similar to that and the resultant at any one spot is the vectorial sum of all the interacting fields.

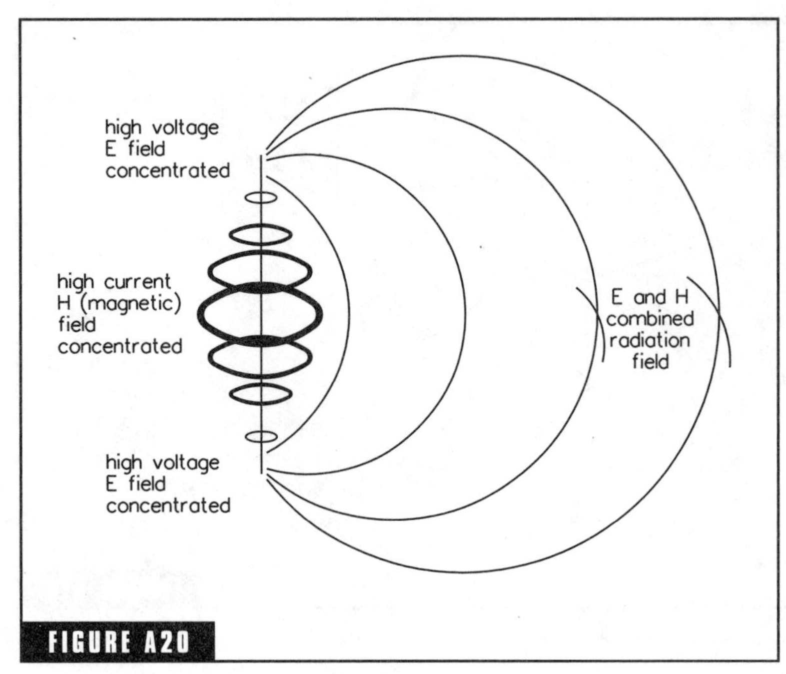

E and H fields around a half-wave dipole.

When a transmitter aerial is attached to the top of a metal latticework tower or mast sections of the tower will absorb and reradiate each of the E and H vector fields independently. The actual strength of the reradiated field will depend in a quite complex manner on the exact length and configuration of the metal section and its degree of illumination by the transmitter aerial. Locations of high-field intensity cannot be predicted by calculation so must be surveyed using a personal hazard radiation meter.

Personal hazard radiation meters use measurement heads which are designed for as near as possible isotropic response in all three dimensions and flat frequency response over a very wide spectrum. Measuring heads for the E field use dipoles which are cut very short so that the self-resonant frequency is well above the intended measuring range. Measuring heads for the H field use small loops also designed so that the self-resonance is well above the intended operating frequency range. Each head has three elements arranged so that all three are at right angles to each other with the transmission line taking signals to the meter being aligned at a 45° angle to all three elements as shown in Figure A21. The detector diode for each element is placed at the center of that element; the signal sent along the transmission line is the DC component. The transmission line must not contribute any signal pickup so it may be formed of a very lossy material such as deposited carbon or the detector diodes may be LEDs with the transmission lines being optical fibers. The sensitivity of these systems as radio receivers is extremely low, but a great degree of loss can be tolerated because the field strength that corresponds to a dangerous power level is in the range of hundreds of volts per meter for the E field ($>10^4$ V^2/m^2) or amps squared per square meter for the H field.

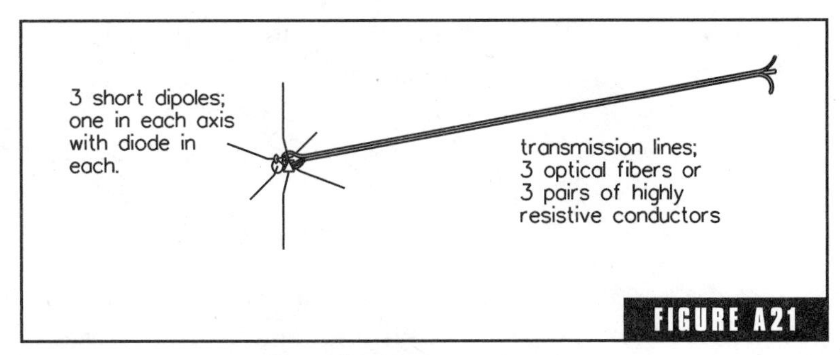

Arrangement of probes for a personal hazard meter.

When the three signals from the three aerial elements arrive at the meter they are combined algebraically and then the result is converted to a square-law output using analog computing techniques. The combining is done in such a way that the final result makes the combined aerial appear as near to a true isotropic radiator as possible. The square-law presentation is done so that the meter can be calibrated directly in volts squared per square meter or amps squared per square meter.

A7 Symbols, Formulas, and Data

Symbols

In all cases in radio technical matters where the following symbols are used they are assumed to have the meanings shown in this list:

α Common base current gain of a transistor; a number with no units close to but slightly less than unity.

B Magnetic susceptibility.

β Common emitter current gain of a transistor; a number with no units usually in the range from 10 to 500.

C Capacitance. Basic unit is farads, usually seen in real circuits as microfarads, nanofarads, or picofarads.

d Diameter of a circle or sphere.

δ Rate of change (or slope of a graph).

E Electromotive force usually expressed in volts but can also be described in ESUs (electrostatic units). In a practical circuit the open circuit voltage at the point of measurement.

e Electron; the subatomic particle which carries a negative electrical charge.

eV Electron-volt; the amount of work done or energy expended when a charge of one electron falls through a potential difference of 1V.

F Depends on the context:

As a unit of quantity (farad) it is the basic unit of capacitance. In radio-frequency circuits microfarads, nanofarads, or picofarads are more common.

As a variable in an equation (frequency) the basic unit is hertz (cycles per second). In radio-frequency circuits kilohertz, megahertz, or gigahertz are more common.

ϕ Phase particularly related to rotating machinery and sine-wave signals; basic unit is degrees (of a circle).

g The acceleration produced by Earth's gravity at sea level.

H Depends on the context:

In a formula dealing with magnetic effects it is called magnetizing force.

In relation to aerials, induction fields, and radiation fields it is the current vector of the electromagnetic field.

I Current; basic unit is amperes. In practical electronic circuits currents expressed in milliamps or microamps are more common.

j A mathematical operator expressing the square root of minus one. For electrical use multiplying by (-1) gives a 180° phase change so multiplying by (j) gives a 90° change. In electronics may also be used to specify reactive impedance in the sense that reactance gives a 90° phase change in the current. Inductive reactance is denoted as $(+j)$ and capacitive reactance as $(-j)$.

k Kilo-; A prefix denoting multiplying by 1000 times as in kilowatt, kilovolt, or kilohm.

L Inductance; basic unit is henry. Inductance of many henrys is seen in mains power circuits; in electronic circuits inductances of millihenrys or microhenrys are more common.

l Length; basic unit is meters.

λ Wavelength, usually denoting the wavelength of a radiated signal in free space in a vacuum.

M Mega-; a prefix denoting multiplied by 1,000,000 times; for example, megahertz and megohms.

m Milli-; a prefix denoting divided by 1000 times; for example, milliamps and millihenrys.

μ Depends on the context:

As a prefix to a unit of quantity it is "micro," which denotes divided by 1,000,000 times; for example, microfarads.

As a stand-alone variable in an equation related to voltages and currents it is the amplification factor of a triode, tetrode, or pentode valve.

As a stand-alone variable in an equation related to magnetic effects it is permeability of a material or magnetic circuit.

n Depends on the context:

As a prefix to a unit of quantity it is "nano," "which means "divided by 10^9 times"; for example, nanofarads.

As a variable it is an algebraic notation for "any number" usually referring to "any integer" as in "the *n*th term of a series."

P Power flow; the rate of transfer of energy; basic unit is watts.

π The ratio of the circumference of a circle to its diameter; numerical value to five decimal places is 3.14159.

Q The ratio of reactance to resistance in a resonant circuit; a number with no units of quantity which for most practical radio circuits is in the range of 10 to 200. The number is never less than unity, and in a regenerative circuit adjusted to exactly the threshold of oscillation the *Q* factor is theoretically infinite but in practice figures higher than about 8,000 to 10,000 can never be measured. The same figure specifies the bandwidth of a tuned circuit in the form of the ratio between the center frequency and the change of frequency required to reduce response by 6 dB.

θ Depends on the context:

In relation to aerial polar diagrams, for example, it is relative direction; in relation to electrical waveforms it is relative phase; the units for both are in degrees (of a circle).

R Resistance or resistive component of a complex impedance; basic unit is ohms; in electronic circuits resistances of kilohms or megohms are more common.

r Radius of a circle or sphere.

ρ Power density in a radiated field.

s Seconds of time.

T or t Time, particularly in the form of time differences; i.e., T_0 = starting time of an action, $(T + 1\ \text{s})$ = a time 1 s after the start.

V Volts; the unit of quantity of electrical pressure.

W Watts; the unit of power flow in both electrical/electronic and mechanics disciplines. In electronic circuits power levels of milliwatts or microwatts are most common.

Ω Ohms; the unit of quantity of resistance, reactance, and impedance. The ratio of volts to amps without reference to phase.

X Reactance; basic unit is ohms.

X_c Capacitive reactance.

X_L Inductive reactance.

Z Impedance (resistance and reactance combined); basic unit is ohms.

Z_0 Characteristic impedance of a transmission line; basic unit is ohms, practical transmission lines are usually in the range from 10 to 600 Ω.

Formulas

The following formulas are worth committing to memory if you are working in a field related to the installation or maintenance of radio systems:

Combining resistances in series:

$$R = R_1 + R_2 + R_3 + \ldots$$

Combining resistances in parallel:

$$1/R = 1/R_1 + 1/R_2 + 1/R_3 + \ldots$$

Combining two resistors in parallel:

$$R = R_1 R_2/(R_1 + R_2)$$

To slightly reduce the value of a resistor by placing a much higher-value resistor in parallel (binomial approximation):

$$\delta R = R/R_{par}$$

For example; to reduce a resistor R in value by 1% place a resistor 100 times the value of R in parallel; to reduce the value by 5% use a parallel resistor 20 times the value. The approximation is not sufficiently accurate for intended changes greater than 11% if better that 1% final accuracy is required.

Inductive reactance (in basic units):

$$X_L = 2\pi f L$$

Inductances in series can be combined as for resistances in series and inductances in parallel can be combined as for resistances in parallel.

Capacitive reactance (in basic units):

$$X_c = 1/(2\pi f C)$$

Capacitances in series can be combined as for resistances in parallel and capacitances in parallel can be combined as for resistances in series. ***Note, however, that capacitive reactances are combined in accordance with the same rules as for resistance and inductance.***

Impedance of resistance and inductance in series (compare with Pythagoras' Theorem):

$$Z = \sqrt{(R^2 + X_L^2)}$$

Can also be written as:

$$Z = (R^2 + X_L^2)^{0.5}$$

Impedance of resistance and inductance in parallel:

$$1/Z = [(1/R)^2 + (1/X_L)^2]^{0.5}$$

Impedance of resistance and capacitance in series:

$$Z = (R^2 + X_c^2)^{0.5}$$

which is derived from:

$$Z^2 = R^2 + X_c^2$$

Impedance of resistance and capacitance in parallel:

$$1/Z = [(^1/R)^2 + (^1/X_c)^2]^{0.5}$$

which is derived from:

$$1/Z^2 = {}^1/R^2 + {}^1/X_c^2$$

Impedance of inductance and capacitance in series:

$$Z = X_L - X_c$$

Impedance of inductance and capacitance in parallel:

$$1/Z = 1/X_L - 1/X_c$$

or if X_c is larger:

$$1/Z = 1/X_c - 1/X_L$$

Wavelength in a vacuum of a radio signal. If f is in kilohertz, λ is in kilometers; if f is in megahertz, λ is in meters; and if f is in gigahertz, λ is in millimeters:

$$\lambda = 299.79/f$$

Length of a resonant conductor:

$$L = 0.5\lambda K$$

Where K is an end effect factor which depends on the $1/d$ ratio of the conductor. For $1/d$ of 100, K is 0.95 and for $1/d$ of 10,000, K is ~0.975.

Voltage standing-wave ratio of a transmission line:

$$\text{SWR} = Z_{(\text{load})}/Z_o$$

This equation only applies if the phase of the load can be regarded as purely resistive; when phase differences must be included a much more complex calculation is required. Convention directs that the result be expressed as a number greater than unity so if $Z_{(load)}$ is lower than Z_o take the reciprocal of the calculated figure.

Ratio between two power levels (P_1 and P_2) converted to decibels:

$$\text{decibels} = 10 \cdot \log (P_2/P_1)$$

Ratio between two voltage levels (V_1 and V_2) at the same impedance converted to decibels:

$$\text{decibels} = 20 \cdot \log (V_2/V_1)$$

The following formulas will also be useful from time to time:

Resonant frequency of a tuned circuit;

$$F = 1/[2\pi(LC)^{0.5}] \quad \text{in basic units.}$$

For scaling to units more commonly used:

- with L in μH and C in μF, frequency is in megahertz.
- with L in henrys and C in μF, frequency is in kilohertz.
- with L in mH and C in nF, frequency is in megahertz.
- with L in henrys and C in pF, frequency is in megahertz.
- with L in μH and C in pF, frequency is in gigahertz.

Power density of a radiated signal (far-field condition):

$$\rho = PG/4\pi R^2$$

Where p = average power at aerial feedpoint

g = numerical gain of aerial in the direction of measurement

r = radius of sphere over which power is distributed

When P is in watts and R is in meters the result will be in watts per square meter.

Capture area of a resonant half-wave dipole in free space:

$$A = 0.09\lambda$$

This applies when the units for the area A are the square of the units for λ; i.e., with λ in meters A is in square meters.

Required impedance of a quarter-wave matching section of transmission line;

$$Z_{(\text{matching})} = [Z_{(\text{load})}\, Z_{\text{o}}]^{0.5}$$

Required length of a quarter-wave matching section of transmission line:

$$l_{(\text{matching})} = 0.25\lambda v$$

Where v is the velocity factor of the transmission line which for most open wire lines is in the range 0.9 to 0.95, for most coaxial cables is in the range 0.65 to 0.68, and for waveguides is a number slightly greater than unity whose exact value depends on the operating frequency of the signal compared to the physical dimensions of the waveguide.

Data

The following data will be useful in many situations:

Gain of a resonant half-wave dipole over an isotropic radiator: 1.6 dB

Coupling loss between two resonant, exactly power-matched half-wave dipoles one wavelength apart in free space: 22 dB

Increase of coupling loss due to greater distance (each time free space path length is doubled): 6 dB

Impedance of free space: 377 Ω

Impedance of an infinitely thin resonant half-wave dipole (in free space, center-fed with no resistive losses): 78 Ω

Power reflected by a 2:1 SWR due to mismatch at the load: 11%

Conductivity of Earth materials:

Sea water: 5000 ms/m

High-conductivity soil: 100 ms/m

Average soil: 10–15 ms/m

Poor-conductivity soil: 1–5 ms/m

Ratio between peak voltage (or current) and RMS voltage (or current) for a pure sine wave: 1.414 times

Ratio between peak-to-peak and RMS voltage for a pure sine wave: 2.828 times

Preferred values for passive components, 20% tolerance series:

10	15	22	33	47	68	100

Preferred values 10% tolerance series:

10	12	15	18	22	27	33
39	47	56	68	82	100	

Preferred values 5% tolerance series:

10	11	12	13	15	16	18
20	22	24	27	30	33	36
39	43	47	51	56	62	68
75	82	91	100			

Preferred values in tolerances closer than 5% require three significant figures for their specification.

Color code for passive components:

Color of band or dot	Digit represented	Color of band or dot	Digit represented
Black	0	Green	5
Brown	1	Blue	6
Red	2	Violet	7
Orange	3	Grey	8
Yellow	4	White	9

There is considerable variation in the way these coded colors are used to specify actual components but if a particular component is known to be a resistance with 5% or wider tolerance and only three or four color bands it is likely that two of the bands will indicate digits in the preferred values series; the third band will be a decimal multiplier; and the fourth, if it exists, will indicate the tolerance range.

For active components only those which have a military specification have any definitely established performance. Usually components with the same major identification as that of a military-spec item have the same characteristics at a particular reference temperature but reduced range of operation with respect to temperature variation. In some cases a semiconductor of a particular design may be marketed in three grades with "mil. spec." offering the widest operating temperature range and narrowest tolerances on other parameters, an "industrial" grade which is adequate for all apart from critical to human life purposes, and a "commercial" or "domestic" grade for mass-produced consumer goods.

INDEX

Note: boldface numbers indicate illustrations; italic t indicates table.